Taiwanese Native Medicinal Plants

Phytopharmacology and Therapeutic Values

T0315217

Taiwanese Native Medicinal Plants

Phytopharmacology and Therapeutic Values

Thomas S.C. Li, Ph.D.

CRC Press
Taylor & Francis Group
Boca Raton London New York

CRC Press is an imprint of the
Taylor & Francis Group, an **informa** business

Published in 2006 by
CRC Press
Taylor & Francis Group
6000 Broken Sound Parkway NW, Suite 300
Boca Raton, FL 33487-2742

First issued in paperback 2019

ISBN-13: 978-0-367-45373-2 (pbk)
ISBN-13: 978-0-8493-9249-8 (hbk)

Library of Congress Card Number 2005026038

Library of Congress Cataloging-in-Publication Data

Li, Thomas S. C.
    Taiwanese native medicinal plants : phytopharmacology and therapeutic values / Thomas S.C. Li.
        p. ; cm.
    Includes bibliographical references and index.
    ISBN-13: 978-0-8493-9249-8
    ISBN-10: 0-8493-9249-7
    1. Materia medica, Vegetable--Taiwan. 2. Medicinal plants--Taiwan. I. Title.
    [DNLM: 1. Plants, Medicinal--Taiwan. 2. Materia Medica--pharma- cology--Taiwan. 3.
Phytotherapy--methods--Taiwan. QV 770 JT2 L693t 2006]

RS180.T35 L5
615'.3210951249--dc22   2005026038

Taylor & Francis Group
is the Academic Division of Informa plc.

Visit the Taylor & Francis Web site at
http://www.taylorandfrancis.com

and the CRC Press Web site at
http://www.crcpress.com

Foreword

Tens of thousands of plant species have been used medicinally, and hundreds of books have been written about them. Most of these are concerned with reviewing the botany or the traditional folk uses of the plants. This is certainly valuable information but insufficient as a basis for prioritizing those species that are especially deserving of research and development. Much more valuable is the approach that Dr. T.S.C. Li has taken in his previous books on medicinal plants and in this one. Dr. Li has concentrated on documenting the chemical components present, and on their therapeutic properties, information that is obviously critical to assessing the potential of these species.

In this book, Dr. Li provides an extremely comprehensive review of the phytopharmacology (dealing with the general science, toxicology, and therapeutics) of more than 1000 medicinal species native to Taiwan, based on his personal familiarity with the plants as well as the very extensive literature on them. As Dr. Li points out, a remarkably large proportion of these have not been addressed in previous analyses of Chinese medicinal plants.

People in every region of the world learned to use local plants for medicinal purposes, and developed "pharmacopoeias" (sets of species and extracts, and associated codified knowledge of their uses) that became the bases of traditional medical systems. With the possible exception of Indian (Vedic) medicine, the Chinese pharmacopoeia is the largest and most impressive ever produced. Western researchers are increasingly acknowledging the importance of the traditional herbal preparations that have been the mainstays of Asian medicine for millennia. Asian medicinal plants, most particularly Chinese herbs, are now enthusiastically being incorporated into Western medical practice. Knowledge of the therapeutic properties of Chinese medicinal herbs is a key to progress in developing more effective medications. This book provides the type of accurate, contemporary information required by those who conduct research on, prescribe, or personally use medicinal plants or their components.

There is no more noble endeavor than the pursuit of promoting human health, but in the realm of medicine this needs to be done carefully, as reflected in the familiar physician's dictum (commonly held to come from the Hippocratic Oath) "First, do no harm." Unfortunately, there are serious hazards and pitfalls associated with medicinal plants, which can only be avoided when authoritative information is available. Dr. Li's monograph is not only useful for indicating how the plants reviewed may be used therapeutically but also points out their potential toxicity.

Although the therapeutic and safety aspects of medicinal plants are of predominant importance to society, one cannot ignore their economic value. Explosive growth is occurring in the multitrillion dollar business of medicinal plants, most evidently in the increasing array of herbal offerings and supplements found in health food stores and supermarkets, but also in the domain of prescription medicine. For the most part, the really lucrative species are restricted in number, with only a few dozen that are highly profitable at a particular time. Those who produce, process, and market medicinal plants and their extracts would be well advised to carefully study Dr. Li's information. It is quite possible that among the species documented, one or more has the potential to become extremely profitable.

Relatively few books are as genuinely scholarly, authoritative, and comprehensive as this present volume. This is a veritable treasure chest of essential and fascinating information critical to all health care professionals who deal in one way or another with medicinal plants. Dr. Li, an internationally renowned scientist with extremely extensive research experience, is to be congratulated on this superb and invaluable synthesis. *Taiwanese Native Medicinal Plants* represents a milestone in educating the world about a gold mine of medicinal knowledge.

Ernest Small, Ph.D.
Principal Research Scientist
Eastern Cereal and Oilseed Research Center
Agriculture and Agri-Food Canada
Ottawa, Ontario, Canada

Preface

The application of medicinal plants to maintain health and treat diseases started thousands of years ago and is still part of medical practice in many countries such as China, Egypt, India, and the developing African countries. Over the centuries, the use of medicinal plants has become an important part of daily life in the Western world despite significant progress in modern medical and pharmaceutical research. Recently, the use of medicinal plants, especially those of Chinese origin, has attracted considerable attention around the world, which has prompted extensive research on their philosophy, principles, and especially the scientific background of the chemical components responsible for their claimed therapeutic value.

Western researchers are increasingly acknowledging the importance of the Chinese medicinal herbs and traditional herbal formulations or preparations that have been the mainstream for centuries in China. Recently, it was found that many of the Taiwanese native medicinal plants with promising therapeutic values have been neglected and are not included in the traditional Chinese herbal pharmacopoeia.

This book is designed to provide researchers, manufacturers, and producers with easy access to information on Taiwanese native medicinal plants compiled from widely scattered literature, including some written in Chinese. This book begins with a general introduction regarding the geographic advantages for growing varieties of medicinal plants, followed by Table 1, which presents current available information on the major constituents and therapeutic values of more than 1000 species. The data are arranged alphabetically by the Latin (generic) names. In addition to an index, three appendices cross-reference major chemical components and their sources as well as the common and scientific names of the medicinal plants cited in Table 1.

The information in this book is primarily for reference and education. It is not intended to be a substitute for the advice of a physician. The uses of medicinal plants described in this book are not recommendations, and the author is not responsible for liability arising directly or indirectly from the use of information in this book.

Acknowledgments

The author thanks Paul Ferguson for his word-processing assistance, Barry Butler for his computer technical assistance, and Lynne Boyd for her efforts in the literature search. I also thank my colleagues, Drs. Tom Beveridge, Dave Oomah, Peter Sholberg, and Ernest Small, and numerous Taiwanese scientists and researchers, for their valuable assistance and contribution. Finally, I would like to thank my wife, Rose, for her encouragement.

Contents

Introduction

Taiwan is 394 km long and 144 km wide, located in the middle of a chain of islands in the west Pacific Ocean stretching from Japan in the north to the Philippines in the southwest. Taiwan's total area, including surrounding islands, is approximately 36,000 km^2. It is known for its eminent mountain features that span from the north to the south of the island. Taiwan has five longitudinal mountain ranges, which occupy almost two-thirds of the land. Agriculture takes place mainly on the remaining 29% (900,000 ha).

There are approximately 800,000 agricultural households with a little over one hectare each. This is too small for efficient agricultural production. The weather is distinct between the north and south. While the south has a tropical, oceanic climate, the north is semitropical or temperate at mountain altitudes, with a touch of snow. Taiwan has every kind of fruit from tropical to temperate available year round due to the wide range of climate. Therefore, it allows cultivation of many varieties of new alternative crops such as Chinese and Western medicinal plants.

The use of medicinal plants for treating human diseases started in China thousands of years ago. Eighty percent of the world population still uses traditional medicine, either because they have no access to Western medicine or because they choose not to use it. Recently, the use of medicinal plants, especially Chinese herbs and their products, has attracted considerable attention around the world. Western researchers are increasingly acknowledging the importance of the traditional herbal preparations. The Taiwan biotechnology industry is poised to cash in on drugs developed from traditional Chinese herbal remedies due to government support for development and utilization of medicinal and health-protective plants. Government-funded companies are trying to extract effective chemical compounds and develop quality control tests and methods of consistent extraction of components from selected Chinese herbs.

Some Taiwanese native medicinal plants with promising therapeutic value have been neglected for years and are not included in the traditional Chinese herbal pharmacopoeia. Recently, the Taiwan government has established agricultural development policy goals to promote research on culture techniques of domestic medicinal plants, to introduce foreign medicinal plants from the West for mass production of high-value chemical components, to develop biotechnology-related industries and research in the fields of agriculture and medicine, and to encourage development of indigenous plants for medicinal uses with high-value-added products.

TABLE 1

Major Constituents and Therapeutic Values of Taiwanese Native Medicinal Plants

TABLE 1 Major Constituents and Therapeutic Values of Taiwanese Native Medicinal Plants

Scientific Name	Major Constituents and Source	Claimed Therapeutical Values
Abelmoschus esculentus (L.) Moench.	(Flowers, seed, root) D-Galactose, L-rhamnose, D-galacturonic acid[1]	Diuretic property; a demulcent; alleviates a hoarse sore throat; treatment for gonorrhea and dysuria
Abelmoschus moschatus (L.) Medicus	(Root, leaf, flowers) A glutinous drug containing araban, lactose, and rhamnosan[1]	Intestinal inflammation; leaves used as a poultice to apply to boils and sores; flowers used for dysentery, blennorrhea
Abrus cantoniensis Hance.	(Whole plant) Abrine, choline[10]	Acute hepatitis, gastric pains, urinary tract infection
Abrus precatorius L.	(Seed) Abrine, glycyrrhizic acid, precatorine, hypaphorine, cycloarternol, squalene, trigonelline, 5-β-cholanic acid[2,10] This herb is toxic[88]	Antiemetic, expectorant, parasiticide, tinea infection, eczema
Abutilon indicum (L.) Sweet *A. taiwanensis* S. Y. Hu	(Whole plant) Flavonoids, amino acids[3]	Flu, fever, diuretic, ringing in the ear, tuberculosis, epidemic parotitis
Acacia confusa Merr. *A. farnesiana* (L.) Willd.	(Peeled branch) D-Catechin, catechutanic acid, epicatechin, gambir-fluorescein, gambirine[4] This herb is toxic[88]	Promotes salivation, resolves phlegm, stops bleeding, treats pyogenic infections
Acalypha australis L. *A. indica* L.	(Whole plant) Acalyphine, flavonoids[3]	Stops bleeding; used for diarrhea, cough, and eczema
Acanthopanax senticosus (Rupr. ex. Maxim.) Harms.	(Root, bark) Eleutherosides, β-sitosterol, glucoside, L-sesamen, syringareinol[5]	Central nervous system activating and antistress function

Acanthopanax trifoliatus (L.) Merr.	(Leaf) Taraxerol[6]	Treats cold, cough, neuralgia, rheumatism
Acanthus ilicifolius L.	(Stem, root) Lignin glucosides, benzoxazinoid glucosides[7]	Treats chronic fever, anticancer
Acer buerferianum Miq. var. *formosanum*	(Whole plant)[583] No information is available in the literature	Free radical–scavenging activity, natural antioxidant
Achillea millefolium L.	(Aerial part) Achillin, flavonoids, betonicine, D-camphor, hydroxycinnamic acids, anthocyanidines, coumarins[7]	Antibacterial; treats menopause, abdominal pain, acute intestinal disorder
Achyranthes aspera L. var. *indica* L. *Achyranthes aspera* L. var. *rubro-fusca* Hook. f.	(Seed) Oleanolic acid, ecdysterone, β-carotene, thiamin, riboflavin, niacin, saponin, ascorbic acid, protein[3,8]	Antispasmodic, diuretic, induces labor, antifertility, antiinflammatory
Achyranthes bidentata Blume	(Root) Inokosterone, ecdysterone, polysaccharide, insect molting hormones[2,652]	Anticancer
Achyranthes japonica (Miq.) Nakai	(Root, leaf) Inokosterone, saponin, oleanolic acid, ecdysterone, calcium oxalate[8]	For antirheumatic, anodyne, amenorrhea, carbuncles, fever, dystocia, urinary ailments
Achyranthes longifolia (Mak.) Mak. *A. ogotai* Yamamoto	(Root) Ecdysterone, ecdysterone, inokosterone, triterpenoid saponin, potassium[27]	Improves blood circulation; diuretic, treats throat infection
Aconitum bartletii Yamamoto *A. fukutomei* Hayata *A. formosanum* Tamura *A. kojimae* Ohwi *A. kojimae* Ohwi var. *lassiocarpium* Tamura *A. kojimae* Ohwi var. *ramosum* *A. yamamotoanum* Ohwi	(Root) Aconitine, aconitum, hypaconitine, carmichaemine, mesaconitine, pseudoaconitine, talatisamine[4,14] This herb is toxic[88]	Cardiotonic, antinociceptive, antiinflammatory, analgesic effect, alleviates pain, improves heart condition, controls fungal infection

TABLE 1 Major Constituents and Therapeutic Values of Taiwanese Native Medicinal Plants (continued)

Scientific Name	Major Constituents and Source	Claimed Therapeutical Values
Acorus calamus L. *A. gramineus* Soland	(Leaf, root) Acoric acid, β-asarone, volatile oil, α-pinene, D-camphene, calamene, calamenol, calamenone[4,15,18]	Anticonvulsant, analgesic, aphrodisiac, carminative, contraceptive, desiccant, diaphoretic
Acronychia pedunculata (L.) Miq.	(Root, wood, leaf, fruit) Acetophenone derivatives[581]	Rheumatic pain, traumatic injury, gastric pain, hernial pain
Actinidia callosa Lindl. var. *formosana* Finet & Gagnep *A. chinensis* Planch	(Whole plant) Metatable acid, iridomyrmecin, allomatatabiol, dihydronepetalactol, isoneomatatabiol, neomatabiol[1,8,16,348]	For esophageal and liver cancer, rheumatoid arthritis, arthralgia, urinary stones, fever; antimutagenic activity
Adenophora stricta Miq. *A. triphylla* (Thunb.) A. DC *A. tetrophylla* Mak.	(Root) Triterpenoid saponins[6]	Hemolyzes blood cells, stimulates myocardial contraction, antibacterial
Adenostemma lavenia (L.) Ktze.	(Root) 11-Hydroxylated kauranic acids[8]	Antiinflammatory, improves lung and liver function, alleviates arthritis pain; a preventative for influenza and measles, infectious hepatitis
Adiantum capillus-veneris L. *A. flabellulatum* L.	(Root) Adipedatol, adiantone, hopadiene, isofernene, fernene, filcene, filicenal, fernadiene, lithium and oleanane flavonoids, astragalin, triterpenoids, kaempferol-3-glucuronide, isoadiantone, hydroxyadianthone[3,16,562] This herb is toxic[88]	Treats cold and grippe, cough; stops bleeding, rheumatism, arthritis pain, dysentery, breast inflammation
Adina pilulifera (Lam.) Franch ex Drake *A. racemose* Lam.	(Whole plant) Nucleoside, secoiridoid glucosides, β-sitosterol, noreugenin, flavonoids, quinoric acid, betulinic acid, morolic acid, cincholic acid, saponin, stimasterol[27,578]	Alleviates pain, fever, cough, throat and liver infections

Species	(Part) Constituents	Therapeutic Values
Agastache rugosa (Fisch. & Mey.) O. Kuntze	(Leaf) Methylchavicol, anethole, anisaldehyde, hexenol, calamene, caryophyllene, β-pinene, octanol, cymene, linalool, elemene[16,17]	Chest congestion, diarrhea, headache, nausea; antipyretic, carminative, febrifuge; stomachache
Ageratum conyzoides L. / A. *houstonianum* Mill.	(Leaf, root) Cyanogenic glucoside, coumarin, agerato-chromene, 7-methoxy-2,2-dimethylchromene, β-caryophyllene[8] This herb is toxic[88]	Digestive disorder, fever, rheumatism, gonorrhea, tetanus, syphilis
Agrimonia pilosa Ledeb.	(Whole plant) Agrimophol, agrimols, agrimonine, agrimonolide, cosmosiin, luteolin-7-β-D-glucoside, apigenin-7-β-glucoside, vitamins C, K[2,16,18,19,645]	Astringent hemostatic in enterorrhagia, hematuria, metrorrhagia, gastrorrhagia, pulmonary, tuberculosis; antiinflammatory, antiplatelet, antimicrobial
Ajuga bracteosa Wall. / A. *decumbens* Thunb. / A. *gray* Thunb. / A. *pygmaea* Thunb.	(Whole plant) Flavone glucoside, lutecolin, ajugasterone, ajugalactone, cyasterone ecdysones, cyasterone, ecdysterone, kiransin, β-sitosterol, cerotic acid, palmitic acid[2,4,8]	Antitussive, antipyretic, antiinflammatory, antiphlogistic, antibacterial; treats bladder ailments, diarrhea, bronchitis
Akebia longeracemosa Matsum. / A. *quinata* Decne	(Stem) Aristolochic acid, triterpenoids, saponin, akebin, akebigenin, caryophyllin[2,8,18,31]	Diuretic, antiphlogistic; for cystitis, edema, goiter, nephritis, urethritis
Alangium chinense (Lour.) Harms.	(Root) DL-Anabasine[2] This herb is toxic[88]	Myocardial stimulation; increases contractility, fibrillation, and blood pressure
Albizzia lebbeck (L.) Benth.	(Bark) Saponins, tannins, amino acids[8,20] This herb is toxic[88]	A tonic, stimulant, anthelmintic, diuretic, piscicidal, vermifuge

TABLE 1 Major Constituents and Therapeutic Values of Taiwanese Native Medicinal Plants (continued)

Scientific Name	Major Constituents and Source	Claimed Therapeutic Values
Aletris formosana (Hayata) Sasaki	(Root) Stigmasterol, β-sitosterol, diosgenin[6]	Antitussive, vermifugal: for ascariasis, marasmus, cough
Aleurites fordii (Hemsl.) A. *moluccana* (L.) Willd. A. *montana* (Lour.) Wilson	(Bark, fruit, seed) Heptadecylic acid, toxalbumin, α-elacostearic acid, phytosterol, pentoson, amino acids[6,8,20] This herb is toxic[88]	Analgesic activity; treats anemia, atrophy, edema; vermicide
Allium bakeri Rogel A. *scorodoprasum* L.	(Leaf) Scorodose, adenosine guanosine, tryptophan, β-sitosterol, β-D-glucoside[3,563]	For stomachache, antiseptic, diuretic, cough with short breath; resolves phlegm; externally used for burns
Allium cepa (L.)	(Whole plant) Coumaric acid, caffeic acid, ferulic acid, sinapic acid, *p*-coumaric acid, protocatechuic acid, quercetin, polysaccharides, quercetin 3,4′-diglucoside, thymine, carotenes[20]	Antibacterial, antimutagenic, anticarcinogenic; lowers plasma cholesterol and lower-density lipoprotein; prevents thrombosis, hypotensive
Allium sativum L. A. *thunbergii* G. Don. A. *tuberosum* Rottler	(Bulb) Allicin, allistatin, glucominol, neoallicin, steroidal saponins, polysaccharides, furostanol saponins, protoisoerubosides, diallyl sulfide[2,49,438,490]	Antibacterial, antimutagenic, anticarcinogenic, carminative, antiarrhythmic; lowers plasma cholesterol and low-density lipoprotein, prevents thrombosis; hypotensive and vessel-protective effect
Alocasia cucullata Schott & Endl.	(Whole plant) Glutelin, fatty acids[8] This herb is toxic[88]	Antirheumatic; for tuberculosis, inflammation, windpipe infection; leaves applied to wounds and ulcers

Species	Constituents	Therapeutic Values
Alocasia macrorrhiza (L.) Schott & Endl.	(Rhizomes) Sapotaxin, ceramide, alocasin[11,501,502] This herb is toxic[88]	Epidemic influenza, high fever, pulmonary tuberculosis
Alpinia speciosa (Wendl.) K. Schum *A. zerumbet* (Pers.) Burtt & Smith	(Seed) Zingiberene, zingiberol[6,643]	Stomachache, abdominal pain, indigestion, vomiting, diarrhea, heartburn
Alternanthera nodiflora R. Br.	(Whole plant)[8] Saponin, coumarin, tannins, flavins[2,638]	For tuberculosis, viral infections, measles, hemorrhagic fever, icteric hepatitis
Alternanthera philoxeroides (Mart) Griseb. *A. sessilis* (L.) R. Brown ex Roem. & Schultes	(Aerial part) Saponin, coumarin, tannins, falvins, 7α-L-rhamnosyl-6-methoxyluteolin[2]	Viral infections, measles, hemorrhagic fever, hepatitis
Alyxia insularis Kaneh. & Sasaki *A. sinensis* Champ. ex Benth.	(Whole plant) Bauereny acetate, scopletin, liriodendrin, pinoresinol-di-0-β-D-glucopyranoside, daucosterol, flaxetin, esculin, aseculin[97,155]	Rheumatic arthralgia, loin pain, diarrhea, amenorrhea
Amentotaxus formosana Li	(Leaf) Lanostanoids[598]	Anticancer
Amorphophallus konjac C. Koch *A. rivieri* Durieu.	(Flower, root) Leviduline, levidulinase, mannose[8]	Febrifuge for aching bones and eye inflammation; treats cancer, ulcers
Ampelopsis brevuoedybcykata (Maxim.) Trautv.	(Root, stem, leaf) Flavonoids, glucosides, amino acids[16,348] This herb is toxic[88]	Antitoxic; alleviates pain and bleeding, treats arthritis; antimutagenic activity
Ampelopsis cantoniensis (Hook. et Arm.) Planch.	(Root, stem, leaf) Flavonoids, glucosides, amino acids[1]	Anodyne, astringent, anticonvulsive, detoxicates; treats tubercular cervical nodes, hemorrhoidal bleeding

TABLE 1 Major Constituents and Therapeutic Values of Taiwanese Native Medicinal Plants (continued)

Scientific Name	Major Constituents and Source	Claimed Therapeutical Values
Ananas comosus (L.) Merr.	(Leaf) Ergosterol peroxide, ananasic acid, 5-stigmautena-3β-7b-diol, 3,4-dihydroxycinnamic acid, 4-hydroxycinnamic acid, bromelin, vitamins[21]	Antioxidant activity; improves digestion, lowers blood pressure; anticancer
Andrographis paniculata (Burm. f.) Nees	(Aerial part) Deoxyandrographolide, andrographolide, neoandrographolide, dehydroandrographolide[2]	Antibacterial, antipyretic, antiinflammatory
Anemarrhena asphodeloides Bunge	(Rhizome) Steroidal saponins, mangiferin, isomangiferin, sarsasapogenin, markogenin, neogitogenin[2,22]	Antipyretic, antiinflammatory, sedative, antibacterial
Angelica acutiloba (Sieb. et Zucc.) Kiitagawa *A. citriodor* Hance	(Root) *n*-Butylidenephthalide, sedanonic acid, *n*-valero-phenones-carboxylic acid, safrol, isosafrole, bergaptene, *p*-cymene, palmitic acid, linoleic acid, oleic acid, dodecanol, tetradecanol, β-sitosterol[6]	Alleviates menstrual pain, regulates menopause periods
Angelica hirsutiflora Liu Chao et Chuang	(Root) Hamaudol, osthol, bergapten, furanocoumarins, xanthotoxin, byakangelicin, byakangelicol, phellopterin, marmesin, psoralen, isopimpinellin, 5-methoxy-8-hydroxypsoralen, stigmasteral[28]	Against fungal infection; alleviates pain, flu, and dizziness; regulates menopause
Angelica keiskei (Miq.) Koitz.	(Whole plant) 4-Hydroxyderricin, 2′,4′4-trihydroxy-3′, chalcone, xanthoangelol, calumbianadin, archangelicin, dihydrooroselol, luteorin, isoquercitrin, rutin, psoralene, angelicin, bergaten, angelic acid, behenic acid, santhotoxin, vitamins[28]	Diuretic; improves blood pressure, diabetes; improves liver function

Plant	Constituents	Therapeutic Values
Anisomeles indica (L.) Kuntze.	(Whole plant) Flavonoids, β-sitosterol, stigmastero[29]	Alleviates pain; treats fever, malaria
Annona muricata L. *A. cherimola* Mill. *A. reticulata* L.	(Seed) Annonaceous acetogenins, annocherine A-B, cherianoine, romucosine H, artabonatine B, acetogenins[503,607,634]	Cytotoxic; acute dysentery, mental depression, antiparasitic, spine disorder, rectal prolapse, swelling
Anoectochilus formosanus Hayata	(Rhizome) Diarylpentanoid, kinsenone, flavonoid glycosides[1,121]	Treats tuberculosis; reduces blood sugar level, blood pressure; diuretic; treats arthritis; antioxidant, hepatoprotective activity, antihyperlipoisis
Aquilaria sinensis (Lour.) Gilg. *A. sibebsus* Gilg.	(Stem wood) Agarospirol, β-agarofuran, benzylacetone, hydrocinnamic acid, *p*-methoxybenzylacetone, α-agarofuran, hydroagarofuran[2,28]	For abdominal pain, abscess, chest pains, high blood pressure; choleretic
Arachis hypogea L. *A. agallocha* Roxb.	(Seed) Amino acids, protein arachine, globulin, biotin, glycyrrhizin, glucosides, thiamin, niacin, riboflavin, carbohydrate[8,16,18]	Stomachache in gastralgia, colic, nervous emesis; nutritive, pectoral, peptic; an emollient
Aralia chinensis L.	(Root) Diterpenoids: (–) pimaradene, (–) kaurene derivatives, L-pimara-8,15-dien-19-oic acid, aralosides, oleanolic acid, taraligenin, β-taralin, α-taralin[8,16,23]	Carminative; treats arthralgia, gastroenteritis, headache; diuretic, antidiabetic; antiseptic
Aralia taiwaniana Liu & Lu ex Lu *A. chinensis* L.	(Stem) β-Taralin, α-taralin, taraligenin, protocatechuic acid, choline, mucilage, saponins, tannins, aralioside[8]	Anodyne, carminative; for arthralgia, gastroenteritis, headache, jaundice, rheumatism
Arctium lappa L.	(Seed, root) Seed: arctin, arctigenin, gobosterin; root: inulin, lappine[18]	Diuretic, antipyretic, expectorant; antiphlogistic in throat infections, pneumonia, scarlet fever, measles, smallpox, syphilis
Ardisia crenata Sims. *A. squamulosa* Presl.	(Root) Ardisic acid, bergenin[27,435]	Treats ulcer, respiratory infection; anti-herpes simplex virus; antiadenovirus activity

TABLE 1 Major Constituents and Therapeutic Values of Taiwanese Native Medicinal Plants (continued)

Scientific Name	Major Constituents and Source	Claimed Therapeutical Values
Ardisia sieboldii Miq.	(Leaf, root, seed) Bergenin, rapanone[3,8,435]	Anticancer, hepatoma; a diuretic, antidote for poison, antiphlegmatic, antiherpes simplex virus; treats arthritis, lymphatic gland infection
Areca catechu L.	(Nut, leaf) Arecoline, arecaine, areca red, catechin, homoarecoline, arecolidine, guvacoline, guvacine; seed contains arecoline, arecaidine, guvacine, guvacoline, isoguvacine arecotidine, leucocyanidin[2,3,565]	Treats taeniasis; insecticide, antifungal, virus; betel leaves have anticarcinogenic agents, nut has cholinomimetic and acetylcholine sterase inhibitory constituents
Arenga engleri Beccari *A. pinnata* (Wurmb.) Merrill *A. saccharifera* Labill.	(Fruit sap) Formic acids, polysaccharides, periodic acid[8,503]	For constipation and dysentery, bleeding, high blood pressure
Arisaema consanguineum Schott *A. erubescens* (Wall.) Schott *A. vulgaris* L.	(Whole plant) Alkaloids, saponin, benzoic acid[2,14,18] This herb is toxic[88]	Treats tetanus, spasms, epilepsy, neuralgia; sedative, anticonvulsive, expectorant
Aristolochia cucurbitifolia Hayata *A. elegans* Mast. *A. manshuriensis* Kom. *A. heterophylla* Hemsl. *A. kaempferi* Willd. *A. kankanensis* Sasaki *A. shimadai* Hayata	(Root, stem) Aristolochic acid, phenanthrene derivatives, aristoliukine C, aristofolin E, aristolochic acid-la methyl ester, madolin-p[6,580,616]	Antinflammatory, diuretic; treats stomachache, alleviates pain; cytotoxicity and antiplatelet activity; used externally for snake bite
Artabotrys uncinatus (Lam.) Merr.	(Whole plant) α-, β-Butenolide alkaloid, uncinine, artabonatine C-F, atherospermidine, squamolone[604]	Against hepatocarcinoma cancer cell lines; antithrombin

Plant	(Part) Constituents	Therapeutic Values
Artemisia capillaris Thunb.	(Shoot) Scoparon, capillene, capillin, capillon, capillarin, capillanol[2]	A choleretic; treats jaundice, acute infectious hepatitis, gallstone-related illnesses
Artemisia indica Willd. A. japonica Thunb.	(Aerial part) Terpinenol-4,β-caryophyllene, artemisia alcohol, linalool, cineol, camphor, borneol, cucalyptol[2]	Antiasthmatic, antitussive; treats chronic bronchitis and hypersensitivity
Artemisia lactiflora Wall. A. princeps Pampanini	(Whole plant) Flavanoid glycoside, coumarin, lactiflorenol, spathulenol, s-guaiazulene, limonene, β-guaienen, trans-β-farnesene, trans-caryophyllene, elemene, copaene, myrcene[21,435]	Diuretic, regulates menstruation, treats headache, high blood pressure; antiherpes simplex virus
Artocarpus altilis (Park.) Fosberg.	(Bark) Triterpenes, β-amyrin acetate, lupeol acetate[1]	Poultice for ulcers
Artocarpus heterophyllus Lam.	(Leaf, seed) Caoutchoue, resin, cerotic acid, protein, minerals[8]	Tonic to treat discomfort from alcohol influences
Asarum hypogynum Hayata A. macranthum Hayata A. hongkongense S. M. Hwang et T. P. Wong A. longerhizomatosum C. F. Liang et C. S. Yang	(Whole plant, root) Asarone, β-sitosterol, 2,4,5-trimethoxybenzaldehyde, 4-(2,4,5-trimethoxyphenyl)-3-en-butylone, 3 β-hydroxystigmast-5-en-7-one[156]	Cough, excessive sputum, rheumatic arthralgia
Asclepias curassavica L.	(Whole plant) Curassavicin, calotropin[11] This herb is toxic[88]	For mastitis, pyodermas, dysmenorrhea
Asparagus cochinchinensis (Lour.) Merr.	(Root, young shoot) Glycolic acid, asparagines, essential oils, methanethiol, (+)-nyasol, asparagine, steroids, β-sitosterol, sarsasapogenin, diosgenin, polysaccharide, oleanen dervatives[4,8,24–26]	Diuretic, laxative, treats cancer; antitumor, antioxidative activity; neuritis, rheumatism, parasitic diseases

TABLE 1 Major Constituents and Therapeutic Values of Taiwanese Native Medicinal Plants (continued)

Scientific Name	Major Constituents and Source	Claimed Therapeutic Values
Aspidistra elatior Blume	(Root, stem, leaf) Aspidistrin (diosgenin 3-0-beta-lycotetraoside)[8,32]	Antifungal: for abdominal cramps, amenorrhea, diarrhea, myalgia, traumatic injuries, urinary stones
Aspidixia articulata (Burm. f.) Van tieghem. *A. liquidambaricala* (Hayata) Nakai	(Whole plant) Oleanolic acid, erythrodiol, inositol, β-amyrin, β-amyrin acetate, betulin, oleanalic acid[20]	Arthritis, diarrhea, cough, high blood pressure, eczema
Asplenium nidus L.	(Whole plant) Dysoxylum gaudichaudianum[8,57]	Estrogenic, spasmolytic; treats fever; infusion alleviates labor pains, asthma, debility, halitosis, sores
Astilbe longicarpa (Hayata) Hayata	(Whole plant) Quercetin, 2-hydroxyphenylacetic acid, bergenin[6]	Alleviates pain, stomachache, cough; treats flu, improves blood circulation, antitoxic
Astragalus sinicus L.	(Whole plant) Canavanine, trigonelline[8]	Used for blennorrhea, unguent for burns
Atalantia buxifolia (Poir.) Oliv.	(Root) Limonoide, N acridone alkaloid[89,90]	Bronchitis, malaria, epigastric pain, abdominal pain, rheumatic arthritis
Atylosia scarbaeoides (L.) Benth.	(Root) Epoxyflavanone[1,46,47]	Antiinflammation, diuretic; alleviates pain, fever, arthritis
Bacopa monniera (Brahmi)	(Whole plant) Bacopaside III, bacopasoponin G, bacopasides A–C[588,589]	Exerts cognitive-enhancing effect in animals; effects on human cognition are inconclusive
Balanophora spicata Hayata	(Whole plant) Taraxasterol, β-amyrin, palmitic acid[20]	Aphrodisiac, antitoxic
Basella alba L.	(Leaf) β-Carotene, thiamin, riboflavin, niacin, ascorbic acid[8]	For urticaria, dysentery, intestinal trouble; diuretic, febrifuge, laxative

Species	Constituents (Part)	Therapeutic Values
Basella rubra L.	(Leaf, berry)[45] Glucan, β-carotene, vitamins A, B, C[20,48]	A demulcent in intestinal troubles, diarrhea, constipation, appendicitis; an emolient; pigmentary addition to facial cosmetics
Bauhinia championi Benth.	(Bark) Kaempferol-3-galactoside, kaempferol-3-rutinoside, flavonoids, stigmasterol, β-sitosterol, β-*p*-glucopyranoside[4,8,630,647]	Oxygen scavenging activity; an astringent, tonic; treats scrofula, skin ailments, leprosy, ulcers, diarrhea
Begonica fenicis Merr. *B. laciniata* Roxb. *B. malabarica* L.	(Whole plant) Luteolin, quercetin, β-sitosterol-3-β-ᴅ-glucopyranoside[11,49,157]	Fever, pneumonia, hemoptysis, ecchymosis, stomachache; influenza, acute bronchitis, internal hematoma, hepatomegaly
Belamcanda chinensis (L.) DC	(Root) Tectoridin, tectorigenin[8,49,564]	Influenza, rheumatic arthritis; antipyretic, antifungus, analgesic; detoxicates; stomachache, cough, wheezing; selective estrogen receptor modulator activities
Bellis perennis L.	(Flower) Flavonol glycosides, apigenin glycosides, triterpenoid glycosides, saponins[58-61]	Inhibits the growth of human-pathogenic yeasts; diuretic, an expectorant agent; hemolytic activity
Berchemia formosana Schneider *B. lineata* (L.) DC	(Root, leaf) Mearnsitrin-3-*O*-α-ʟ-rhamnoside, myricitrin, narcissin, flavones; phenolic, carboxylic acids; flavanones, flavanonols, (−)-catechin[10,158]	Chronic bronchitis, peptic ulcer bleeding, schizophrenia
Bidens pilosa L. var. *minor* (Blume) Sherff. *B. racemosa* Sieb et Zucc.	(Leaf) Polyacetylenes, phenytheptatriyne[4]	Antibiotic; treats bug bites, diarrhea, snakebite; bactericidal, fungicidal
Biota orientalis (L.) Endl.	(Twig) Quercitrin, pinipicrin, thujone[2]	Hemostatic, shortens blood clotting time, antitussive
Bischofia javanica Blume	(Leaf, root) β-Amyrin, ursolic acid, β-sitosterol, friedelin, methyl betulin, friedelan-3α-yl-acetate, ellagic acid, epifriedelanol acetate[8,20]	Astringent, for ulcers, diuretic, nocturnal emission

TABLE 1 Major Constituents and Therapeutic Values of Taiwanese Native Medicinal Plants (continued)

Scientific Name	Major Constituents and Source	Claimed Therapeutical Values
Bixa orellana L.	(Seed, leaf, unripe fruit) Ethereal oil, resin, tannins, cellulose, orlean, palmitin, phytosterol, vitamin A, stearic and oleic acids[1]	Astringent, antipyretic, rubefacient, emollient; treats febrile catarrh
Blechnum orientale L.	(Root, shoot) Panasterone A, inokosterone, woodwardic acid, chlorogenic acid[1,3,8,29]	For urinary complaints; young shoot is applied as a compress to swellings and boils; treats bleeding, headache, flu; antitoxic
Blechum pyramidatum (Lam.) Urban. *B. orientale* L. *B. amabile* Makino	(Leaf) Chlorogenic acid, cholestanes[1,3,504]	Treats blennorrhea, vulnerary, intestinal parasites, vomiting with blood, flu, measles
Bletilla formosana (Hayata) Schltr.	(Rhizome) Mucilage, essential oil, glycogen[1,18]	Lung tonic for tubercular patients vomiting blood; it has demulcent properties
Bletilla striata (Thunb.) Reichb.	(Tuber) Gelatin, stilbenoids, blespirol, blestrianol, phenanthrene, bisphenanthrene[2,33-38]	Hemostatic, promotes leukocyte and platelet aggregation; treats hematuria, blood splitting, primary hepatic carcinoma; antimicrobial
Blumea aromatica DC *B. lacera* (Burm. f.) DC	(Whole plant) Flavonoids, β-sitosterol, volatile oil[3,435]	Antileukemia activity; arthritis, headache; externally for skin eczema
Blumea balsamifera (L.) DC var. *microcephala* Kitamura	(Leaf, shoot) Borneol, camphor, cineole, limonene, palmitic acid, myristic acid, sesquiterpene alcohol, dimethyl ether, limonene, flavonoids, pyrocatechic tannins[16,20,39]	Treats itch, sores, wounds, stomachache; sudorific, diaphoretic, anticatarrhal
Blumea laciniata (Roxb.) DC	(Whole plant) Carotene, vitamin C, cineole, citral, fenchone, camphor[8]	Insect repellant, insecticide, juice deobstruent, febrifuge, stimulant
Blumea lanceolaria (Roxb.) Druce.	(Whole plant)[1,20] Flavonoids, β-sitosterol, volatile oil[3]	For arthritis, malaria, influenza, beri-beri, headache, flu; antiswelling

Species	Constituents (Part)	Therapeutic Values
Blumea riparia (Blume) DC var. *megacephala* Randeria	(Leaf, root) Polysaccharide, protein[1,20]	Treats headache, alleviates colic; diuretic
Boehmeria densiflora Hook. et Arn.	(Leaf, root) β-Carotene, thiamin, lignin, riboflavin, niacin, ascorbic acid[16]	Astringent, antiabortifacient, drooling, demulcent, diuretic, resolvent, uterosedative, antihemorrhagic, styptic
Boehmeria nivea (L.) Gaud. var. *tenacissima* (Gaudich.)	(Root) Chlorogenic acid, flavonoids, 5-hydroxytryptamine[16,18,97]	Diuretic; treats stomachache, arthritis; antibiotic, antipyretic, antiinflammatory
Bombax malabarica DC	(Root, bark, flower, fruit) Naphthaquinone, lactone, 3-methy-1-5-isopropyl, 1-6-hydroxy-7-methoxynaphthalene, 1-8 lacone, lupeol, catechutannic acid, stearin, arabinose, galactose, protein, mucilage[4]	Cholera, pneumonia, pleurisy, neuralgia, leprosy, smallpox, bleeding gums, toothache, rheumatism
Bothriospermum tenellium (Hornemann) Fischer & Meyer	(Whole plant)[20] No information is available in the literature	Bleeding, cough, vomiting blood, liver diseases
Botrychium daucifolium (Wall.) Hook & Grev. *B. lanuginosum* (Wall.) Hook & Grev.	(Whole plant)[1,3] No information is available in the literature This herb is toxic[88]	Lymphatic gland inflammation; tonic; antitoxic; leaf used for eyewash
Bougainvillea spectabilis Willd.	(Root, flower, vine) D-Pinitol (3-O-methyl-chiroinositol), pinitol[3,91]	Antidiabetic; flower used as a tonic; vine used to treat hepatitis
Boussingaultia gracilis Miers var. *pseudobaselleoides* Bailey	(Leaf) Isoproterenol[154,159,435]	Antiherpes simplex viruses, antiadenoviruses activity; treats constipation; antiinflammatory
Bredia oldhamii Hooker *B. scandens* (Ito et Matsum.) Hayata	(Root, whole plant)[48] No information is available in the literature	Arthritis; alleviates pain, restores normal menstruation
Bredia rotundifolia Y. C. Liu et C. H. Ou	(Whole plant)[48] No information is available in the literature	Vomiting blood, bleeding stomach ulcer, cough, throat inflammation, arthritis

TABLE 1 Major Constituents and Therapeutic Values of Taiwanese Native Medicinal Plants (continued)

Scientific Name	Major Constituents and Source	Claimed Therapeutical Values
Breynia accrescens Hayata *B. fruitcosa* (L.) Hook f.	(Root) Phenolic compounds[8,21,49]	Antipyretic, antitoxic, antiswelling, antipruritic; fever, headache, hemorrhage, mumps, puerperium, stomachache; antiseptic for cuts and sores, bruises, syphilis, abscesses, suppurating sores, lactagogue
Breynia officinalis Hemsley	(Root, leaf) Terpenic glucosides (turpinionoside B and betulalbuside A), phenolic glycosides, megastigmane glucosides (breyniaionosides A-D), isorobustaside A, breyniosides A and B[6,55,56,160]	Antiinflammatory, toxic, treats cancer, improves blood circulation, treats contusions, heart failure, venereal diseases, growth retardation, conjunctivitis; it causes hepatocellular liver injury
Broussonetia kazinoki Kazinoki Sieh & Zucc.	(Root, bark, fruit) Kazinol B (isoprenylated flavan), pyrrolidine alkaloids, broussonetines R-T, V-X, M1, U1, J2–3[1,50,51,53,54]	A tonic for curing sexual impotence and insomnia and increasing vision; antitumor effect *in vivo*; kazinol B inhibits nitric oxide synthesis
Broussonetia papyrifera (L.) L'Herit.	(Leaf, root, bark, fruit) Papyriflavonol A, amylase, cerotic acid, chymase, lipase, protease[1,8,52]	Antiinflammatory; leaf sap for bug bite, eczema, epistaxis; for dysentery, gonorrhea; diuretic, invigorating, ophthalmic, stimulant; for stomachache, impotence, kidney trouble
Brucea javanica (L.) Merr.	(Fruit) Yatanoside, yatanine, bruceines, bruceolide, brusatol, oleic acid[2,10] This herb is toxic[88]	Treats amebic dysentery, malaria; antiamebial, anticancer, antiprotozoan
Bryophyllum pinnatum (Lam.) Kurz.	(Leaf) Bryophylline; citric, lactic, malic, and succinic acids; allelopathic compounds; *p*-coumaric-, cinnamic-, ferulic-, and caffeic acids; glucosides A, B, and C[8]	Treats intestinal troubles caused by bacteria; antiseptic, bactericidal, cicatrizant, diuretic, emollient, hemostat, soporific, vulneraria

Species	Constituents	Therapeutic Values
Buddleja asiatica Lour. *B. formosana* Hatushima	(Flower bud) Buddleoglycoside[2,18,39]	Improve visual acuity, prescribed as ophthalmic in nyctalopia, asthenopia, cataract
Bupleurum chinensis DC *B. falcatum* L.	(Root) Triterpenoid saponins, sapogenins, saikosaponins, bupleuran, L-arabinose, D-glucose, arabinan polymer[2,40,43,642]	Antitumor, antipyretic, treats chronic hepatitis, nephrosis, autoimmune diseases; antiulcer; treats inflammation of inner organs, immunopharmacological activities
Bupleurum kaoi Liu, Chao & Chuang *B. chinense* DC	(Root) Saponins such as saikogenin, bupleurumol, furfurol; lignoceric, linoleic, oleic, palmitic, and stearic acids[8]	For amenorrhea; analgesic, antipyretic, antitoxic; for catarrh, diarrhea, malaria, dysmenorrhea, dyspepsia, fever, hepatitis; hemostat, sedative; for cholecystitis, cough, gallstones, gastritis, hypertension, insomnia, nervousness, rhinitis, tuberculosis
Buxus microphylla Sieb et Zucc.	(Root) Cyclovirobuxine D, C, cycloprotobuxamine A, C, buxtamine, cyclohoreanine B, busamine E, buxpine, cyclovirobuxine D, buxtauine, cycloprotobuxamine A, C[27] This herb is toxic[88]	Treats heart conditions, alleviates pain, blood vomiting; a detoxicant
Caesalpinia pulcherrima (L.) Sw.	(Stem wood) Alkaloid, gallic acid, resins, tannins[1] This herb is toxic[88]	Febrifuge, for stomachache, diuretic, astringent, anticholeric
Callicarpa formosana Rolfe *C. japonica* Thunb.	(Flower, root) Volatile substances, gamma-caryophyllene, 1-octen-3-ol, 2-hexenal, germacrene B, aromadendrene II[6,161]	Diuretic; arthritis, nerve pain, gonorrhea; emmenagogue
Callicarpa longissima (Hemsl.) Merr. *C. loureiri* Hook. et Arn.	(Leaf) Flavonoids, tannins[49]	Hemoptysis, hematemesis, epistaxis, rheumatism, vomiting blood

TABLE 1 Major Constituents and Therapeutic Values of Taiwanese Native Medicinal Plants (continued)

Scientific Name	Major Constituents and Source	Claimed Therapeutical Values
Callicarpa nudiflora Hook. et Arn. *C. pedunculata* R. Br.	(Leaf, root) Tannins[2,49]	Treats suppurative skin infections and burns, uterine bleeding, upper respiratory tract infection, gastrointestinal bleeding, hemoptysis, epistaxis
Camellia japonica L. var. *hozanensis* (Hayata) Yamamoto	(Flower bud) Camelliagenins, D-catechol, L-epicatechol, leucoanthocyanin, arabinose, camellin, rhamnose, theasaponin[8,18]	For hemoptysis, epistaxis, gastrointestinal hemorrhage, metrorrhagia
Camellia oleifera Abel. *C. sinensis* L.	(Leaf) Caffeine, theophylline, tannic acid, theobromine, xanthine, polyphenols[2,66,67,68]	Diuretic, increases renal blood flow, stimulates central nervous system, antitumor, prevents lung cancer
Camptotheca acuminata Decne	(Fruit) Camptothecine, venoterpine, hydroxyleamptothecin, methoxyl-camptothecin, irinotecan, 10-hydroxycamptothecin[2,62-64] This herb is toxic[88]	Treats breast cancer, carcinoma of the stomach, rectum, colon, and bladder, chronic leukemia
Canavalia ensiformis (L.) DC	(Seed) Canavaline, canavanine, urease, gibberelin A_2, gibberelin A_{22}, canavalia gibberelin I-II, canavalia[2]	A tonic, bactericidal, fungicidal; treats stomachache
Canna flaccida Salisb. *C. indica* L.	(Root) Molluscacides, zingiberales, acetylcholinesterase, adenosine triphosphatase[20,49,162]	Stops bleeding, adjusts monthly period, diuretic, treats acute hepatitis
Capsella bursa-pastoris (L.) Medic.	(Whole plant) Bursic acid, alkaloids, vitamin A, choline, citric acid[2]	Hemostatic, anthypertensive; chyluria, nephritis, edema, hematuria
Capsicum frutescens L.	(Fruit) Capscin, capsaicin, solanine[4]	Carminative, digestive, hypnotic, stomachic, for cholera, diarrhea, dysentery

Cardiospermum halicacabum L.	(Whole plant) Eicosenoic acid, 1-cyano-2-hydroxy methylprop-2-ene-1-ol, 1-cyano-2-hydroxy methylprop-1-ene-3-ol[20] This herb is toxic[88]	Diuretic, antitoxic; pneumonia, diabetes
Carex baccans Nees	(Root, fruit)[154] No information is available in the literature	Root for stopping menses bleeding, fruit to treat cough and as a diuretic
Carpesium divaricatum Sieb. et Zucc.	(Whole plant) Sesquiterpene lactone, thymol derivatives[21,163,164]	Antitoxic; treats fever, headache, throat inflammation, antipyretic, analgesic, vermifugic, antiinflammatory
Carthamus tinctorius L.	(Flower) Cartharmin, neocarthamin, safflower yellow, quinochalone, safflomin[2]	Promotes blood circulation, removes blood stasis, restores normal menstruation
Caryopteris incana (Thunb.) Miq.	(Whole plant) Flavonoids, phenolic compounds, amino acids[11]	Chronic bronchitis, whooping cough, rheumatic arthralgia, gastroenteritis, dysmenorrhea
Casearia membranacea Hance	(Leaf, twig) Clenodane diterpenoids[574]	Cytotoxic
Cassia fistula L.	(Seed) Chrysophanein, 5-(2-hydroxyphenoxymethyl)furfural, (2'S)-7-hydroxy-5- hydroxymethyl-2-(2'-hydroxypropyl)chromone, benzyl 2-hydroxy-3,6-dimethoxybenzoate, benzyl 2beta-O-D-glucopyranosyl-3,6-dimethoxybenzoate, 5-hydroxymethylfurfural, (2'S)-7-hydroxy-2-(2'-hydroxypropyl)-5-methylchromone, chrysophanol[591]	Treats hypertension, hepatitis, swelling liver, eye infections
Cassia mimosoides L.	(Whole plant) Protein, fatty acids, tannin, aloe-emodin, emodin[20]	Improves liver, stomach, and kidney functions, inflammation; diuretic

TABLE 1 Major Constituents and Therapeutic Values of Taiwanese Native Medicinal Plants (continued)

Scientific Name	Major Constituents and Source	Claimed Therapeutical Values
Cassia occidentalis L. *C. torosa* Lloydia	(Seed, root) Anthraquinones, torosachrysone, *n*-methylmorpholine, apigenin, galactomannan, cassiollin, xanthoria, dianthronic, heteroside, helminthosporin[2,69] This herb is toxic[88]	Mild purgative, lowers blood pressure, antioxidative, antiasthmatic, antitoxic, antimalarial, antibacterial; hepatoprotective activities
Cassia tora (L.) Roxb.	(Seed) Anthraquinones such as emodin, chrysophanol, physcion, rhein, aurantio-obtusin, obtusifolin, chryso-obtusin, naphthopyrones, obtusin, rubrofusarin, nor-rubrofusarin, toralacton[2,70,71]	Purgative, treats ophthalmia, hypercholesterolemia, vaginitis
Catharanthus rosens (L.) G. Don	(Whole plant) Vinblastine, vincristine, carosine, vinrosidine, lenrosine, lenrosivine, rovidine, pervine, perividine, vindolinine, pericalline[2] This herb is toxic[88]	Anticancer in chronic lymphocytic leukemia, Hodgkin's disease, and acute lymphocytic leukemia
Cayratia japonica (Thunb.) Gagnep.	(Whole plant, root) Araban, nitre, potassium nitric acid[6]	Antitoxic; treats inflammation, arthritis, bloody urine, hepatitis
Celastrus kusanoi Hayata *C. hypoleucus* (Oliv.) Warb.	(Leaf, stem, fruit) Pristimerin, celastrol, sesquiterpene esters[154,505,506]	Improves blood circulation, antiinflammatory, alleviates pain
Celastrus orbiculatus Thunb. *C. punctatus* Thunb. *C. paniculatus* Wild.	(Whole plant) Kaempferitrin[11,13,507]	Treats rheumatic arthritis, traumatic injury; intestinal relaxant effect; neurasthenia, palpitation, insomnia
Celosia argentea L.	(Whole plant, seed) Celosiaol, nicotinic acid[6,11,166]	Treats acute conjunctivitis, chronic uveitis, hypertension; improves wound healing

Celosia cristata L.	(Whole plant, seed) Protein, glycoproteins, asparagine, asparagine-linked glycon[20,165,167]	Vomiting with blood, dysentery; antiviral; treats sores, ulcers, skin eruptions; uterotonic action
Centella asiatica (L.) Urban	(Leaf) Glucoside, asiaticoside, sitosterol, tannin, hydrocotyin, vallarine, pectic acid[4]	Leprosy, epilepsy
Cephalotaxus wilsonianer Hayata	(Shoot, stem) Cephalotaxine, tetraflavonoid, cephalotaxinone, alkaloids, acetycophalotaxine, wilsonine, demethylcephalotaxine, epicephalotaxin, harringtonine, hormoharringtonine, c-3epi-wilsonine, biflavone, hydroxyeephalotaxine, isoharringtonine[29,571,619] This herb is toxic[88]	Cytotoxic, antitumor, anticancer; treats lymphatic gland swelling, improves digestion, an insecticide
Ceratopteris thalictroides (L.) Brongn.	(Whole plant) DL-(3-14C) Cysteine, organoids, pepsin A, pronase, cysteine, trypsin[6,508]	Improves blood circulation, antitoxic; treats cough, throat infection
Chaenomeles japonica (Thunb.) Lind.	(Fruit) Vitamin C, malic acid, tartaric acid, citric acid, hydroxyanic acid[18]	Treats arthralgia, diarrhea, cholera, gout, arthritis
Chamaecyparis formosensis Matsumura C. *obtusa* Sieb. & Zucc. var. *filicoides* Bei C. *obtusa* Sieb. & Zucc. var. *formosana* Bei	(Wood, root) Turpentine, benihiol, benihinol, benihinal, D-α-pinene, pinene, cadinane-type sesquiterpenes, camphene, α-terpineol, allylpyrocatchin, 1-α-pinene[27,601]	Controls infection, bacterial; treats fever; diuretic; alleviates gasp, headache
Chamaesyce hirta (L.) Millsp. C. *thymifolia* (L.) Millsp.	(Stem, leaf) Flavonoids[3]	Dysentery, intestinal infection, diarrhea, hemorrhoidal bleeding

TABLE 1 Major Constituents and Therapeutic Values of Taiwanese Native Medicinal Plants (continued)

Scientific Name	Major Constituents and Source	Claimed Therapeutical Values
Chenopodium album L.	(Whole plant) Palmitic acid, carnaubic acid, oleic acid, linoleic acid, nonacosane, oleyl alcohol, sitosterol, betaine, amino acids, sterol, ferulic acid, vanillic acid, aleanolic acid, L-1-leucine, ferulic acid, vanillic acid[28]	Lowers blood pressure, improves heart function; treats diarrhea, fever, dysentery, skin infection
Chenopodium ambrosioides L.	(Leaf) Volatile oil, ascaridol, geraniol, saponin, L-limonene, *p*-cymene, D-camphor, kaempferol-7 shannoside, ambroide[1,4]	An anthelmintic to treat ascaris, ancylostomiasis; vermifuge, carminative
Chichorium endivia L.	(Flower, bud)[1,54] No information is available in the literature	Antiinfection, eye discomfort
Chloranthus oldham Solms.	(Leaf, stem) Essential oils, flavonoids, pelargonidin-3-rhannosylglucoside[6] This herb is toxic[88]	Treats bone fractures, vomiting, contusions, lung infection; an astringent; antitumor; improves immune system; alleviates arthritis pain
Chloranthus spicatus (Thunb.) Mak.	(Whole plant) Volatile oil[21]	Relaxes muscles; controls pain, bleeding, arthritis
Chlorophytum comosum (Thunb.) Baker	(Leaf) Flavonoids, pelargonidin-3-rhannosylglucoside[6]	Inflammation; improves blood circulation; treats pneumonia, enteritis, cancer
Chrysanthemum indicum L.	(Flower) Carvone, cineol, camphor, borneol, yejuhualactone, chrysanthinin, limonene, chrysanthemaxanthin, α-pinene[2]	Antibacterial; alleviates headache, insomnia, and dizziness due to high blood pressure
Chrysanthemum morifolium Ramat.	(Flower) Borneol, chrysanthemin, camphor, stachydrine, choline, acacetin-7-rhamnoglucoside, cosmosiin, acacetin-7-glucoside, diosmetin-7-glucoside, adenine[2]	Antipyretic, antitoxin; remedy for common cold, headache, dizziness, red eye, swelling, hypertension

Species	Constituents (part)	Therapeutic values
Chrysanthemum segetum L.	(Whole plant) Borneol, camphor, adenine, chrysanthenone, stachydrine, monobornylphthalate, chrysanthenol, bornyl acetate, flavonoids, luteolin-7-glucoside, cosmosiin, choline, chrysanthemin[3,28]	Diuretic, a tonic
Cibotium barometz (L.) J. Sm.	(Root) Palmitic acid, linoleic acid[8,72]	Tonic, digestive, laxative, analgesic in rheumatism, lumbago, myospasm
Cibotium cumingii Kunze	(Root) Phenolic compounds, starch, tannins[11]	Cold-caused bone pain, lumbago, sore extremities, hemiplegia, leukorrhea
Cichorum endivia L.	(Root) Flavonoids, kaempferols, sesquiterpene lactones[168-170]	Phytoprotection, prevention of UVB-induced erythema, pyrimidine dimer formation, IL-6 expression
Cinnamomum cassia Presl.	(Root) Cinnamic aldehyde, cinnamyl acetate, cinnamic acid, eugenol, phellandrene, coumarin, phenylpropyl alcohol, orthomethylcoumaric aldehyde[2,11,18,73,74]	Treats stomachache, diarrhea, cough, wheezing
Cinnamomum camphora (L.) Presl.	(Root, branch, leaf) D-Camphor, eucalyptole, cineole, pinene, camphene, aromadendrene, azulenen, α-camphorene, laurolitsine, limonene, safrole, terpineol, carvacrol, eugenol, cadinone, bisabolene, cumaldehyde, pinocarveol, L-acetyl-4-isopropylidenecyclopentene[2,6,11,39]	Treats stomachache and distention, rheumatic bone pain, flu; alleviates pain
Cinnamomum insulari-montanum Hayata *C. kotoense* Kaneh. & Saski *C. micranthum* Hayata	(Leaf, stem) Polysaccharides, dehydrosulfurenic acid, 15α-acetyl-dehydrosulfurenic acid[3,154,171]	Used to treat headache by drinking the sap or crushing the leaves and putting on the forehead; improves blood circulation; antiinflammatory
Cirsium albescens Kitamura	(Whole plant) Essential oil, rutin, acacetin-7-rhomnoglucoside, protocatechuic acid, caffeic acid, chlorogenic acid[16,18]	Hemostat, diuretic; stops bleeding; used externally for wound infections

TABLE 1 Major Constituents and Therapeutic Values of Taiwanese Native Medicinal Plants (continued)

Scientific Name	Major Constituents and Source	Claimed Therapeutical Values
Cirsium japonicum DC *C. japonicum* DC var. *australe* Kitam.	(Leaf, stem) α-Amyrin, β-amyrin, β-sitosterol, stigmasterol, taraxsteryl acetate, inulin, labenzyme, pectolinarin[1,8]	Hemostat, diuretic; treats intestinal ulcer bleeding
Cissus repens Lam. *C. sicyoides* L.	(Root) Steroidal sapogenins, coumarin glycoside, hecogenin, diosgenin[20,509,510]	Antitoxic, alleviates inflammation, cystitis, lymphatic gland infection
Citrus maxima (Burm. f.) Merr. *C. sinensis* (L.) Osbeck var. *sekken* Hayata	(Leaf) Pyranoanthocyanins, nobiletin, flavonoids, anthocyanins[29,172,173]	Used to treat headache by putting the leaves on the forehead; treats coronary heart disease, cholesterol concentrations, prevents atherosclerosis
Citrus medica L. var. *gaoganensis* (Hayata) Tanaka	(Fruit) Pinene, α-lemonene, citropten, limettin, diosmin, hesperidin[29]	Alleviates pain, stomachache, arthritis, headache
Citrus medica L. var. *sarcodactylis* Swingle	(Fruit) Volatile oil[97]	Abdominal distention, gastric pain, anorexia, vomiting, productive cough
Citrus tangerina Hort. ex Tanaka	(Leaf) D-Limonene, citral, nobiletin, hesperindin, vitamin B_1[3]	Headache; adjusts blood pressure, improves blood circulation, lowers cholesterol
Claoxylon polot (Burm. f.) Merr.	(Root)[49,174,677] Triterpenoids	Rheumatic arthritis, lumbago, beri-beri
Clausena lansium (Lour.) Skeels.	(Leaf) Flavonoids, amino acids[29]	Alleviates pain, inflammation, stomachache; diuretic; treats flu, windpipe infection
Clausena excavata Burm. f.	(Root, leaf) Limonoid, coumarins, carbazoles, claulactones A-J, clauszoline M. umbelliferone, clauseactones A-D[20,175-177]	Alleviates pain, antitoxic; treats arthritis, stomach; antimycobacteria, antifungal, antinociceptive; immunomodulatory activities

Species	Constituents (Plant part)	Therapeutic values
Cleistocalyx operculatus (Roxb.) Merr. et Perry	(Flower bud, root) Flavonoids, phenolic compounds, amino acids[10]	Alleviates colds, fever, summer heat; treats indigestion, acute gastroenteritis, bacillary dysentery, hepatitis
Clematis chinensis Osbeck *C. florida* Thunb.	(Root) Anemonin, anemonol, saponins, ckenaogebik A, dihydro-4-hydroxy-5-hyroxymethy-2(3H)-furanone[2,18,75,178,179,641]	Hepatic protective, analgesic, diuretic, carminative, antitumor, antiinflammatory; treats arthritis, backache, headache
Clematis gouriana Roxb. ex DC subsp. *lishanensis* Yang & Huang	(Stem, leaf) Protoanemonin[3]	Antiinfection, alleviates pain, inflammation, skin disorders; antitoxic; resolves extravasate blood, rheumatic pain
Clematis grata Wall.	(Root) Sitosterol, iresenin, hederagenin, fatty acids, protoanemonin, anemonin[29,48]	Rheumatism, numbness of extremities, rigidity in joints
Clematis henryi Oliv.	(Whole plant, root)[29] No information is available in the literature	Alleviates pain, antioxic; used for headache, muscle pain, arthritis
Clematis lasiandra Maxim	(Vine)[29] No information is available in the literature	Improves blood circulation; diuretic; alleviates pain
Clematis montana Buch.	(Stem, root) Clemontanos-C, saponin[6,180,181]	Urinary pain, menstruation disorder; diuretic; improves blood circulation
Cleome gynandra L.	(Seed) Cleomin, lactone, tannins, volatile oils[8]	Treats dysentery, gonorrhea, malaria, rheumatoid arthritis
Clerodendrum calamitosum L.	(Twig, leaf) Pheophorbide-related compounds, methyl ester, (10S)-hydroxypheophytin[28,182]	Diuretic, lowers blood pressure, alleviates kidney stones, treats lung carcinoma, breast adenocarcinoma, malignant melanoma, ovarian and kidney carcinoma

TABLE 1 Major Constituents and Therapeutic Values of Taiwanese Native Medicinal Plants (continued)

Scientific Name	Major Constituents and Source	Claimed Therapeutical Values
Clerodendrum cyrtophyllum Turcz.	(Leaf, root) Indirubin, ingigo, tryptanthrin, isatan B, glucobrassicin, 3-indolylmethylgluco-sinolate, neoglucobrassicin, isoindigo, indican, lacerol, pheophorbide-related compounds[39,182,608]	Antipyretic, detoxicant, diuretic, preventative for epidemic meningitis, cytotoxic
Clerodendrum inerme (L.) Gaertn.	(Leaf) (24S)-Ethylcholesta-5,22,25,3β-o1 apigenin-7-0-glucurondes, scutellarein-7-0-glucuronides, 4'-methyl scutellarein, pectolinarigenin, 3-epicaryoptin[27]	Antitoxic; used for inflammation, arthritis, flu, stomachache, hepatitis, malaria
Clerodendrum japonicum (Thunb.) Sweet *C. kaempferi* (Jacq.) Siebold ex Stend.	(Root, leaf) Flavonoid, *iso*-prenepolymer, galactitol, stigmasterol, cyrtophyllin, melissyl alcohol, γ-sitosterol, *n*-pentacosane, clerosterol, picein, friedelin, epifriedelinol, clerodendrin[11,27]	Rheumatic bone pain, lumbago, pulmonary tuberculosis with cough, hemoptysis, hemorrhoids, dysentery
Clerodendrum paniculatum L.	(Root) β-Epimer poriferasterol, α-epimer stigmasterol[29]	For gonorrhea, skin diseases; diuretic, regulates menses
Clerodendrum petasites (Lour.) Moore	(Root) Flavonoid hispidulin, bronchodilator flavonoid[11,183,184]	Hepatitis, cough, colds, fever, pulmonary tuberculosis, hemoptysis, bacillary dysentery
Clerodendrum philippinum Schauer	(Root, leaf)[12] Flavonoids, phenolics, tannins	Rheumatic arthritis, lumbago, beri-beri, edema, leukorrhea, bronchitis, hemorrhoids, prolapsed rectum, scrofula, chronic ostemyelitis
Clerodendrum trichotomum Thunb. *C. trichotomum* var. *fargesii* (Dode) Rehder	(Leaf, stem, root) Glycosides, clerodendrin, acacetin-7-glucurono-(1,2)-glucuronide, mesoinositol, clerodolone, apigenin-7-diglucuronide, friedelin, epifriedelin[2,16,76]	Treats hypertension, arthritis pain, malaria, diarrhea; externally for skin eczema, infection, hemorrhoids
Cleyera japonica Thunb.	(Whole plant)[583] No information is available in the literature	Free radical–scavenging activity

Species	Constituents	Therapeutic Values
Clinopodium laxiflorum (Hayata) Matsum	(Whole plant) α-Spinasterol, stigmast-7-en-3β-ol, oleanolic acid, fatty acid, ursolic acid, isosakurannetin, 4-hydroxybenzaldehyde, 3,4-dihydroxybenzoic acid, caffeic acid, rosmarinic acid, narirutin, epi-narirutin, nesperidin, besperidin, betula acid[28]	Controls bleeding, antitoxic, antiinflammatory; blood in urine, uterine bleeding, flu, headache
Clinopodium umbrosum (Bieb.) C. Koch	(Whole plant) Dydimin, hesperidin, siosakuranetin, apigemin, ursolic acid[16]	Hemostatic, stimulates uterine contractions, antibacterial
Cocculus orbiculata (L.) DC	(Root)[3] Trilobine, isotrilobine, homotrilobine, trilobamine, normenisarine, epistephanine[3]	Abdominal pain
Cocculus sarmentosus (Klour.) Diels.	(Root, stem) Menisarine, trilobine, *iso*-trilobine[27]	Antinfection, alleviates pain, lowers blood pressure
Cocculus trilobus (Thunb.) DC	(Root) Trilobine, *iso*-trilobine[12]	Rheumatic arthralgia, gastric pain, urinary tract infection, dysmenorrhea, sore throat, nephritis
Codonopsis kawakami Hayata	(Root) Saponin[28]	A tonic, relieves cough, improves spleen and stomach function
Coix lacryma-jobi L.	(Seed, root) Coixenolide, coixol, protein, myristic acid, palmitic acid, stearic acid, oleic acid, linoleic acid, polysaccharides, triglycerides, phospholipids, benzoxaxinones, adenosine, benzoxazinones[8,18,77-82]	For intestinal or lung cancers and warts; antitumor, antirheumatic, diuretic, refrigerant
Coleus scutellarioides (L.) Benth. var. *crispipilus* (Merr.) Keng C. *parvifolius* Benth.	(Whole plant) Luteolin 5-0-β-D-glucopyranoside, luteolin, luteolin 7-methyl ether, luteolin 5-0-β-D-glucuronide, rosmarinic acid, daucosterol, β-amyrin[21,511]	Antitoxic; for cough, eye infection, hepatitis; improves digestion; inhibitory activities against HIV-1

TABLE 1 Major Constituents and Therapeutic Values of Taiwanese Native Medicinal Plants (continued)

Scientific Name	Major Constituents and Source	Claimed Therapeutical Values
Colocasia antiquorum Schott var. *illustris* Engler C. *esculenta* (L.) Schott	(Rhizome, leaf) Digalactosyl, monogalactocyl diacylglycerols, starch 70%, protein, fatty acids, vitamins B_1, B_2 [13,656] This herb is toxic [88]	Antihyperlipemia activity; rhizome: scrofula, furuncles, carbuncle; leaf stalk: urticaria, diarrhea, ulcer
Colocasia formosana Hayata	(Leaf) [3] No information is available in the literature	Headache, carbuncle, inflammation
Commelina benghalensis L. C. *communis* L.	(Aerial part) Awobanin, commeinin, delphin, delphinidin, flavocommelitin [2,10,20]	Antibacterial, antipyretic, diuretic, antiedematic, antitoxic; for epidemic influenza, upper respiratory tract infection
Conyza canadensis (L.) Crong C. *dioscoridis* Desf.	(Aerial part) Essential oils, matricaria ester, dehydromatricaria ester, sterols, linoleyl acetate, limonene, linalool, dephynyl methane-2-carboxylic acid, cumulene, *O*-benzoylbenzoic acid, triterpenes, tannins, flavonoids, glycosides [16,512]	Alleviates swelling, itchiness; treats intestine and liver infection; a detoxicant; externally for skin eczema, wounds, pain caused by arthritis, toothache
Conyza sumatrensis (Retz.) Walker C. *blinii* L.	(Aerial part) Sphingolipid, friedelinol, *n*-triacontanol, daucosterol, triterpenoid daponins, conyzasaponins I-Q [185-187,513,654]	Antinflammatory, acetic acid-induced abdominal contractions, alleviates formalin-induced pain
Coptis chinensis Franch	(Root) Berberine, coptisine, urbenine, worenine, palmatine, jatrorhizine, columbamine, lumicaerulic acid, berberine hydrochloride [1,2,83] This herb is toxic [88]	Antiarrhythmic, antibacterial, antiviral, antiprotozoal, anticerebral ischemic

Corchorus aestuand L.	(Whole plant) Quercetin[20]	Antitoxic, stops bleeding, measles
Corchorus capsularis L. *C. olitorius* L.	(Leaf, flower) Glycosides, capsularin, corchorin, corchoritin, aglycone, strophanthidin, digitoxigenin, coroloside, glycovatromonoside, oleic acid, erysimoside, olitoriside, linoleic acid, corchoroside, helveticoside, corchotoxin, palmitic acid, stearic acid[1, 84-86] This herb is toxic[88]	Treats dysentery, consumptive cough, epistaxis, bladder diseases; inhibitory effect on lipopolysaccharide-induced NO production in cultured mouse peritoneal macrophage
Cordyline fruticosa (L.) Goeppert	(Leaf) Phenolic compounds, amino acids[29]	Stops bleeding, vomiting with blood, blood in urine, cough, stomachache
Coriandrum sativum L.	(Leaf) Acetone, borneol, coriandrol, cymene, decanal, decanol, decylic aldehyde, dipentene, geraniol, rutin, limonene, linalool, malic acid, nonanal, oxalic acid, phellandrene, tannic acid, terpinene, terpinolen, umbelliferone, scopoletun, coumarins, quercetin, kaempferol, aflatoxins[4,8]	Eruptions of pox and measles
Coriaria japonica A. Gray ssp. *intermedia* (Matsum.) Huang & Huang *C. intermedia* Matsumura	(Leaf) Coriamyrtin, coriose, tutin[29] This herb is toxic[88]	Antiinflammatory: alleviates pain, uterine cancer
Corydalis pallida (Thunb.) Pers.	(Whole plant) Alkaloids, long-chain carboxylic acid[188,568] This herb is toxic[88]	Antibacterial activity
Costus speciosus (Koen.) Sm.	(Whole plant) Diogenin, tigogenin, corticosteroids, 3-(4-hydroxyphenyl)-2 (E)-propenoate[4,8,87]	For fever, anasarca, asthma, bronchitis, cholera; antifungal

TABLE 1 Major Constituents and Therapeutic Values of Taiwanese Native Medicinal Plants (continued)

Scientific Name	Major Constituents and Source	Claimed Therapeutic Values
Crassocephalum crepidioides (Benth.) S. Moore	(Whole plant) Dihydroisocoumarin, carrageenan[21,514,515]	Antiinflammatory; antimalaria, antitoxic, improves stool movement, flu, dysentery
Crateva adansonii DC subsp. *formosensis* Jacobs *C. nurvala* Ham.	(Root bark) Triterpenes[6,49,516]	Antitoxic, flu, hepatitis, malaria, diarrhea, rheumatic activities
Cratoxylon ligustrinum (Spach.) Blume.	(Root, bark, leaf) Flavonoids, phenolics, amino acids[49]	Prevention of heatstroke and dysentery; colds, fever, enteritis, diarrhea, cough, hoarseness
Crawfurdia fasciculata Wall.	(Whole plant)[11] No information is available in the literature	Pulmonary abscess, nephritis, urinary tract infection, high fever in children, bronchitis, pulmonary tuberculosis with hemoptysis, pneumonitis
Crossostephium chinense (L.) Makino	(Whole plant) Taraxerol, taraverone, taraxeryl acetate[6]	Antitoxic, arthritis bone and joint pain, flu, cough, windpipe infection
Crotalaria albida Roth	(Seed) Croalbidine, monocrotaline[13,28] This herb is toxic[88]	Treats urinary tract infection; diuretic; for bronchitis, wheezing; antiinflammatory, anticancer
Crotalaria pallida Ait.	(Seed) Mucronatine, usaramine, nilgirine, crotastriatine, β-sitosterol, luteolin, vitexin, vitexin-*o*-syloside, pterocarpanoid[28,586]	Liver diseases, liver cancer; antitoxic, antiinflammatory; for dysentery, mammary gland infection

Crotalaria sessiliflora L.	(Whole plant) Monocrotalines, relroncine, platynecic acid, flavonoids, tetrahydraxyflavone, trihydroxyisoflavone, dihydroxy flavone, isovitexin[2,28,348,517] This herb is toxic[88]	Anticancer; leukemia, uterine cancer, skin cancer; antimutagenic activity
Crotalaria similis Hemsl.	(Whole plant)[28] No information is available in the literature	Flu, fever, cough, hepatitis, gastritis, enteritis
Croton crassifolius Geisel	(Whole plant) Amino acids[11] This herb is toxic[88]	Gastric, duodenal ulcer; chronic hepatitis, rheumatic arthralgia, hernial pain
Croton lachnocarpus Benth.	(Root, leaf) Phenolic compounds[49]	Rheumatic arthritis, antiinflammatory; for pruritus, pyodermas, ringworm
Croton tiglium L.	(Seed) Croton oil, crotonic acid, tiglic acid, crotin, crotonoside, phorbol diester, croton resin, phorbol[2,14,20] This herb is toxic[88]	Anticancer, antiinflammatory; for diarrhea, purgative, wound healing property
Cryptocarya chinensis (Hance) Hemsl.	(Wood) Alkaloids, isoquinoline alkaloids, β-phenylethylamines[599,600]	Electrophysiological effect and antiarrhythmic activity, against ischemia/reperfusion arrhythmia
Cryptotaenia canadensis (L.) DC *C. japonica* Hassk	(Whole plant) Cryptotaenen, kiganen, kiganol, petroselic acid, isomesityl oxide, mesityl oxide, methyl isobutyl ketone, trans-β-ocimene, terpinolene[6,8,16]	For diarrhea, dysmenorrhea, rheumatism, tubercular glands, antiinflammatory, pneumonia

TABLE 1 Major Constituents and Therapeutic Values of Taiwanese Native Medicinal Plants (continued)

Scientific Name	Major Constituents and Source	Claimed Therapeutic Values
Cucumis melo L. subsp. *melo*	(Pedicel) Melotoxin, cucurbitacin B, cucurbitacin E, sterol[2,92] This herb is toxic[88]	Induces vomiting for drug intoxication, treats toxic and chronic hepatitis and cirrhosis of the liver
Cucurbita moschata Duchesne ex Poir.	(Seed, fruit) Cucurbitine, sterol, amino acids, rarotenoids, vitamins[2,29,92,93]	Treats taeniasis, antiinflammatory, stops pain, diuretic
Cudrania cochinchinensis (Lour.)	(Whole plant, root) Cudraxanthone S, B, toxyloxanthone C, wighteone, benzophenones[154,189,190]	Treats wounds, antiinflammatory, antilipid peroxidation, against *Candida*, *Cryptococcus*, *Aspergillus* species; hepatoprotective effects
Cunninghamia korishii Hayata	(Wood, whole plant) Diterpenoids[623]	Chronic bronchitis, epigastric pain, rheumatic arthralgia, impotence, nocturnal ejaculation
Curculigo capitulata (Lour.) O. Kuntze	(Rhizome) Calcium oxalate, resin, tannins[8,20]	Improves immunity, stimulates endocrine system, antiinflammatory, for arthritis
Curculigo orchioides Gaertn.	(Root, stem) Tannin, fatty acids, resin[20]	Tonic, improves immune system, aphrodisiac
Curcuma domestica Valet	(Tuber) L–Curcamene, sequiterpene, camphor, camphene, curmarin, curzerenone, curzenene, curcumol, zederone, furanodienone, furanodiene, diol, curcolone, procurcumenol, curdione, curcumin[2,10,94,95]	Anticancer, antiinflammatory, antitumor, antiinfectious properties, antioxidative activity, activates blood flow, removes blood stasis, hematemesis, infectious hepatitis
Curcuma longa L.	(Whole plant) Curcumin, curcumol, tumerone, phellandrene, cineole, sabinene, borneol, zingiberen, flavonoids[13,20]	Anticancer; treats numbness of arm and shoulder, dysmenorrhea, amenorrhea; antitoxic in liver, alleviates pain

Curcuma zedoaria (Berg.) Rose	(Rhizome) Curzerenone, curzerene, zederone, zerumbone, furanodiene, curdione, furanodienone, curculone, curcumin, diol, procurcumenol, sesquiterpene alcohols, turmerone, zingiberene, 3-4-hydroxyphenyl-2 (E)-propenoate, α-curcumene, curcumol, curcumenal, isocurcumenol, stigmasterol[2,87,96,348] This herb is toxic[88]	Inhibits mutagenesis and tumor promotion, antiinflammatory, antitumor, antiinfections, antifungal, anti-HIV, antimutagenic activity
Cyathea lepifera (Hook.) Copel. C. *podophylla* L.	(Stem) Dryocrassy formate, sitostanyl formate, 12α-hydroxyfern-9(11)-ene[20,191]	Antitoxic, antiinflammatory; abdominal pain; stops bleeding
Cyathula prostrata (L.) Blume	(Leaf, root) Ecdysterone, cyaterone[3,8,27]	Laxative, dysentery, antitoxic, alleviates pain, flu, cough, rheumatism, syphilis
Cycas revoluta Thunb.	(Stem, leaf, fruit) Cycasin, neocycasin A-G, β-carotene, gryptoxanthine, zeaxanthine, diazomethane, sotelsulflavone, hinokiflavone, amentoflavone[2,29] This herb is toxic[88]	Promotes blood circulation, anticancer, antiinflammatory, antitoxic, treats cough
Cyclea insularis (Makino) Hatusima C. *barbata* (Wall.) Miers	(Leaf, root) 1-Curine, dimethyl, dimethiodide[3,8]	Alleviates pain, arthritis joint pain, abdominal and stomach pain
Cyclobalanopsis stenophylla (Makino) Liao.	(Whole plant)[583] No information is available in the literature.	Free radical–scavenging activity

TABLE 1 Major Constituents and Therapeutic Values of Taiwanese Native Medicinal Plants (continued)

Scientific Name	Major Constituents and Source	Claimed Therapeutical Values
Cymbopogon citratus (DC) Stapf.	(Leaf, root) Elemicin, cymbopogonol, citral, dipentene, methylheptenone, β-dihydropseudoionone, linalool, methylheptenol, α-terpineol, myrceus, 1-borneol, 1,8-*p*-menthadien-5-ol, geraniol, nerol, farnesol, caprylic, citrogellol, citronellal, decanal, farnesal, isovaleric acid, geranic acid, citronellic acid[1,2,78]	Treats blood in the urine, fever; antiseptic, preservative, antiinflammatory; for stomachache, windpipe infection
Cymbopogon nardus (L.) Rendle	(Whole plant) Piperitone, citronellal, citronellol, geraniol, terpenen, camphane, dipentene, limonene, methylheptenone, borneol, linalool, nerol, eugenol, chavicol, γ-cadinene, citral, myrcene, dipentene, citronellic acid[2,27,96]	Antagonizes muscle contraction; antitussive, antibacterial; helps digestion, stops vomiting, flu, windpipe infection, cough
Cynanchum paniculatum (Bunge) Kitagawa	(Root) Paeonol, paeonin, tomentogenin, deacylcyanchogenin, sarcostin, deacylmetaplexlgenin[2,8,97]	Sedative, analgesic; effect on the cardiovascular system, lowers plasma cholesterol level; for insufficient lactation, neurasthenia, chronic nephritis, pulmonary tuberculosis
Cyperus alternifolius L.	(Aerial part)[8,29] No information is available in the literature	Antiinflammatory, flu, diaphoretic, diuretic, emmenagogue, litholytic, sedative, stomachic, vermifuge; treats cervical cancer
Daemonorops margaritae (Hance) Beccari	(Aerial part) Dracoalban, dracoresene, dracoresinotannol, benzolacetic ester[1]	Astringent
Dalbergia odoriferer T. Chen.	(Wood) Flavonoids[626]	Antiallergic, antiinflammatory

Species	Constituents (part)	Therapeutic values
Damnacanthus indicus Gaertn. f.	(Root, stem) Rubiadin, rubiadin-1-methyl ether, physcie, 1,4-dihydroxy-2-methylanthraquinone, 1-hydroxyanthraquinone, 1-hydroxy-2-methylanthraquinone, 1,6-dihydroxy-2,4-dimethoxyanthraquinone[154,192]	Treats rheumatism, recover from tiresome, antiinflammatory
Daphne arisanensis Hayata; *D. odora* Thunb.	(Root, stem) Flavan, 5,7,4'-trihydroxy-8-ethoxycarbonyl flavan[28,518]	Alleviates pain; antitoxic, antiinflammatory; arthritis pain, headache
Daphniphyllum calycinum Benth.	(Root, leaves) Calycine, glaucescine[11]	Treat colds, fever, tonsillitis, rheumatic arthralgia
Daphniphyllum glaucescens Blume spp. *oldhamii* (Hemsl.) Huang	(Leaf) Daphniglaucins A-B, polycyclic quaternary alkaloids[3,193]	Antitoxic; improves blood circulation; flu, fever, tonsil infection
Datura metel L.; *D. metel* L. f. *fastuosa* (L.) Degener; *D. tatula* L.	(Leaf, seed, flower) Scopolamine, hyoscyamine, daturodiol, daturolone, hyoscine[2,4,14]; This herb is toxic[88]	Spasmolytic, analgesic, antiasthmatic, antirheumatic agent; a general anesthetic for major operations
Davallia mariesii Moore ex Bak.	(Root) Hesperidin, starch, glucosides[3]	Improves kidney function, blood circulation
Debregeasia edulis (Sieb. et Zucc.) Wedd.; *D. salicifolia* L.	(Root, twig) Triterpene, pomolic acid, uvaol, ursolic acid, pomolic acid methyl ester, tormentic acid[6,194]	Stops bleeding, improves blood circulation; arthritis, cough with blood, malaria; antimicrobial activity
Dendranthema indicum (L.) Des Moul.	(Leaf) Buddleoglucoside, yejuhualactone, chrysanthemin, chrysanthemaxanthin, α-pinene, limonene, carvone, cineol, camphor, borneol[3]	Flu, high blood pressure, hepatitis

TABLE 1 Major Constituents and Therapeutic Values of Taiwanese Native Medicinal Plants (continued)

Scientific Name	Major Constituents and Source	Claimed Therapeutical Values
Dendrobium moniliforme (L.) Sw.	(Stem) Phenanthraquinones[606]	Ejaculation can occur spontaneously during sleep; improves stomach, kidney, and liver function; improves strength after sickness
Dendropanax pellucidopunctata (Hayata) Merill	(Whole plant) Limonene, carvone, cineal, camphor, chrysanthemin, chrysanthemaxanthin, α-pinene, borneol, buadleglucoside, acacetin-7-rhamnosidoglucoside. [3,48]	Treats flu, high blood pressure, hepatitis, rheumatic arthritis, hemiparesis, migraine, brachial plexus neuritis, irregular menstruation
Derris elliptica Benth.	(Root) Rotenoids, rotenone, deguelin, elliptone, 12a-hydroxy-, 6a-, 12a-dehydro-analogs, amorphigenin, rotenone acid, dalpanol, munduserone[195,196] This herb is toxic[88]	Rheumatic arthralgia, pruritus, eczema
Derris trifoliata Lour.	(Whole plant) Triterpenoid, taraxerol-3-β-*O*-tridecyl-ether, 6-α, 12α-12a-hydroxyelliptone, β-cariiebem deguelin, α-toxicarol[48,197] This herb is toxic[88]	Anticancer, skin tumor
Desmodium capitatum (Brum. f.) DC	(Whole plant) Phenolics, swertism, canavanine[6]	Vomiting with blood, inflammation with water, abdominal pain
Desmodium caudatum (Thunb.) DC	(Whole plant, root) Flavonoids, phenolic compounds[48]	Gastroenteritis, dysentery, infantile malabsorption, rheumatic arthralgia
Desmodium laxiflorum DC D. gangeticum DC D. multiflorum DC	(Leaf, fruit, root) Flavone, isoflavonoid, dipheny picyl, glucosides, hydrazyl, nitric oxide, ferryl-bipyridyl, hypochlorous acid, morphine[3,200,201,519–521]	Stomachache, diarrhea, parotitis, dysentery; antiinflammatory, antinociceptive, antioxidant

Desmodium pulchellum (L.) Benth. *D. sequax* Wall.	(Aerial part) Bufotenine, higerine, donoxime[2,3,97]	Antimalarial, antipyretic, antischistosomiasis, antitoxic, antiinflammatory; hepatitis, parotitis
Desmodium triquetrum (L.) DC	(Leaf) Potassium oxide, silicic acid, tannins[1,8]	A tonic for dyspepsia, hemorrhoids, infantile spasms; insecticide, germicide
Desmodium triflorum (L.) DC	(Whole plant, root) Desmodium alkaloids[20,198,199]	Dysentery, antitoxic, hepatitis, red eye with inflammation, lymph infection, anthelmintic action against *Ascaris lumbricoides*
Deutzia cordatula Li *D. taiwanensis* (Maxim) Schneidr. *D. corymbosa* R. Br. *D. gracilis* Sieb & Zucc.	(Root, stem, berry) Saponins, hydroxyleucine, arabinopyranosyl, rhamnopyranose, umbellferone, sitosterol, hydroxyleucine[27,202,203,522,523]	Diuretic, flu, windpipe infection, urination during the night; antitoxic; high blood pressure, malaria
Dianella ensifolia (L.) DC ex Red. *D. chinensis* (L.) DC *D. longifolia* (L.) DC	(Root) Antgictabubs, anthocyanidin, anthraquinone, chrysophanic acid[97,204,205] This herb is toxic[88]	Lymphangitis, tuberculous lymphadenitis, tinea infection, antiinflammatory; external use in furunculosis
Dianthus chinensis L.	(Root)[3] Triterpenoid saponins[674]	Bladder infection, hepatitis
Dichondra micrantha Urban	(Root)[3] No information is available in the literature	Antitoxic, diuretic, improves blood circulation, dysentery, hepatitis, malaria, abdominal pain
Dichrocephala bicolor (Roth) Schlechtendal	(Leaf) 4,5-Dicaffeoyl quinic acid, 3,4-dicaffeoyl quinic acid, 3,5-dicaffeoyl quinic acid, ethyl 4,5-dicaffeoyl quinate, methyl 3,5-dicaffeoyl quinate, 5-caffeoyl quinic acid, caffeic acid, quercetin-3-O-rutinoside[3,29,206]	Antitoxic, diuretic; stops bleeding, pneumonia, throat infection, diabetes, high blood pressure; immunomodulatory

TABLE 1 Major Constituents and Therapeutic Values of Taiwanese Native Medicinal Plants (continued)

Scientific Name	Major Constituents and Source	Claimed Therapeutical Values
Dichrocephala integrifolia (L.f.) Kuntze	(Leaf, flower) Essential oils[12,673]	Promotes circulation; for irregular menses, sprains; antiinflammatory, antiswelling
Dichroa febrifuga Lour.	(Root) Dichroines, dichroidine, 4-quinazolone, dichrins[2] This herb is toxic[88]	Antiamebial, antipyretic; for use against chicken malaria
Dicliptera chinensis Juss. D. *riparia* Nees.	(Leaf) Glycosides, dicliriparisides A, C, β-sitosterol, 2,5-dimethoxy-*p*-benzoquinone, vanillic acid, daucosterol, lugrandoside, poliumonside, amino acid[3,11,207]	Flu, fever, cough, hepatitis, conjunctivitis, epidemic encephalitis, enteritis, pneumonitis, acute appendicitis
Dicranopteris dichotoma (Thunb.) Bernh.	(Root, leaf) Polysacchrides, rare earth elements, La, Ce, Nd, Sm, Eu, Tb, Yb, Lu, clerodane glycosides, flavonoids[49,208]	Antipyretic, diuretic, expectorant, hemostatic, urinary tract infection, leukorrhea, bronchitis
Dicranopteris linearis (Burm. F.) Under.	(Leaf, stem) Quercitrin, afzelin, nonacosane, hepacosane, nonacosan-10-one, nonacosan-10-ol[6]	Anthelmintic, a poultice for fever; improves blood circulation, diuretic
Digitalis purpurea L.	(Whole plant) Digitoxigenin, gitoxigenin, gitanin, gitaloxigenin, digitoxin, gitoxin, gitaloxin, digicoside, strospeside, digipurin, digicirin, digifolein, digitonin, purpureal glycosides[1] This herb is toxic[88]	For gonorrhea, sclerosis of the breast
Dioscorea bulbifera L.	(Rhizome) Saponins, dioscorecin, iodine, dioscoretoxin, saponins, diosgenin, giosbulbin, tannins, campesterol, β-sitosterols, stigmasterol, diosbulbines[2,10,16]	Treats cancer of gastrointestinal tract, goiter, tuberculous lymphadenitis, hematemesis, hemoptysis, uterine bleeding

Dioscorea opposita Thunb.	(Leaf, tuber, root) Allantoin, arginine, choline, glutamine, leucine, tyrosine, diosgenin, sinodiosgenin[8]	Leaf sap for snakebite, root for asthma, cachexia, cough, debility, diarrhea, neurasthenia, polyuria; tuber is anthelmintic
Diospyros eriantha Champ. ex Benth.	(Leaf, root) Astragalin, myricitrin, L-eitrulline, carotenoids, flavonoids, phenolics, tannins, dioscorine, cocaine[3,20]	Alleviates pain
Diospyros angustifolia L.	(Bark) Coumaroyl triterpene lactone, phenolic and naphthalene glycoside, dispyrosooleans, friedelin, diospyrososide, β-amyrin, betulinic acid, lupeol, diospyrosonaphthoside[209]	Astringent, stomach discomfort, respiration problems; treats diarrhea, enterorrhagia, hemorrhoids; antifebrile, antivirous, demulcent
Diospyros kaki L.	(Seed, leaf) Oleanolic acid, betulinic acid, acetylcholine, choline, shibuol, ursolic acid; seed oil contains fatty acid, leaf contains pectic polysaccharide, flavone[3,8,210-212]	An astringent, styptic, antitussive, laxative, nutritive, stomachic; for constipation, hemorrhoids, diarrhea, bronchial complaints, dry cough, hypertension
Diplazium megaphllum (Bak.) Christ. *D. subsinuatum* (Wall. ex Hook & Grev.) Tagawa	(Whole plant) Hopane-triterpene lactone glycosides, hydroxymethylene, monoacetyl derivative, diplazioside, diplaziosides[3,213]	Antitoxic, hepatitis, liver diseases
Diplocyclos palmatus (L.) C. Jeffrey.	(Leaf, seed) Punicic acid, fatty acids[3,678,679] This herb is toxic[88]	Headache, tumor, inflammation
Dipteracanthus repens (L.) Hassk. *D. prostratus* Nees.	(Seed, whole plant) Fatty acids[214]	Improves lung function, relieves body temperature
Dodonaea viscosa (L.) Jacq.	(Leaf, bark) β-Sitosterol, stigmasterol, isorhamnetin, alkaloid, glucoside, tannins, resins[1,6] This herb is toxic[88]	Shoulder pain; remedy for fever; astringent to treat eczema

TABLE 1 Major Constituents and Therapeutic Values of Taiwanese Native Medicinal Plants (continued)

Scientific Name	Major Constituents and Source	Claimed Therapeutical Values
Dolichos lablab L.	(Flower, seed) Glucokinin, plant insulin, tryptophane, arginine, lysine, tyrosine[48,98]	Treats menorrhagia, leucorrhea, diarrhea, leukorrhea, chronic nephritis
Drynaria cordata (L.) Willd.	(Whole plant) Codeline phosphate, nitrates, phytosteroids, potassium compounds, succinates, succinic acid, potassium nitrate, stigmasterol[10,215,216]	Antiinflammatory; treats cough; antipyretic, diuretic, antiswelling; acute hepatitis, jaundice, pterygium, antitussive, indigestion
Drynaria diandra Blume. D. fortunei (Kunze) J. Smith	(Root) Drymaritin, diandraflavone, drymarin A, B, c-glycoside flavonoid[3,154,217,218]	Headache, antitoxic, hepatitis, malaria; improves kidney function, blood circulation; alleviates pain
Duchesnea indica (Andr.) Focke	(Whole plant) Emodin, chrysophanic acid, phytosterol, volatile oil, calcium[1,10,348]	Insecticide, antidote; treats whitlow, burns, snakebite, hepatitis, antipyretic, influenza, dysentery, diphtheria, stomach and lung cancer, nasopharynx disorder; antimutagenic activity
Dumasia bicolor Hayata	(Fruit, leaf)[3] No information is available in the literature	Relaxes tight muscles, alleviates pain
Dumasia villosa DC D. truncata Sieb et Zucc.	(Fruit) Triterpenoidal saponins[3,219]	Alleviates muscle pain
Dumasia pleiantha (Hance) Woodsen	(Stem) Podophyllotoxin, dehydropodophyllotoxin, deoxypodophyllotoin, astragalin, β-sitosterol[3]	Antitoxic; resolves phlegm; lowers blood pressure, blood sugar level
Duranta repens L.	(Fruit) Methyl p-methoxycinnamate, scutellarein, pectolinangenin, durantoside-1,4-oleanolic acid, ursolic acid, β-carotene[3,6] This herb is toxic[88]	Improves blood circulation, resolves extravasated blood; antiinflammatory, antitoxic, insecticide

Species	Constituents	Uses
Dysosma pleiantha (Hance) Woodson	(Root) Podophyllotoxin, etoposide, peltutin, dehydropodophyllotoxin, hyperin, astragalin, deoxypodophyllotoxin[2,3] This herb is toxic[88]	Antitoxic, antiinflammatory; lowers blood pressure, blood sugar level; treats condyloma acuminata, exophytic wart
Ecdysanthera rosea Hook. & Arn. E. utilis Heyne	(Leaf) Proanthocyanidins, epicatechin, procyanidin B2, proanthocyanidin A1, A2, aesculitannin C[3,13,220]	Antiinfection, antibacterial; throat infection, enteritis, rheumatic bone pain; has immunopharmacological activity
Echinochloa colonum (L.) Link.	(Whole plant) Phenolic compounds[3,524]	Diuretic, stops bleeding, inflammation; antioxidant
Echinops grilisii Hance	(Root, flower stalk) Echinopsine[16]	Anthelmintic, galactagogue, depurative; treats tumors, swellings, leukorrhea, gout
Eclipta alba (L.) Hassk. E. prostrate L.	(Aerial part) Alkaloids, nicotine, ecliptine[1,638]	Hemostatic effect, antimyotoxic, antihemorrhagic; for dysentery, epistaxis, hepatitis, neurasthenia, premature graying of hair
Ehretia acuminata R. Br. E. dicksonii Hance E. resinosa Hance	(Leaf, root) (10E,12Z,15Z)-9-Hydroxy-10,12,15-octadecatrienoic acid, methyl ester, antiinflammatory compound[3,221]	Treats tooth pain, antiinflammatory; stops diarrhea, treats intestinal infection
Eichhornia crassipes (Mart.) Solms.	(Whole plant) Delphinidin-3-diglucoside, carotenoids, SiO_2 and other minerals[29]	Antitoxic, diuretic, antiinflammatory
Elaeagnus macrophylla Thunb. E. glabra Thunb. E. lanceollata Warb.	(Whole plant) Flavonol glycosides, epigallocatechin[29,525]	Dysentery, treats sores
Elaeagnus obovata Li E. loureirli Champ. E. bockii Diels.	(Whole plant) Flavonol glycosides[12,526]	Asthma, bronchitis; antinociceptive; haemolysis, gastric pain, diarrhea, chronic hepatitis, osteomyelitis, acute orchitis

TABLE 1 Major Constituents and Therapeutic Values of Taiwanese Native Medicinal Plants (continued)

Scientific Name	Major Constituents and Source	Claimed Therapeutical Values
Elaeagnus oldhamii Maxim. *E. thunbergii* Serv. *E. wilsonii* Li	(Root) Sitosterol, muslinic acid, sitosteryl glucopyranosid, arjunolic acid[27,29]	Alleviates arthritis pain, treats asthma; antiinflammatory; improves blood circulation, rheumatism pain, arthritis
Elaeagnus morrisonensis Hayata *E. angustifolia* L.	(Whole plant, seeds) Epigallocatechin, flavonoids, phenolics, amino acids[29,527]	Antiinflammatory, alleviates pain, improves blood circulation, cough, malaria, diarrhea
Elatostema lineolatum Wight var. *majus* Wedd. *E. edule* C. Robinson	(Whole plant)[3] No information is available in the literature	Bacterial dysentery, rheumatism, arthritis; antiinflammatory, alleviates pain
Elephantopus mollis Kunth. *E. scaber* L.	(Root, leaf) Epifriedelinol, elephantin, lupeol, dotriacantan-1-ol, stigmasterol, triacontanol-ol, deoxyelephantopin, elephantopin, isodeoxyelephantopin, molephantin, phantomolin, dotriacontanol, lupeol acetate[3,10,27,622,629,640]	Hepatoprotective effects, diuretic, hepatitis, antitoxic, antiinflammatory; furunculosis, eczema, influenza, tonsillitis, pharyngitis, conjunctivitis, epidemic encephalitis B; iceteric hepatitis, chronic nephritis; antibacterial activity against *Streptococcus mutans*, carrageenan- and adjuvant-induced paw edema in rats
Emilia sonchifolia (L.) DC	(Whole plant) Senecionine, flavonoid glycoside, phenolic compounds, alkaloids[3,99]	Windpipe infection, sore throat, pneumonia, intestinal infection; for dysentery, phthisis, coughs; a detoxicant, diuretic, febrifuge
Emilia sonchifolia (L.) DC var. *javanica* (Burm f.) Mattfeld	(Leaf) Alkaloids[101]	For dysentery, phthisis, coughs; a detoxicant, diuretic, febrifuge
Entada phaseoloides (L.) Merr.	(Stem, seed) Entageric acid[2]	Antirheumatic, promotes collateral flow, alleviates blood stasis, hernial pain, gastric pain, rheumatic arthritis
Epimeredi indica (L.) Rothon	(Whole plant) Flavonoids, phenolic compounds, tannins[10]	Epigastric pain, rheumatic arthritis, cold, fever

Plant	Part / Constituents	Uses
Epiphyllum oxypetalium (DC) Haw.	(Flower, stem)[97,222,528] No information is available in the literature	Pulmonary tuberculosis with cough, hemoptysis, uterine bleeding, pharyngitis
Epipremnum pinnatum (L.) Engl.	(Whole plant)[21] No information is available in the literature	Cough, stomachache, antitoxic, stops bleeding, encephalitis
Equisetum ramosissimum Desf.	(Whole plant) Equisetonin, equisetrin, articulain, isoquereitrin, galuteolin, populnin, kaempferol-3,7-diglucoside, astragalin, palustrine, grossypitrin, 3-methoxypyridine, herbacetrin[16] This herb is toxic[88]	Antihemorrhagic, anodyne, carminative, diaphoretic, diuretic
Erechtites valerianaefolia (Wolf.) DC	(Whole plant)[154] No information is available in the literature	Improves blood circulation, diuretic, antiinflammatory
Erigeron canadensis L.	(Whole plant) Matricaria ester, dehydromatricaria, limonene, linalool, gallic acid, dipentene, methylacetic acid, terpeneol, lacnophylium, matricaria, crigeron, tannic acid, hexahydromatricaria, diphenylmethane-2-carboxylic acid[3,8]	Antitoxic, tooth pain, arthritis pain, mouth cavity infection; for hemorrhage, diarrhea, dysentery, internal hemorrhage of typhoid fever
Eriobotrya japonica (Thunb.) Lindl.	(Leaf, flower, fruit) Levulose, sucrose, malic acid, citric acid, tartaric acid, succinic acid, amygdalin, crytoxanthin, carotenes, phenyl ethyl alcohol, pentosans, essential oils[8] This herb is toxic[88]	Antitussive, expectorant; treats bronchitis, cough, fever, nausea; externally applied to epistaxis, smallpox, ulcers
Erycibe henryi Prain	(Leaf, stem, root) Scopoline, erycheline, scopoletin[29]	Leaf poultices applied to sores and to the head to treat headache; treats arthritis, swelling, pain
Eryngium foetidum L.	(Whole plant) α-Cholesterol, brassicasterol, campesterol, stigmasterol, β-sitosterol, delta-5-avenasterol, delta(5)24-stigmastadienol, delta-7-avenasterol[10,223,224]	Colds, chest pain, indigestion, diarrhea, enteritis

TABLE 1 Major Constituents and Therapeutic Values of Taiwanese Native Medicinal Plants (continued)

Scientific Name	Major Constituents and Source	Claimed Therapeutical Values
Eucalyptus robusta Smith	(Leaf) Essential oils, cineol, thymol, gallic acid, phenolic compounds, sitosterol[2,97]	Antibacterial, antimalarial, upper respiratory tract infection, intestinal candidiasis, influenza, pharyngitis; externally, treats *Trichomonas vaginalis*
Euchresta formosana (Hayata) Ohwi	(Root with stem) Tectorigenin, 3′,4′,5′-trihydroxyisoflavone, euchretin F, arachidonic acid, quercetin, euchretin M, formosanatin C, coumaronochromones, flavanones, formosanatins A-D, euchrenone, euchretins[20,225,226]	Stops fungal infection, antiinflammatory; alleviates pain, stomachache, skin diseases
Eucommia ulmoides Oliver	(Bark) Pinoresinol-di-β-D-glucoside, resin, aucubin, ajugoside, reptoside, harpagide acetate, encommiol[2]	Improves liver and kidney function, lowers blood pressure
Euonymus echinatus Wall. *E. laxiflorus* Champ. *E. chinensis* Champ.	(Root, stem, bark, leaf) Sesquiterpenes, triterpene, laxifolone A, ebenifoline, carigorinine, euojaponine, emarginatine, triterpenoids, putranjivadione[13,227,228]	Low back pain, fractures, euonymus, laxiflorus, chronic nephritis, cytotoxicity
Eupatorium amabile Kit. *E. lindleyanum* DC	(Root, stem) Sesquiterpenoids, eupachinilides[154,229,230]	Diuretic, alleviates phlegm
Eupatorium cannabinum L. subsp. *asiaticum* Kitam.	(Leaf) Eupaformonin, eupatolide[3]	Anticancer, leukemia, diuretic, pneumonia, antiinflammatory
Eupatorium clematideum (Wall. ex DC) Sch. Bip.	(Root) Flavonoids, phenolics, amino acids[49]	Diphtheria, tonsillitis, pharyngitis, cold, fever, measles, pneumonitis, bronchitis, rheumatic arthritis, furunculosis
Eupatorium formosanum Hayata	(Whole plant) Sesquiterpine lactones, eupatolide, eupaformonin, eupaformosanin, parthenolide, michelenolide, costunolide, santamarine[2,100]	Anticancer

Eupatorium tashiroi Hayata	(Whole plant) Odoratin, α-sitosterol, β-sitosterol, linalool, eupatol, coumarin, hyperin, fumaric acid, eupatene, succinic acid, tataxasterol, euparin, eupaformosanin, eupatolide[3,28] This herb is toxic[88]	Diuretic, antiinflammatory, antitoxic; alleviates pain, flu, fever, cough
Euphorbia atoto Forst. f.	(Whole plant) Taraxerol, taraxerone, friedelan-3α-ol, friieddan-3β-ol, epifriedelanol, flavonoids, euphorbol, euphal, cydoartinal[6,154] This herb is toxic[88]	Improves menses; antitumor
Euphorbia formosana Hayata	(Whole plant) Ellagic acid, dimethylether[6,154] This herb is toxic[88]	Antitoxic, skin infection; alleviates phlegm, rheumatism, skin ulcer; externally for snakebite
Euphorbia heterophylla L.	(Root, seed) *N*-Acetylgalactosamine-specific lectin, Euphorbiaceae lectins[154,232,233] This herb is toxic[88]	Regulates menses, stops bleeding, antiinflammation
Euphorbia hirta L.	(Stem) Camphol, leucocyanidol, quercitol, quercitrin, rhamnose, euphorbon, chlorophenolic acid, taraxerol, taraxerone, gallic acid[8,11] This herb is toxic[88]	For asthma, bronchitis; externally for athlete's foot; bacillary dysentery, acute enteritis, phyllitis, chronic bronchitis, nephritis
Euphorbia jolkini Boiss.	(Whole plant) Putranjivain A[154,234] This herb is toxic[88]	Antiviral, inhibiting viral attachment and penetration, viral replication

TABLE 1 Major Constituents and Therapeutic Values of Taiwanese Native Medicinal Plants (continued)		
Scientific Name	Major Constituents and Source	Claimed Therapeutical Values
Euphorbia lathyris L.	(Seed) Euphorbiasteroid, betulin, 7-hydroxylathyrol, lathyrol diacetate benzoate, lathyrol diacetate nicotinate, euphol, euphorbol, euphorbetin, esculetin, daphnetin[2,14,39] This herb is toxic[88]	Treats bronchitis, antiinflammation
Euphorbia milli Ch. des Moulins *E. neriifolia* L.	(Root) Euphorbin, lectin[20,235,236] This herb is toxic[88]	Alleviates diarrhea, antitoxic, eliminates pus
Euphorbia thymifolia L.	(Seed) Euphorbiasteroid, betulin, 7-hydroxylathyrol, lathyrol, diacetate, benzoate, diacetate nicotinate, euphol, euphorbol, euphorbetin, esculetin, daphnetin[2,11,14,39] This herb is toxic[88]	Diuretic to remove edema, eliminate blood stasis and resolve masses, antitumor; bacillary dysentery, bleeding hemorrhoids
Euphorbia tirucalli L.	(Stem bark) Lectin, latex[237] This herb is toxic[88]	For mosquito control
Euphoria longana Lam.	(Fruit) 2-Amino-4-hydroxymethylhex-5-ynoic acid, 2-amino-4-hydroxyhept-6-ynoic acid, dihydrosterulic acid, quercetin, quercetrin, friedelin, 16-hentriacontanol, epifriedelinol, sigmateryl-D-glucoside[3]	Tonics, insomnia, memory problems, stops bleeding, alleviates pain

Plant	Part (constituents)	Uses
Euryale ferox Salish. *E. chinese* R. Brown	(Seed) Protein, starch[13,39]	Treats diarrhea, spontaneous emission, and leukorrhagia; prevention of epidemic influenza
Evodia meliaefolia (Hance) Benth.	(Leaf, root bark, fruit) Terpenoids[3,13,238]	Stomachache, gastric discomfort, vomiting, headache, tuberculosis
Evolvulus alsinoides L.	(Whole plant) Flavonoids, phenolic compounds, β-sitosterol[13]	Bronchial asthma, cough, gastric pain, indigestion, dysentery, urinary tract infection
Excoecaria orientalis Pax et Hoffn *E. agallocha* L. *E. kawakamii* Hayata	(Whole plant) Huratoxin, resin[28]	Improves spleen function, antitoxic, alleviates pain, cough, indigestion, hepatitis
Farfugium japonicum (L.) Kitamura	(Whole plant) 4-Butyrolactone, naphthalenes, terpenes, furans, farfuomolide A and B, sesquiterpenes, furanosesquiterpenes[12,239-242]	Cold, influenza, pharyngitis, tonsillitis, hemoptysis, amenorrhea
Fatoua pilosa Gaud.	(Leaf)[3] No information is available in the literature	Sore throat infection, parotitis
Fatsia polycarpa Hayata *F. japonica* (Thunb.) Decne. & Planch.	(Bark, leaf) Triterpene glycosides, saponins[29,243,244]	Improves blood circulation, alleviates pain, arthritis
Ferula assa-foetida L.	(Gum, resin) Vanillin, asarensinotannol, ferulic acid, farnesiferols[2]	Anthelmintic; treats ascites, dysentery, malaria
Ficus carica L. *F. benjamina* L.	(Leaf, fruit) Bergaptin, cerotinic acid, ficusin, glutamine, papain, pepsin, psoralen, guaiaxulene, amyrin, lupeol, retin, octacosane, guaiacol, queritin, rhamnose, sitosterol, tyrosine, urease[2,20,97,98,245]	Antitumor, antibacterial; treats respiratory disorders, skin diseases; for warts, stomachache; externally for swollen hemorrhoids, corns; fruit is laxative, digestive, treats pharyngitis, anthelmintic; hypolipidemic and hypotriglyceridemic activities; treats hoarseness, asthma, constipation, hemorrhoids

TABLE 1 Major Constituents and Therapeutic Values of Taiwanese Native Medicinal Plants (continued)

Scientific Name	Major Constituents and Source	Claimed Therapeutical Values
Ficus erecta Thunb. var. *beecheyana* (Hook. & Arn.) King	(Root)[29] No information is available in the literature	Improves blood circulation; antiinflammatory; arthritis, rheumatism
Ficus formosana Maxim.	(Root, twig, leaf)[29] No information is available in the literature	Improves blood circulation, alleviates pain, improves lung function; antiinflammatory, antitoxic
Ficus hispida L.	(Root, leaf, fruit) Tannins[97]	Cold, bronchitis, indigestion, dysentery, rheumatic arthritis, axillary carbuncle
Ficus microcarpa L. f.	(Root, leaf) Phenolic compounds, amino acids, flavonoids, tannins[3,12,20]	Tonsil gland infection, flu, fever, persistent cough, acute enteritis, bronchitis; prevention of influenza, tonsillitis
Ficus pedunculosa Miq. var. *mearnsii* (Merr.) Corner. *F. religiesa* L.	(Root) β-Sitosterol-D-glucoside[21,154]	Lowers blood sugar; gall bladder infection; alleviates fever; treats cough with blood, vomiting
Ficus pumila L. var. *awkeotsang* (Makino.) Corner	(Whole plant) Meso-inositol, taraxeryl acetate, β-sitosterol, latex, β-amyrin acetate[8,20,49]	Carbuncle, dysentery, hematuria, hemorrhoids, hernia, oligogalactia, amenorrhea, nocturnal ejaculation, impotence, chyluria, bladder inflammation
Ficus sarmentosa Buch. et J.E. Sm. var. *nipponica* (Fr. & Sav.)	(Root, stem)[154] No information is available in the literature	Antitoxic, blood clearance
Ficus septica Burm. f. *F. superba* (Miq.) Miq. var. *japonica* Miq.	(Root, leaf, fruit) Tylophorine, tylocrebrine, septicine, antofine[20,154] Root is toxic[154]	Anticancer; antifood toxin; leaf and fruit for diarrhea and vomiting
Ficus virgata Reinw. ex Blume	(Root, twig)[27] No information is available in the literature	Itchiness caused by skin cancer, antiinflammatory, abdominal pain, diarrhea, arthritis, antitoxic

Plant	Constituents	Uses
Ficus wightiana Wall.	(Root)[20] No information is available in the literature	Antitoxic, kills worms, treats ulcer
Flemingia macrophylla (Willd.) Merr.	(Root) Phenols, coumarin, amino acids[28,154]	Alleviates rheumatism; treats arthritis, joint infections
Flemingia prostrata Roxb.	(Whole plant) Phenolics, coumarin, amino acids[28]	Improves spleen function, arthritis, rheumatism pain
Foeniculum vulgare Mill.	(Fruit) Anethole-D-fenchone, anisaldehyde, methylchavicol, fenicularin, vitamin A[2,3]	Stomachache, hernia pain; restores normal stomach function, treats disease caused by schistosome, pain caused by menses, pain caused by hernia
Galium echinocarpum Hayata	(Whole plant) Rivalosides C-E, momordin lib, rivalosides A-B, monotropein, scandoside, deacetylasperulosidic acid[28,246]	Antiinflammatory; improves blood circulation; antitoxic, anticancer, urine with blood, dysentery
Gardenia angusta (L.) Merrill var. *kosyunensis* Sasaki *G. oblongifolia* Champ.	(Fruit, flower, bark) Gardenin, α-crocetin, volatile oil, chlorgenin, glycosides, mannit[99,247]	Emetic, stimulant, febrifuge, diuretic, hemostatic, antihemorrhagic, emmenagogue
Gardenia jasminoides Ellis.	(Fruit) Shonzhiside, gardonin, β-sitosterol, carotenoid, jasminoidin, geniposide, crocin, genipin-1-β-gentiobioside[3,10]	Fever, vomiting with blood, hepatitis, inflammation, bloody urine, buccal ulcer, hepatitis, insomnia, conjunctivitis, epistaxis
Gelsemium elegans Benth.	(Whole plant, root) Gelsemine, gelsemidine, koumine, kouminicine, sempervirine, kouminine, douminidine[2,8,11] This herb is toxic[88]	Treats eczema, tinea corporis, hemorrhoids, scrofula, pretibial ulcer, boils and pyodermas, leprosy
Gendarussa vulgaris Nees.	(Root) Justicin, volatile oil[3]	Antiinflammatory; alleviates pain, arthritis pain

TABLE 1 Major Constituents and Therapeutic Values of Taiwanese Native Medicinal Plants (continued)

Scientific Name	Major Constituents and Source	Claimed Therapeutical Values
Gentiana arisanensis Hayata	(Whole plant) Oleanolic acid, mangiferin[29]	Stomach infection, stomachache, hepatitis
Gentiana scabrida Hayata *G. scabrida* Hayata var. *horaimontana* (Masam.) Liu et Kuo *G. lutea* L.	(Whole plant, root) Triterpenoid, (S)-(+)-and (R)-(−) gentiolactones[8,29]	Improves stomach function, hepatitis, urinary tract infection, arthritis, cancer, carbuncle, fever, epilepsy
Gentiana atkinsonii Burk *G. campestris* L. *G. flavo-maculata* Hayata	(Whole plant) Xanthones, bellidin, bellidifolin, swertianolin, norswertianolin, swertiamarin, gentiopicroside[29,248,249]	Antitoxic, improves stomach function, hepatitis, throat inflammation
Geranium nepalense Sweet var. *thunbergii* (Sieb. & Zucc.) Kudo *G. suzukii* Masamune	(Whole plant) Gallic acid, quercetin, succinic acid, tannin[6,8,250]	Antitoxic, stops diarrhea, alleviates arthritis pain, intestinal infection, dysentery, antirheumatic, bacillary diseases
Glechoma hederacea L. var. *grandis* (A. Gray) Kudo	(Aerial part) 1-Pinocamphone, 1-menthone, isomenthone, 1-pulegone, α-pinene, β-pinene, 1,8-cineol, isopinocamphone, limonene, menthol, α-terpineol, linalool, *p*-cymene[16]	Febrifuge, anodyne; treats earache, fever, toothache; diuretic, decoagulant, arthritis
Glehnia littoralis Schmidt et Miq.	(Leaf, root) Stigmasterol, β-sitosterol, imperatorin, psoralen, osthenol-7-*o*-β-gentiobioside, petroselenic acid, petroselidinic acid, polyine, polysaccharides, falcalindiol, anthocyanin, furanocoumarin[101–106]	Anthelmintic; for chronic bronchitis, cough and hoarseness; antiproliferative activities; antimycobacterial, immune-suppressive activities
Glochidion eriocarpum Champ. *G. acuminatum* Muell. *G. zeylanicum* A. Juss.	(Root, leaf) Glochidiolide, isoglochidiolide, acuminaminoside, megastigmane glucosides, glochidacuminoside A-D[49,251,252]	Urticaria, eczema, enteritis, dysentery, contact dermatitis, pruritus, desquamative dermatitis, gum inflammation

Plant	Constituents (part)	Uses
Glochidion lanceolarium (Roxb.) Veigt.	(Whole plant) Friedelan-3-ol, glochidonol, β-sitosterol[11,13]	Antinflammatory, jaundice, stomatitis
Glochidion puberum (L.) Hutch.	(Root)[13] Phenolic compounds, amino acids	Influenza, bone pain, glaciation puberun, gastroenteritis, dysentery
Glochidion rubrum Blume	(Leaf) Glochidone, glochidonol, β-sitosterol[3]	Arthritis, nerve pain
Glossogyne tenuifolia (Kabukk.) Cass.	(Whole plant) Oleanolic acid, luteolin-7-glucoside[97,253]	Acute tonsillitis, pyorrhea, bronchitis, enteritis, diarrhea, urinary tract infection; antipyretic, antiinflammatory
Glycine javanica L. G. tabacina (Labill.) Benth. G. tomentella Hayata	(Seed, root) Sitosterols[154,675]	Tonic; treats rheumatic arthritis and joint infection
Glycosmis citrifolia (Willd.) Lindl.	(Root, leaf) Dimeric acridone alkaloids[575]	Treats cough, flu; stomachic, improves digestion; relieves pain caused by hernia
Glycyrrhiza uralensis Fisch.	(Outer cortex of root) Glycyrrhiza, triterpenoid saponin, flavonone glucoside, liquirtin, aglycone, liquiritigenin, chalcone, glucose, isoliquiritin, isoliquiritigene, glycyrrhizic acid, β-glycyrrhetinic acid[2,107–110]	Antinflammatory, anticonvulsant, carminative, antidote, antitumor, antispasmodic, antiulcer
Gnaphalium affine D. Don G. luteoalbum (L.) ssp. affine (D. Don.) Koster	(Whole plant, flowers) Fat, resin, phytosterol, xylose, essential oil, carotene, glucose, arabinose, flavonoids, galactose, sitosterol, polysaccharide, vitamin B_1[8,11,16,18,529,530]	Remedy for lung disease, antifebrile, antimalarial, reduces blood pressure; for stomach and intestinal ulcers, chronic bronchitis, asthma, acute hemolysis, rheumatism
Gnaphalium hypoleucum DC G. adnatum Wall. ex DC	(Leaf) Butein, cardamunin, luteolin 4'-β-D-glucoside, gnaphalin[3,27]	Cold, flu, cough, shortness of breath; dysentery, mouth cavity inflammation

TABLE 1 Major Constituents and Therapeutic Values of Taiwanese Native Medicinal Plants (continued)

Scientific Name	Major Constituents and Source	Claimed Therapeutical Values
Goldfussia formosanus (Moore) Hsieh et Huang *G. psilostachys* C. B. Clarke & W. W. Smith.	(Whole plant)[27,254] No information is available in the literature	Antitoxic; flu, parotitis, sore throat, hepatitis; antimitotic
Gomphrena globosa L.	(Flower) Saponins, β-cyamines, gomphrenin, amaranthin, isoamaranthin[2,11]	Treats chronic bronchitis, whooping cough, dysentery, pertussis, pulmonary tuberculosis with hemoptysis, infantile fever
Goniothalamus amuyon (Blanco) Merr.	(Whole plant) Styrylpyrones[585]	Cytotoxic activity
Gonostegia hirta (Blume) Miq. *G. pentandra* (Roxb.) Miq.	(Whole plant)[3,29] No information is available in the literature	Eliminates pus, sore; antiinflammatory; stops bleeding, dysentery, skull itch
Goodyera procera (Ker-Gawl.) Hook *G. schlechtenda* Liana. *G. nankoensis* Fukuyama	(Whole plant) Flavonol glycoside, goodyerin, goodyeroside A, kinsenoside, rutin, kaempferol-3-D-rutinoside, isorhamnetin-3-D-rutinoside[3,13,255-257]	Rheumatism, arthralgia, hemiplegia, bronchitis, asthma, hypertension, flu, cough
Gossampinus malabarica (DC) Merr.	(Flower, root) Daucosterol, oleanolic acid, hesperidin, potassium nitrate, 2-*O*-methylisohemigossylic acid, lactone, sesquiterpene[13,258,259]	Tuberculous, enteritis, dysentery, hematoma, rheumatism, contusion, epigastric pain
Graptopetalum paraguayense E. Walther	(Whole plant) Phenol, anthocyanin[28, 671]	Treats hypertension, hepatitis, flu, sore throat, arthritis; antioxidant activity, reduces radical scavenging and lipid peroxidation inhibition
Grevillea robusia A. Cunn.	(Leaf) Robustol[29]	Used externally for wounds

Plant	(Part) Constituents	Uses / Effects
Gynostemma pentaphyllum (Thunb.) Makino	(Leaf) Panaxatriol, panaxadiol, saponin, glypenosides, sterol, gypenocide, ginsenosides Rb$_1$, Rb$_3$, Rd, Rf, flavonoids[2,3,111–114,121]	Antitoxic, eliminates infection, stops cough, hepatitis, enteritis, gastritis; regulating effect on lymphocyte transformation; protective effect against myocardial and cerebral ischemia, relaxes ischemic heart ventricles
Gynura bicolor (Willd.) DC	(Whole plant) Flavonoids[6]	Improves blood circulation, stops bleeding; a detoxicant; alleviates swelling, cough with blood
Gynura formosana Kitamura G. elliptica Yabe & Hayata	(Root) p-Hydroxyacetophenone–like derivative, (+)-gynunone, chromane, 6-acetyl-2,2-dimethylchroman-4-one, vanillin[21,618]	Flu, diuretic, antitoxic, antiinflammatory, encephalitis; antiplatelet aggregation activity
Gynura japonica Juel. var. flava (Hayata) Kitamura	(Root, leaf) Saponins, quinonoid terpenoid, gynuraone, steroids, caryophyllene oxide, vanillin, benzoquinone, benzoic acid[18,260]	Hemostat, furunculosis, hemorrhage, hemorrhea; externally for bruises and wounds, insect bites, snakebites; antiplatelet aggregation activity
Habenaira dentata (Sw.) Schltr. H. repens Nutt.	(Root) Habenariol, bis-p-hydroxybenzyl-2-isobutylmalate[3,670]	Diuretic; stops infections, cough; treats hepatitis, enteritis, gas trite
Haraella retrocalla (Hayata) Kudo.	(Leaf, root)[154] No information is available in the literature	Improves lung, liver function, fever; alleviates alcohol effects; antitoxic
Hedychium coronarium Koenig	(Root) Volatile oils[48]	Cold, body aches, headache, rheumatism
Hedyotis corymbosa (L.) Lam.	(Whole plant, seed) Borneol, bornyl acetate, l-camphor, linalool, nerolidol, sitosterol, phenolic compounds, flavonoids[10,18,641]	Tonsillitis, pharyngitis, bronchitis, malignant neoplasm, malaria, stomachache; hepatic protective; mouthwash to alleviate toothache, as a poultice to heal wounds, small sores
Hedyotis diffusa Willd.	(Whole plant) β-Sitosterol, acyl flavonol, di-glycoside, iridoid glucosides, anthraquinone, essential oils, p-vinylphenol, p-vinylguaiacol, linalool[8,10,115–119,649]	Malignancy, bronchitis, tonsillitis, appendicitis, hepatitis; antitumor, immunopotentiation activity; antibacterial, antipyretic; a detoxicant, diuretic, anticancer; externally applied as lotion

TABLE 1 Major Constituents and Therapeutic Values of Taiwanese Native Medicinal Plants (continued)

Scientific Name	Major Constituents and Source	Claimed Therapeutic Values
Hedyotis pinifolia Wall.	(Whole plant) β-Sitosterol, ursolic acid[13]	Treats snakebites, infantile malabsorption, abscesses
Hedyotis uncinella Hook. & Arn.	(Whole plant) Usnic acid, β-sitosterol, β-sitosterol-D-glucoside[6,27]	Antiinflammatory, arthritis
Helianthus annuus L.	(Whole plant, seed) Fatty acids, linoleic acids, palmitic acid[97,261,262]	Hypotensive, antiinflammatory, diuretic, antitussive, analgesic, antimalarial, treats wounds
Helicteres angustifolia L.	(Whole plant) Flavonoids, phenolic compounds, tannins, lupane-type triterpenoids[49,603] This herb is toxic[88]	Influenza, high fever, tonsillitis, pharyngitis, measles, dysentery
Heliotropium indicum L.	(Whole plant) Indicine[11]	Treats pneumonitis, lung abscess, buccal ulcer, sore throat
Helwingia formosana Kanehira et Sasaki H. *japonica* (Thunb.) Dietr. subsp. *formosana* (Kaneh. & Sasaki) Kara & Kurosawa	(Root, aerial part) Tritepenoids[6]	Improves blood circulation, alleviates pain, cough with shortness of breath, arthritis, irregular menstruation
Hemerocallis fulva L.	(Root) D-Glucoside, chrysophanol, rhein, asparagine, friedelin, obtusifolin, β-sitosterol, vitamin C, obtusifolin, jervine, colchicine, hemerocallin, trehalase, protoveratrine, pseudojervine[1,6,11] This herb is toxic[88]	Epistaxis, hemoptysis, hepatitis, cystitis, oliguria, hematuria, sternutative, anthelmintic, evacuant properties, mastitis, cervical lymphadenitis

Hemerocallis longituba Miq.	(Root, aerial part) r-Hydroxyglutamic acid, β-sitosterol, lysine, succinic acid[3,28]	Parotitis, hepatitis, cystitis, urine with blood
Hemiboea bicornuta (Hayata) Ohwi	(Whole plant)[21] No information is available in the literature	Antitoxic, stops bleeding, diuretic; for high blood pressure, toxic inflammation
Hemiphragma heterophyllum Wall var. *dentatum* (Elmer) Yamazaki	(Whole plant) Phenylpropanoid, iridoid glycosides[20,669]	Alleviates pain, stops bleeding, improves blood circulation, antitoxic, cough, vomiting with blood, and mouth cavity infection
Heterostemma brownii Hayata	(Root, whole plant) Puriniums, pyrimidines, heteromines D, E[21,263]	Antitoxic, malaria
Heterotropa hayatanum F. Maekawa *H. macrantha* (Hook. f.) Maekawa ex Nemoto	(Root) Chamazulene, bisabolol constituents[3,154]	Alleviates pain; treats rheumatic diseases, flu; fungicidal, antiinflammatory, antibacterial
Heterotropa taitonensis (Hayata) Maekawa ex Nemoto	(Root) Volatile oils[3,264]	Alleviates pain, antitoxic
Hibiscus mutabilis L.	(Leaf, flower, root) Isoquercitrin, hyperoside, rutin, quercetin-4-glucoside, quercetin, quercimeritirin[3,8,97]	Cough, mammary gland infection, pulmonary empyema, lung ailments, dysuria, menorrhagia, leukorrhea; applied to swellings, burns, ulcers
Hibiscus rosa-sinensis L.	(Flower, leaf, stem) Quercetin, cyanidin glucoside, kaempferol gossypetin, thiamin, riboflavin, niacin, cyandidin-3-sophoroside[3,8]	Antiinflammatory: used as poultice on cancerous swellings and mumps
Hibiscus sabdariffa L.	(Leaf, flower, bark) Saponin, saponaretin, vitexin[8]	Stomachache; diuretic, expectorant; hematochezia, vertigo, gas
Hibiscus syriacus L.	(Root bark) Lignans, hibiscuside, syringaresinol, feruloyltyramines, acetyldaidzin, acetylgenistin, hydroxydaidzein[154,265]	Antitoxic, alleviates fever, rheumatic diseases, itchiness; antioxidant

TABLE 1 Major Constituents and Therapeutic Values of Taiwanese Native Medicinal Plants (continued)

Scientific Name	Major Constituents and Source	Claimed Therapeutic Values
Hibiscus taiwanensis W.Y. Hu	(Root, stem) Phenylpropanoid esters, secoisolaricinresinol, demethylcarolignan E, hibiscuwanin A-B[20,267,533]	Treats infection, antitoxic, antiinflammatory, infection in rib membrane area; cytotoxic activity against lung and breast carcinoma cell lines *in vitro*
Hibiscus tillaceus L. H. esculentus L.	(Leaf, bark, flower) Sterol, sitosterol, campesterol, stigmasterol, 5-avenasterol, cholesterol[13,531,532]	Bronchitis, cough, fever, cassava poisoning
Hippeastrum equestre (Ait.) Herb. H. regina (L.) Herb.	(Stem, root) Lycororine, lycoramine, galanthamine, tazettine[3,21]	Antiinflammatory, antitoxic, diuretic, treats hernia, resolves phlegm, forces vomiting
Hippobroma longiflora (L.) G. Don.	(Whole plant) Volatile constituents[154]	Maintains vigilance, alleviates pain; antitoxic, antiinflammatory
Houttuynia cordata Thumb.	(Whole plant) Essential oil, houttuynium, decanoylacetaldehyde, quercitrin, isoquercitrin[2,3,49]	Controls fungus infection, alleviates pain; treats malaria, pulmonary empyema, mastitis, cellulitis, pneumonitis, bronchiolitis, encephalitis, conjunctivitis, tonsillitis
Hoya carnosa (L.f.) R. Br.	(Leaf) Condurangin, hoyin, phytosterindigitonid[8]	To hasten maturation of anthrax and furuncles
Humulus scandens (Lour.) Merr.	(Aerial part) Humulone, lupulone, asparagine, choline, luteolin[2]	Inhibits tubercle bacillus; antipyretic, diuretic
Hydrangea chinensis Maxim.	(Root) Carbonyl compounds, β-sitosterol[3]	Diuretic, antiinflammatory, headache, malaria
Hydrangea macrophylla (Thunb.) Ser.	(Whole plant) Febrifugin, hydrangeic acid, hydrangenol, rutin, daphnetin[8,13] This herb is toxic[88]	Treats malaria, fever, anxiety, sore throat; antitussive, diuretic

Plant	Constituents	Uses
Hydrocotyle formosana Masamune *H. asiatica* L.	(Whole plant)[20,534]	Cardioprotective activity, diuretic, antitoxic, flu, sore throat, intestinal infection, kidney stone, encephalitis
Hydrocotyle nepalensis Hook.	(Leaf) No information is available in the literature[3]	Flu, cough, tuberculosis, asthma, windpipe infection
Hydrocotyle sibthorpioides Lam.	(Whole plant) Hydrocotylosides I-VII, udosaponin B, glycosidic constituents, oleanane-type triterpenoid saponins, hydrocotylosides I-VII, udosaponin B[3,268,269,535]	Hepatitis, bladder stones, flu, cough
Hylocereus undatus (Haw.) Br. & R.	(Flower, stem) β-Sitosterol[97]	Bronchitis, tuberculous lymphadenitis, pulmonary tuberculosis
Hypericum chinense L. *H. patulum* Thunb. ex Murray	(Whole plant) 1,1-Diphenyl-2-picrylhydrazyl, carboxylic acid[27,270]	An antioxidant, free-radical scavenger; antitoxic, stops diarrhea; diuretic; flu, hepatitis, hernia, dysentery, tonsil infection
Hypericum japonicum Thunb.	(Whole plant) Quercetin, quercitrin, isoquercitrin, sarolactone, hypericin, usigtoercin, protohypericin, kaempferol[1,2,4,39,120]	Chronic hepatitis, hepatic cirrhosis, conjunctivitis, tonsillitis; antipyretic, antibacterial, a detoxicant, treats acute icteria, hepatitis, lowers blood pressure, dysmenorrhea, gonorrhea, skin ailments
Hypericum geminiflorum Hemsley	(Stem, leaf, root) Xanthones, prenyl chalcone, gemichalcone C, xanthones-6,7-dihydroxy-1,3-dimethoxyxanthone, 4-hydroxy-1,2-dimethoxyxanthone, gemixanthone A[28,271,272]	Improves blood circulation, alleviates pain, antinflammatory; for vomiting with blood, uterine bleeding, hepatitis
Hyphea kaoi Chao	(Whole plant) 5,7-Dihydroxy flavonone 7-glucoside, pentahydroxy flavone[20]	Overbleeding during menses, sickness after giving birth
Hypoestes purpurea R. Br.	(Whole plant) Furanolabdane diterpenes, hypopurin A-D, lignans, hinoguinin, helioxanthin, 7-hydroxyhinokinin, dehydroxycubebin, justicidine E, lupeol, betulin[3,273]	Tuberculosis, windpipe infection, diabetes; cytotoxic toward the KB cell line with an IC(50) value of 9.4 µM

TABLE 1 Major Constituents and Therapeutic Values of Taiwanese Native Medicinal Plants (continued)

Scientific Name	Major Constituents and Source	Claimed Therapeutical Values
Hypolepis tenuifolia (Forst.) Bernh.	(Whole plant) Ptaquiloside[3,536]	Carcinogenic and antitumor activity; treats flu, fever
Hypoxis aurea Lour.	(Whole plant) Phytosterols, β-sitosterol, hypoxoside, glycone, ruoperol[3,668]	Hypoxoside activity, anticancer, HIV-AIDS, antiinflammation, antimutagenic, a tonic, relieves hernia pain
Hyptis rhomboides Mart. & Gal.	(Whole plant) Butulinic acid[6,20]	Antiinflammatory, improves blood circulation, flu, heatstroke, tuberculosis, shortness of breath
Hyptis martiusii Benth. *H. suaveolens* (L.) Poir.	(Leaf, root) Butulnic acid, abietane diterpenoids[3,6,10,274,275]	Headache, gastrointestinal distention, antitoxic; alleviates pain; cytotoxic activity against tumor cell lines
Ilex asprella (Hook & Arn.) Champ.	(Root)[6] β-Sitosterol, caffeine, theophylline, oleanolic acid, hederagenin, ursolic acid, glucose, rhamnose	Improves heart function, blood circulation; antitoxic; for flu, dizziness
Ilex cornuta Lindl.	(Root, leaf, fruit) Volatile oil, caffeine, tannins[49]	Acute hepatitis, pulmonary tuberculosis, headache, fever
Ilex pubescens Hook. et Arn.	(Leaf, root) Flavone, ursolic acid, scopoletin, 3,4-dihydroxyacetophenone, hydroquinone, vomifliol[2]	Treats angina pectoris, acute myocardial infarction, central angiospastic retinitis, cerebral thrombosis, thrombophlebitis
Ilex rotunda Thunb.	(Root, stem) Flavonoids, phenolics, ursolic acid, amino acids, sitosterol[21]	Antitoxic; improves blood circulation, alleviates pain; diuretic, antiinflammation, flu, cough, throat inflammation
Illicium arborescens Hayata	(Fruit) Sikimin, shikimintoxin, skimmianine, hananomin, illicin, shikimetin, safro eugenol, chavicol[28] This herb is toxic[88]	Increases blood sugar level, concretizes blood

Impatiens balsamina L.	(Root) Anthocyanins, cyanidin, kaempferol, quercetin, monoglycoside, delphinidin, pelargonidin, malvidin[6]	Improves blood circulation, alleviates pain, arthritis
Imperata cylindrica (L.) Beauv. var. *major* (Nees) C.E. Hubb. ex Hubb. & Vaughan	(Root, stem) Anemonin[6]	Diuretic, stops bleeding, vomiting with blood, shortness of breath, inflammation, hepatitis, kidney infection
Indigofera suffruticosa Mill.	(Leaf) Indirubin, malospicine[276] This herb is toxic[88]	Emetic, antiinflammatory, alleviates pain, reduces fever
Indigofera tinctoria L.	(Whole plant, root) Indimulin, indican, indoxyl, indigotin[3]	Antitoxic, removes extravasation, encephalitis, parotitis, inflammation caused by sores
Indigofera zollingeriana Miq. *I. longeracemosa* Boiv. ex Baill.	(Whole plant) Abietane diterpenoid, indigoferabietone[3,537,538]	Mouth cavity infection: antitoxic, antiinflammatory; sore throat, hepatitis
Ipomoea batatas (L.) Lam. *I. obscura* (L.) Ker-Gawl. *I. stans* Cav.	(Root) Anthocyanins, tetrasaccharide glycosides, stansins 1-5, vitamins A, B_1, B_2, C, alcorhic acid, ipomarone[29,539,540] This herb is toxic[88]	Improves blood circulation, stomach and intestine function; treats constipation; diuretic
Ipomoea pes-caprae (L.) Sweet subsp. *brasiliensis* (L.) Oostst.	(Whole plant) Behenic acid, melissic acid, myristic acid, sitosterol, volatile oils[29,97]	Antiinflammatory; flu, arthritis, alleviates pain, hemorrhoid pain, rheumatic arthralgia, hemorrhoidal bleeding, antiswelling
Ipomoea quamoclit L.	(Whole plant, root) No information is available in the literature[154]	Alleviates fever, seeds used to treat diarrhea
Iris tectorum Maxim.	(Root) Flavonoids[97]	Traumatic injury, rheumatic pain, sore throat, indigestion and abdominal distention; antiinflammatory

TABLE 1 Major Constituents and Therapeutic Values of Taiwanese Native Medicinal Plants (continued)

Scientific Name	Major Constituents and Source	Claimed Therapeutical Values
Ixeris chinensis (Thunb.) Nakai	(Whole plant) Luteolin-7-glucoside[3,20,277,435,633]	Treats infection, antitoxic, alleviates pain, stops bleeding; pneumonia, breast cancer, flu; stomachache; inhibits the proliferation of K562 cells; hepatoprotective activity
Ixeris laevigata (Blume) Schultz-Bip. ex Maxim. var. *oldhamii* (Maxim.) Kitamura	(Whole plant) α-Naphthyl-isothiocyanate, carbon tetrachloride[3,20,122,123,278]	Improves blood circulation; antitoxic; flu, shortness of breath, cough, windpipe infection, hepatitis, pneumonia; hepatoprotective effect
Ixeris tamagawaensis (Makino) Kitamura	(Leaf, root) Lupenol acetate, tricosyl, pentacosyl, bauenyl acetate, luteolin, luteolin-7-0-glucoside, strearyl palmitate, strearyl strearate[3]	Antiinflammatory, pneumonia, stone in urinary tract
Ixora chinensis Lim.	(Leaf) Phenolics, amino acids[6]	Liver clearance, improves blood circulation, high blood pressure, calms uterus movement
Jasminum hemsleyi Yamamoto	(Root) α-Tropolone, β-tropolone, nootkatin, tropolone, cryptozaaponol, sugiol, 7-α-methoxy-deoxyptojaponol, 7-β-methoxy-deoxyptojaponol, isocedrolic acid, detuydrosugiol, cedrol, β-sitosterol, δ-cardinol, chamaecin, emodin, detetrahydroconidendrin[28]	Diabetes, kidney infection, inflammation, arthritis, liver diseases
Jasminum sambac (L.) Ait.	(Flower, root) Jasmine, linalool, benzoic acid, benzylalcohol, linalyl benzoate, formic acid, acetic acid, anthranilic acid, sesquiterpene, sesquijasmine[1,27]	Sedative, anesthetic, vulneraria properties; for congestion, headache; lactifuge, alleviates pain, insomnia

Plant	Parts and constituents	Medicinal uses
Jatropha curcas L.	(Bark, seed) Taraxerol, β-amyrin, α-amyrin, β-sitosterol-3-0-β-D-glucoside, *n*-1-triacontanol, campesterol, β-sitosterol, 7-deto-β-sitosterol, stigmast-5-ene-3β, 7α-diol, stigmast-5-ene-3β, 7β-diol, palmitic, palmitoleic, stearic acids, linoleic acid, linolenic acid, oleic acid, dulcitol acid, myristic acid[20] This herb is toxic[88]	Alleviates convulsions, itchiness, and pain; stops vomiting and bleeding; antiinflammatory; acute gastritis and enteritis
Juncus effusus L. var. *decipiens* Buchen.	(Whole plant) Tripeptide, *r*-glutamyl-valyl-glutamic acid, apigenin, juglandic acid, juglandic acid, barium, luteolin-7-glucoside, luteolinidin, oxalic acid, arsenic, vitamins[4,16]	Diuretic, sexually transmitted diseases; antiinflammation
Juniperus formosana Hayata	(Leaf, fruit) Volatile compounds, wood lignans[3,279,280]	Kidney infection, inflammation; gallbladder, liver disease; spleen diseases
Justicia gendarussa Burm. f. *J. procumbens* L. *J. procumbens* L. var. *hayatai* (Yeman.) Ohwi.	(Root) Gentianine, gentlanidine, gentianol, lignans[2,595]	Treats rheumatism and fever, antipyretic, effects on nitric oxide and tumor necrosis, antiinflammatory, antihypersensitivity, and antihistaminic effects
Kadsura japonica (L.) Dunal.	(Vine) Kadsuric acid, kadsurin, kadsurarin A, germacrene[2,124]	Against hepatitis B, alleviates pain, a detoxicant, improves blood circulation, alleviates arm and leg numb feelings
Kaempferia galanga L.	(Whole plant) Cineole, borneol, ethyl cinnamate, ethyl-*p*-methoxycinnamate, camphene, haempferol, kaempferide, flavonoids[13,27]	For stomachache, gastric pain, acute gastroenteritis, diarrhea; stops infection, treats gastritis, enteritis, arthritis, cholera
Kalanchoe spathulata (Poir.) DC *K. pinnata* (Lam.) Pers. *K. gracillis* Hance *K. crenata* Raym-Hamet *K. tubiflora* (Harvey) Hamet	(Leaf) Bufadienolides, acetic acid, β-amyrin, β-sitosterol, bryophylin, caffeic acid, citric acid, ferulic acid, kaempferol[4,8,20,27,154,281,282,283]	Antitoxic, stops bleeding; antiinflammation, ulcer, encephalitis, hypertension, ear infection; leaves applied as paste on burns

TABLE 1 Major Constituents and Therapeutic Values of Taiwanese Native Medicinal Plants (continued)

Scientific Name	Major Constituents and Source	Claimed Therapeutical Values
Kalimeris indica (L.) Schultz-Bip	(Whole plant) Triterpene saponin, shimadoside A, oleanolic acid[29,284]	Flu, fever, hepatitis, antiinfection, antitoxic, sore throat, windpipe infection, vomiting with blood, night soil with blood
Kyllinga brevifolia Rottb.	(Whole plant) Volatile oils, oligoglycosidic compounds, flavonoid glycosides, quercetin, triglycoside[3,20,285,286]	Diuretic, antiinflammation; alleviates pain, cough, throat infection, flu, headache; antiviral, abdominal pain, appendix, alleviates stress; a sedative agent
Lactuca indica L.	(Whole plant, seed) β-Amyrenl, germanicyl, dimeric guianolides, lignan glycoside, lactucin, pectic compound, oxalic acid, malic acid, citric acid, ceryl alcohol, ergosterol, vitamin E[8,13]	Treats tonsillitis, mastitis, cervicitis; anodyne, lactogogue; for genital swelling, hemorrhoids, lumbago; antidiabetic activity
Lagenaria siceraria (Molina) Standl. var. *microcarpa* (Naud.) Hara	(Fruit) Lagenaria D[48]	Ana sarda ascites, beri-beri; antiswelling; abdominal swelling, swelling feet
Lagerstroemia subcostata Koehne	(Flower, root) Tannins, ellagic acid[6]	Abdominal pain, antitoxic
Lantana camara L.	(Whole plant) Lantadene A, B, lantic acid, humulene, lantanotic acid, tannins, β-caryophyllene, γ-terpinene, α-pinene, *p*-cymene[6] This herb is toxic[88]	Improves blood circulation; arthritis, flu; antiinflammation, antitoxic; abdominal pain, vomiting, and diarrhea
Laportea pterostigma Wedd. *L. moroides* Wedd.	(Leaf, flower)[541,542] No information is available in the literature This herb is toxic[88]	Antiinflammatory, treats poison caused by swelling
Laungusa galanga (L.) Stuntz.	(Root, stem) Tannins, volatile oils, cineole, eugenol, pinene, galangin[6]	Diuretic, antiinflammation, pneumonia, sore throat, bladder stone

Species	(Part) Constituents	Uses
Leea guineensis G. Don	(Root, leaf, wood) Volatile constituents[21,667]	Arthritis, antitoxic, antiinflammation
Lemmaphyllum microphyllum Presl.	(Whole plant) Vitamins, luteolin-7-β-D-glucopyranoside, flavonoids, D-apiose, protein, resin[16]	A poultice for animal bites, itchiness; a lotion for smallpox; alleviates headache
Leonurus artemisia (Lour.) S. Y. Ha	(Whole plant, seed) Leonurine[97]	Menstrual irregularities, amenorrhea, postpartum hematoma and hypogastric pain, nephritis, edema, oliguria, hematuria
Leonurus sibiricus L. f. *albiflora* (Miq.) Haieh.	(Seed) Essential oil, leonurine[18]	Emmenagogue, diuretic, vasodilator
Lepidagathis formosensis Clarke ex Hayata L. *hyalina* Nees L. *cristata* Willd.	(Leaf) Triterpenoid, cristatin A, cycloartenol, stigmasta-5, 11(12)-diene-3β-ol[3,287,288]	Pneumonia, flu, mouth and lip infection
Lespedeza cuneata (Dumont d. Cours.) G. Don	(Root, whole plant) Pinitol, flavonoids, β-sitosterol[2,11]	For chronic bronchitis; antiasthmatic, antiphlogistic, antibacterial, gastrointeritis, antitussive
Leucas aspera (Willd.) Link. L. *chinensis* (Retz.) R. Br.	(Whole plant) Lignans, flavonoids, nectandrin B, meso-dihydro-guaiaretic acid, macelignan, acacetin, apigenin[48,289,543]	Antiinflammatory, analgesic, prostaglandin inhibitor, and antioxidant activities; cough, sore throat, chronic phyllitis, furuncles, mastitis
Leucas mollissima Wall. var. *chinensis* Benth. L. *lavandulaefolia* Rees.	(Whole plant) Flavonoids, lignans, psychopharmacological properties[20,544]	Antitoxic; treats cough, chest pain, intestinal infection; appendix, uterus infection; mammary gland infection, dysentery
Ligustrum lucidum Ait. L. *pricei* Hayata	(Fruit, root, bark) Nuzhenide, oleanolic acid, ursolic acid[2,28,125]	Increases leukocyte count, a cardiac tonic, diuretic, treats urological tumors; alleviates pain, cough, windpipe infection, irregular menses
Ligustrum sinense Lour.	(Root) Glycosides, dihydrochloride[28,290]	Stops bleeding, cough, pain; diuretic, antitoxic, antiinflammation, acute hepatitis, mouth cavity infection; protects red blood cell membrane to resist hemolysis

TABLE 1 Major Constituents and Therapeutic Values of Taiwanese Native Medicinal Plants (continued)

Scientific Name	Major Constituents and Source	Claimed Therapeutic Values
Lilium formosanum Wallace *L. speciosum* Thunb.	(Stem, bulb) Phenolic glycoside, steroidal saponin[28,291,292]	Pneumonia, throat infection; stops coughing, headache, abdominal pain; improves lung function
Limonium sinense (Girald) O. Kuntze.	(Aerial part) Samarangenin B, flavonols, flavonol glycosides, flavonol glycoside gallates, flavones, flavanones, flavan-3-ols, galic acid[29,293,294]	Stops bleeding, antitoxic; alleviates pain, irregular menstruation, urinary disorders with blood, arthritis, diabetes, bladder infection; suppresses herpes simplex virus type 1 replication
Lindera communis Hemsl.	(Fruit) Fatty acids[20]	Alleviates swelling, pain, bleeding; treats infection
Lindera glauca (Sieb. & Zucc.) Blume	(Fruit) Essential oils, cineole, limonene, caryophyliene, bornylautate, fatty acids, camphene, β-pinene[20]	Carminative properties, treats arthritis joint pain
Lindera aggregata (Sims) Kosterm. *L. okoensis* Hayata	(Root tuber, leaf) Linderane, acetyllindenanolide B-1, B-2, kaempferol, β-sitosterol, linderalactone, dehydrolindestrenolide, hydroxylinderstrenolide[295,297]	Antibacterial, antiinflammatory
Lindera strychifolia (Sieb. et Zucc.) Villar.	(Root) Linderane[3,296]	Root extract induces apoptosis in lung cancer cells
Liparis cordifolia Hook. f. *L. loeselii* (L.) L. C. Rich.	(Whole plant) Alkaloids[298]	Treats cough, infantile malabsorption, diarrhea
Liparis keitaoensis Hayata	(Whole plant) Pyrrolizidine alkaloids[3,28]	Improves lung function, stops coughing, headache, treats sore throat, abdominal pain, stops bleeding, cough, high blood pressure
Liquidambar formosana Hance	(Bark, leaf, root) Balsam (resin), cinnamic alcohol, cinnamic acid, 1-borneol, camphene, dipentane, terpene[1,126]	Antihemorrhagic, externally as antiphlogistic and astringent in skin diseases

Liriope spicata Lour.	(Root) Mucilage;[18] this herb is used to produce ophiopogon[1]	Antitussive, expectorant, emollient
Litsae acutivena Hayata	(Whole plant) Butanolides[602]	Cytotoxic
Litsea cubeba (Lour.) Persoon	(Bark, fruit) Isocorydine, magnocurarin, methyl heptenone, D-sabinene, linalool, citronellal laurotetannine, citral[2,6,11,299]	Rheumatic bone pain, headache, gastric pain; treats chronic bronchitis and bronchial asthma, protects against hypersensitization shock
Litsea japonica (Thunb.) Juss. *L. hypophaea* Hayata	(Root, leaf) Lactones, litsealacton A-B, hamabiwalactone A-B, akolactone B, akolacton B hamabiwalactone B[21,300]	Stops pain, antitoxic; improves stomach function, treats heart diseases, hernia
Lobelia chinensis Lour.	(Whole plant) Lobeline, lobelanine, pyrrolidine alkaloids, radicamines A-B, lobelanidine, isolobelamine[2,8,76,301] This herb is toxic[88]	Diuretic, increases respiration via stimulation of carotid chemoreceptors; treats snakebites; insecticide, reduces swelling, depurative, antirheumatic, and antisyphilitic
Lobelia nummularia Lam. *L. laxiflora* L.	(Leaf, stem, flower) Piperidine alkaloids[3,302]	Malaria, improves blood circulation, antitoxic, rheumatic pain, antinflammatory
Lonicera japonica Thunb. *L. japonica* Thunbery var. *sempervillosa* Hayata *L. apodonta* Ohwi	(Flower bud, whole plant) Luteolin, inositol, loganin, lonicerin, syringin, saponin, tannins, chlorogenic acid, luteolin-7-rhamnoglucoside[2,16,20,545]	Antibacterial, cytoprotective, antilipemic, antiphlogistic
Lonicera kawakamii (Hayata) Masamune *L. confusa* DC	(Buds, flower, stem, leaf) Rutin, quercetin, luteilin-7-0-β-D-galactoside, lonicern, chlorogenic acid, β-sitosterol, tetratriacontane[546]	Treats flu, intestinal infection, pneumonia, appendicitis, mammary gland infection
Lonicera macrantha DC	(Flower, stem, leaf) Lonicern, luteolin-7-rhamnoglucoside, luteolin, inositol[10,20]	Lobar pneumonia, lung empyema, mastitis, appendicitis, influenza, upper respiratory tract infection, acute conjunctivitis, ententes bacillary dysentery, rheumatic arthritis, ryodermas

TABLE 1 Major Constituents and Therapeutic Values of Taiwanese Native Medicinal Plants (continued)

Scientific Name	Major Constituents and Source	Claimed Therapeutic Values
Lonicera shintenensis Hayata	(Stem) Luteolin, inositol, tannins, lonicerin, luteolin-7-rhamnoglucoside[20]	Antitoxic, blood in night soil, arthritis, dysentery; treats infection, lowers cholesterol
Lophatherum gracile Brongn.	(Aerial part) Arundoin, cylindrin, friedelin, taraxevol, phenolics[2,6]	Antipyretic, diuretic, antibacterial, improves lung and stomach function, treats lung disease and coughing
Loropetalum chinense (Rbr.) Oliver	(Whole plant) Flavone, quercitrin, isoquercitrin[2]	Antipyretic, a detoxicant, hemostatic, treats angina pectoris, bronchitis, bleeding, alimentary indigestion
Ludwigia octovalvis (Jacq.) Raven	(Whole plant, seed) Oleanane-type triterpenes, carbohydrates, flavonoid glycoside, phenols, amino acids, fatty acids, linolenic acid[28,576,640]	Diuretic, lowers blood pressure, antiinflammation, urinary tract infection, gallbladder infection; cytotoxic activity against human cancer cell lines; antibacterial activity against *Streptococcus mutans*
Luffa cylindrica (L.) Roem.	(Fruit, fibers) Xylose, mannosan, galactan, saponins, acetic acid, valeric acid, pinenes, limonene, cineole, sterol, menthone, linalool, bourbonene, caryophyllene, menthol, carvone, vitamins A, B, C[8,18,92]	Hemostatic, analgesic in enterorrhagia, dysentery, metrorrhagia, orchitis, hemorrhoids
Lycium chinense Mill.	(Root, bark) Amino acids, zeaxanthin, cinnamic acid, betaine, peptides, acyclic diterpene glycosides, kukoamine, polysaccharide[2,13,390]	Treats oligospermia, sexual neurasthenia, dizziness; lowers blood sugar; antipyretic, antibacterial; for type I pneumocystis
Lycopersicon esculentum Mill.	(Root, leaf) Protein, vitamin A, thiamin, nicotinic acid, riboflavin[8]	Alleviates numb feelings, arthritis pain, sexually transmitted diseases

Lycopodium cunninghamioides Hayata	(Whole plant) Alkaloids, saponins[21,303,304]	Relaxes muscles, improves blood circulation
Lycopodium salvinioides (Hert.) Tagwa	(Whole plant) Complanatine, tohogenol, lycopodine, nicotine, α-obscunine, serratenediol[3]	Diuretic, rheumatic pain, muscle pain; numbs feeling, lymphatic disease
Lycopus lucidus Turcz. var. formosana Hay.	(Aerial part) Resin, lycopose, raffinose, glucose, stachyose[16]	For abdominal distention, abscesses, congestive edema, blood extravasation
Lygodium japonicum (Thunb.) Sw.	(Leaf with sporangia) Fatty oil[18]	Diuretic, antirheumatic against venereal diseases, disorders of the urinary tract
Lysimachia ardisloides Masam. L. capillipes Hemsl. L. davurica Ledeb.	(Whole plant) Triterpenoid saponins, oleanolic acid, triacontanic acid, palmitic acid, β-amyrin, stigmasterol, soyacerebroside[305-307]	Treats inflammation, irregular menses, discharge; improves blood circulation, flu, cough, windpipe infection, shortness of breath; stomachache; pain caused by arthritis
Lysimachia mauritiana Lann. L. davurica Knuth. L. simulans Hemsl.	(Whole plant) Triterpenoid saponins[3,21,666]	Treats liver and stomach diseases, hernia, stomachache, irregular menses, flu, vaginal discharge; antiinflammatory
Macaranga tanarius (L.) Muell.	(Root, fallen leaf) Tannins, prenyl flavanones[20,608]	Induces vomiting, stops cough with blood, malaria; allelopathic activity
Machilus kusanoi Hayata M. zuihoensis Hayata	(Root) Alkaloids, di-coclaurine[21,308,583]	Antiinflammation, antitoxic, cholera; free radical-scavenging activity
Maesa perluria var. formosana (Mez.) Yuen P. Yang M. japonica Moritzi.	(Root) Triterpenoid saponins[3,309]	Antiinflammation; improves stomach and spleen function; alleviates stomachache, headache, and lumbago
Maesa tenera Mez. M. lanceolata Forssk. M. laxiflora Benth.	(Whole plant) Maesaquinone, triterpenoid saponins, alkylated benzoquinones[27,310-312]	Stomachache, hepatitis; lowers cholesterol level, treats cold, headache; antioxidant activities

TABLE 1 Major Constituents and Therapeutic Values of Taiwanese Native Medicinal Plants (continued)

Scientific Name	Major Constituents and Source	Claimed Therapeutical Values
Magnolia liliflora Desr.	(Bark) Alkaloids, magnocurarine, magnoflorine, β-eudesmol, neo-lignans, magnolol, konokiol, liriodenine, crytomeridiol[127]	Central nervous system depressant action, sedative, anticovulsant, muscle relaxation
Mahonia japonica (Thumb.) DC *M. oiwakensis* Hayata	(Root, stem, leaf) Volatile compounds, aromatic compounds, benzldehyde, benzyl alcohol, indole, linalool[154,313]	Improves lung, kidney, and liver function
Mallotus apelta (Lour.) Muell-Arg.	(Root, leaf) Pentacyclic triterpenoids, benzopyran derivatives, 3α-hydroxyhop-22(29)-ene, hennadiol, friedelin, friedelanol, opifriedelanol, taraxerone, epitaraxerol[10,314-316,547]	Chronic hepatitis, hepatosplenomegaly, leukorrhea, enteritis diarrhea, prolapse of uterus and rectum, edema in pregnancy
Mallotus paniculatus (Lam.) Muell. *M. japonicus* (Thunb.) Muell.	(Stem, leaf) Amino acids, bergenin, tannin, rutin, malloprenol[6,21]	Treats wounds, improves liver function, stops bleeding, hepatitis, spleen inflammation, vaginal discharge
Mallotus repandus (Willd.) Muell.	(Stem, leaf) Mallorepine, bergenin, repandusinin, repandusinic acids, mallotinim[1,128,129,630,642]	Oxygen-scavenging activity, antihepatotoxic actions; an insecticide; alleviates itching; antiinflammatory, ulcer
Mallotus tiliaefolius (Blume) Muell.	(Leaf, root) Rottlerin, isorottlerin, 4-hydroxyrottlerine, 3,4-dihydroxyrottlerine, phloroglucinal, bergenin[3,20]	Stomach ulcer, gastritis, enteritis, skin infection
Malvastrum coromandelianum (L.) Garcke.	(Whole plant) Acetylsalicylic acid[21,317]	Antiinflammatory, hepatitis, liver infection, enteritis, diarrhea, arthritis, sore throat, cough
Manihot utilissima Pohl.	(Root) Hydrocyanic acid[101]	Used for dressing ulcerous sores
Maranta arundinacea L.	(Root) Crude protein, fat, starch, dextrin, sugars, crude fiber, ash[28,318]	Diuretic, heatstroke, flu, sore throat; improves lung function; used for oral hydration

Plant	Components	Uses
Mariscus cyperinus Vahl.	(Whole plant) Flavans, flavanones[3]	Antiinflammatory, removes accumulated blood from infection, regulates menses; headache, malaria
Marsdenia formosana Masamune	(Leaf) Triterpenoids[3,319]	Alleviates pain, arthritis pain, muscle pain
Marsilea crenata Presl. *M. minuta* L.	(Whole plant) Fatty acid, sitosterol[3]	Kidney infection, diuretic, antitoxic, vomiting with blood, blood in urine, hepatitis, vaginal discharge, irregular menses
Maytenus diversifolia (Gray) Hou	(Leaf, bark) Dulcitol, maytansine, succinic acid, syringic acid, 3-oxykojie acid, loliolide[8]	Antitumor; bark is used for cancer of the liver and stomach
Maytenus emarginata (Willd.) Hou *M. serrata* (Hochst. et A. Rich) Wilez.	(Fruit, bark, rhizome) Maytansine, maytanprine, maytanbutine, maytanvaline, maytanacine, maytansinol[2]	Treats lung cancer, breast and ovarian cancer, acute lymphocytic leukemia, colon carcinoma, kidney carcinoma
Medicago polymorpha L.	(Whole plant) Lucernol, sativol, coumesterol, formonetin, daidzein, tricin, citrulline, canaline, dicoumarol, methylene-bishydroxy-coumarin, medicagemic acid, ononitol, petunidin, malvidin, delphinidin, linalool, myrecene, limonene, serine, leucine, phenylalanine, vitamins A, B_1, C, E, K^{16}	Depurative, diuretic, stomachache; treats intestinal and kidney disorders, kidney stones, poor night vision; improves spleen, stomach function; hepatitis, gall bladder stones; diuretic; swelling
Melanolepis multiglandulosa (Reinw.) Reich. f. et Zoll.	(Root) Friedelin, triterpene[20]	Diuretic, antiinfection; used as insecticide
Melastoma candidum D. Don	(Whole plant) Flavonoids, castalagin, procyanidin B2, helichrysoside[6,320,321]	Antitoxic, antiinfection; stops diarrhea, indigestion, stomachache, stool with blood, overdischarge during menses
Melastoma dodecandrum Lour	(Whole plant)[49] Phenolics, amino acids, tannins	Prevents cerebrospinal meningitis, bacillary dysentery, rheumatism, anemia of pregnancy, menorrhagia

TABLE 1 Major Constituents and Therapeutic Values of Taiwanese Native Medicinal Plants (continued)

Scientific Name	Major Constituents and Source	Claimed Therapeutical Values
Melastoma septemnervium Lour.	(Whole plant)[49] Phenolics, amino acids, tannins	Diarrhea, melaena, menorrhagia, wound bleeding, antitoxic, antiinflammation
Melia azedarach L.	(Stem, root, bark) Toosendanin, nimbin, kulinone, methylkulonate, melianol, gedunin, melianodiol, melianotriol, melialactone, azadarachtin, nimbolins, fraxinella, palmitic acid, lauric acid, valerianic acid, butyric acid, stearic acid, cycloencalenol[2,14,18]	Treats intestinal parasites, antibacterial, anthelmintic
Melicope semecarpifolia (Merr.) T. Hartley	(Whole plant) Quinoline alkaloids[605]	Antiplatelet aggregation activity
Melissa officinalis L.	(Whole plant) Polyphenols, essential oil[154,322–324]	Antioxidant and antitumor capacity; sedative, spasmolytic, antibacterial; for incense and spice
Melodinus angustifolius Hayata	(Fruit) Melodinus[21]	Meningitis, improves blood circulation, lung function; rheumatic heart disease
Mentha canadensis L.	(Leaf) Menthol, menthylacetate, camphene, limonene, isomenthone, pinene, menthenone, rosmarinic acid[3]	Flu, headache, sore throat, skin infection
Mentha haplocalyx Briq.	(Aerial part) Menthol, menthone, menthyl acetate[2]	Stimulates gastrointestinal tract motility and central nervous system, dilates peripheral blood vessels, increases sweat gland secretion
Mesona chinensis Benth.	(Whole plant) Phenolic compounds, tannins[1,10]	Heatstroke, colds, antipyretic, antioxidant; hypertension, muscle and joint pains, diabetes, hepatitis; a remedy for gonorrhea and kidney diseases

Species	Part / Compounds	Uses
Mesona procumbens Hemsley	(Whole plant) Carbonhydre compounds[20]	Lowers blood pressure; heatstroke, flu, muscle pain, arthritis, high blood pressure, diabetes; alleviates thirst
Messerschmidia argentea (L.) Johnston	(Root)[21] No information is available in the literature	Diuretic, antitoxic, antiinflammatory; rheumatic bone pain
Michelia alba DC	(Flower bud) Acetic acid, linalool, michelabine, methylethylacetic ester, methyl eugenol, oxoushinsunine, salcifoline, ushinsunine[8]	For sapremia following miscarriage
Microcos paniculata L.	(Leaf, stem bark) Piperidine alkaloid[49,325]	Colds, heatstroke, indigestion, dyspepsia, diarrhea, hepatitis
Microglossa pyrifolia (Lam.) O. Kuntze	(Root, stem) 1-Acetyl-6-E-geranyl, acetylenic glucosides, anthocyanidins, diangelate, dihydrochalcones[20,664]	Treats malaria, improves blood circulation; irregular menses, colic; antiinflammation, vaginal discharge, stops bleeding; antiinfection, antitoxic
Mikania cordata (Burm. f.) B. L. Rob	(Whole plant) Sesquiterpene dilactone, mikanolide, dihydromikanolide, scandenolide, deoxymikanolide[27,326,327]	Antitoxic, antiinflammatory, alleviates pain; pneumonia, lung disease, windpipe infection, flu, excess white corpuscles
Millettia nitida Benth.	(Vine) Rotenone, anhydroderrid, rotenoid, phenolics[21,49]	Anemia, amenorrhea, irregular menses
Millettia pachycarpa Benth.	(Root) Rotenone, rotenoids[3]	Shortness of breath, numb feeling, abdominal pain, dizziness
Millettia speciosa Champ.	(Root)[10] No information is available in the literature	Low back pain, rheumatism, chronic bronchitis, dry cough, pulmonary tuberculosis, chronic hepatitis, nocturnal ejaculation, leukorrhea
Millettia taiwaniana (Matsum.) Hayata	(Leaf) Rotenone, anhydroderrid, isoflavonoids, millewanin A-E, rotenoids, auriculasin[1,328]	Cancer chemopreventive activity, antitumor promoter, insecticide

TABLE 1 Major Constituents and Therapeutic Values of Taiwanese Native Medicinal Plants (continued)

Scientific Name	Major Constituents and Source	Claimed Therapeutical Values
Mimosa pudica L.	(Whole plant) Minosine, flavonoids, phenolics[20,130] This herb is toxic[88]	Treats neurosis, trauma wounds, and hemoptysis; antitoxic, enteritis, insomnia, gastritis, cough; it has a tranquilizing effect
Mirabilis jalapa L.	(Root) Amino acids, betaxanthins[6]	Diuretic, improves blood circulation, antitoxic; vomiting with blood, arthritis, stomach ulcer
Miscanthus floridulus (Labill.) Warb. ex Schum. & Laut.	(Whole plant) Pentose, hexose, tricin, miscathoside[27]	Diuretic, stops diarrhea; flu, hernia
Miscanthus sinensis anders. var. *condensatus* (Hack.) Makino	(Stem) Pentose, hexose, flavonoids, misrathoside, prunia, tricin, diphenhydramine[27,548]	An antioxidant; diuretic, antitoxic, cough, vaginal discharge; lowers blood pressure; it has inhibitory mechanisms for glycoprotein fraction
Mollugo pentaphylla L.	(Whole plant) Nitre, saponin, saltpeter[1,20]	Antitoxic, antifungal, antidote, abdominal pain, diarrhea, dyspepsia
Momordica charantia L.	(Seed) Anti-HIV protein MAP 30, sterol, momordicine, elaterin, charantin, β-sitosterol-β-ᴅ-glucoside, 5, 25-stigmastardien-3β-ol-β-ᴅ-glucoside, momordicine, β-elaterin[2,20,92,131,132] This herb is toxic[88]	For immune disorders and common infections; capable of inhibiting infection of HIV-1 in T lymphocytes and monocytes; antitumor, antitoxic; stomachache, blood in stool, diarrhea
Monochoria vaginalis (Burm.) f. Presl.	(Root) Trigonelline[6]	Antitoxic, alleviates pain, tonsil infection, sore throat, vomiting, enteritis, diarrhea

Plant	Part / Constituents	Uses
Morinda citrifolia L.	(Root) Dihydroxy methyl anthraquinone, glucoside, morindin, moridon, trihydroxy anthraquinone monomethylether, morindanigrin, rubichloric acid, alizarin, α-methyl ether, rubiadin-1-methyl ether, tannins, morindadiol, masperuloside, soranjudiol, nordamnacanthal[8,132]	Treats beri-beri, cancer, lumbago, cholecystitis, increases leukocyte count, stimulates endocrine system; tuberculosis, diarrhea
Morinda umbellata L.	(Root, stem) 2-Hydroxyanthraquinone, tectoquinone, β-sitosterol, 1-methoxy-2-methylanthraquinone, alizarin, morindin, damnacanthal, rubiadin, purpuroxanthin, alizarin-1-methyl ether, rubiadin-2-methylether, munjistin, lucidin, stigmasterol[28]	Rheumatic stomachache, antitoxic, antiinflammatory, alleviates pain, improves blood circulation; flu, headache, hepatitis
Morus alba L.	(Young twig) Morin, dihydromorin, maclurin, dihydrokaempferol, mulberrin, 2,4,4',t-tetrahydroxybenzophenone, mulberrochromene, cyclomulberrochromene[2]	Antirheumatic, antihypertensive, diuretic, removes obstructions of the intestinal tract
Morus australis Poir	(Leaf) Carotenoids, adenine, choline, isoquercitrin[3,640]	Antibacterial activity against *Streptococcus mutans*, lowers blood sugar level, lowers blood pressure, diuretic; flu, coughs
Mosla punctulata (J. F. Gmel.) Nakai	(Stem) Essential oil, carvacrol, thymo-hydroquinone, *p*-cymene, phellandrene, terpinene[1,154]	Alleviates fever, thirsty feeling; lowers blood pressure; diuretic
Mucuna macrocarpa Wall. M. *nigricans* (Lour.) Steud. M. *puriens* Bits & Pieces.	(Stem, root, seed) Tetrahydroisoquinoline alkaloids[154,549]	Treats rheumatism, alleviates back pain
Muehlenbeckia platychodum (F. V. Muell.) Meisn. M. *hastulata* (Sm.) Johnst.	(Aerial part, root) Epicatechin, emodin-8-glycoside, rutin[154,329]	Antitoxic, resolves extravasated blood, antiinflammation

TABLE 1 Major Constituents and Therapeutic Values of Taiwanese Native Medicinal Plants (continued)

Scientific Name	Major Constituents and Source	Claimed Therapeutical Values
Murdannia keisak (Hassk.) Hand. *M. loriformis* (Hassk.) R. Rao et Kammathy.	(Whole plant) Quinoline, pyridine, imidazole, acrylamide[154]	Treats hepatitis; antimutagenic activity, cancer chemopreventive activity
Murraya paniculata (L.) Jack.	(Leaf, twig) L-Cadinene, methylanthranilate, bisabolene, β-caryophyllene, geraniol, carene, 5-gualzulene osthol, paniculatincoumurrayin, carene-3, eugenol, osthole, paniculatin, coumurrolin, exoticin, 8-isopentyl-limettin, 5-7-dimethoxy-8-(2,3-dihydroxyisopenty) coumarin, L-cadinene, methyl anthranilate, methyl salicylate, citronellol, S-quaiazulene, scopoletin, semicarotenoides[2,6]	Alleviates pain, removes toxic substances, an antispasmodic, antagonizes muscular spasms, rheumatic pain, skin eczema, improves blood circulation, alleviates pain
Musa insularimontana Hayata *M. paradisiaca* L.	(Whole plant) Pectin, banana lectin[3,550,551]	Diuretic, alleviates pain, flu, cough, high blood pressure
Musa sapientum L. *M. formosana* (Warb.) Hayata *Musa basjoo* Siebold var. *formosana* (Warb.) S. S. Ying	(Root, trunk juice, fruit, flower) Serotonin, norepinephrine, dopamine, musarin, vitamins A, B, C, E[1,3]	Carbuncles, all kinds of tumors, swellings, measles, headache with fever, sunburn; stimulates the smooth muscle of the intestine, treats certain forms of heart collapse, antitoxic, diuretic, antiinflammatory; stroke, high blood pressure
Mussaenda parviflora Miq.	(Leaf, root) Triterpenoid saponins, mussaenoside, shanzhiside, methyl ester, β-sitosterol, arjunolic acid, phenolics[3,133]	Treats malarial fever, diarrhea, windpipe infection, throat infection; antitoxic, improves blood circulation; enteritis, tonsillitis
Mussaenda pubescens Alt. f.	(Vine leaf) Polyphenolic compounds, triterpens, triterpenoid saponins[10,331–333]	Bronchitis, tonsillitis, influenza, colds, heart attack, pharyngitis, nephritis, edema, enteritis, diarrhea, uterine bleeding; antirespiratory syncytial virus
Myrica adenophora Hance	(Root) Myricitrin, cannabiscitrin, myricetin, gum, taraxerol, lupeol, mycinositol, α-amyrin, β-amyrin, anthocyanidin[27]	Stops bleeding, stomachache; alleviates pain, diarrhea, hemorrhoid bleeding

Plant	Constituents	Uses
Myrica rubra (Lour.) Sieb. et Zucc.	(Fruit) Myricetin, prodelphinidin, B-2, 3,3'-di-*O*-gallate[2,552]	Treats gastric pain, diarrhea, dysentery; antiproliferative activity
Myristica cagayanensis Merr. *M. fragrans* Houtt.	(Seed) Lauric acid, myristic acid, stearic acid, hexadeceronic acid, oleic acid, linoleic acid, amylodextrins, pectins, resins, campherene, cymene, dipentene, eugenol, geraniol, isoeugenol, linalool, myristicin, peinene, safrole, terpineol[8]	For hysteria, hypochondria, agoraphobia, cramps, crying jags, dysmenorrhea, amnesia; improves stomach function
Nandina domestica Thunb.	(Fruit, bark, leaf) Domesticine, nandinine, cyanic acid, nandazurine, berberine[18] This herb is toxic[88]	Antitussive
Nelumbo nucifera Gaertn.	(Leaf) Nuciferine, roemerine, anonaine, *O*-nornuciferine, liriodenine, anneparine, dihydronuciferine, pronuciferine, *N*-methylcoclaurine, *N*-methylisococlaurine[2]	Relaxing effect on smooth muscles, increases essential body energies
Neolitsea acuminatissima (Hayata) Kanehira et Sosaki	(Bark) Alkaloids[596,597]	Cytotoxic
Nephrolepis auriculata (L.) Trimen	(Stem) Sequoyitol[3]	Hepatitis, diarrhea, hernia, mammary gland infection, lymphatic gland infection
Nerium indicum Mill.	(Leaf, stem, flower, root) Oleandrin, oleandrose, neriodorin, nerioderin, karabin, scopoletin, scopoline, neriodin, ursolic acid, adynerin[2,4,11] This herb is toxic[88]	Treats psychosis, congestive heart failure; analgesic, emmenagogue; for epilepsy, asthma, externally for paronychia
Nervilia taiwaniana Ying *N. purpurea* (Hayata) Schltr.	(Whole plant) Cyclonerviol, cyclomonerviol, stigmasterol, dihydrocyclonerviol, ergosterol, epibrassicasterol, nervisterol, cyclonervilol[27]	Improves lung and liver function, cough, high blood pressure, diabetes; as a protective medicine postpartum; treats throat infection, pneumonia, high blood pressure, diabetes

TABLE 1 Major Constituents and Therapeutic Values of Taiwanese Native Medicinal Plants (continued)

Scientific Name	Major Constituents and Source	Claimed Therapeutical Values
Nicotiana tabacum L.	(Leaf) Nicotine, nicotimine, nicoteine, nicotelline[1] This herb is toxic[88]	Treats soreness in the joints, numbness, hemicrania, poisonous snakebites; insecticide, antidysenterica, emetic
Nothapodytes foetida (Wight) Sleumer *N. nimmoniana* (Graham) Mablerley	(Root, stem) 9-Methoxycamptothecine, trigonelline, scopoletin, acetylcamptothecin, linoleic acid, hydroxybenzaldehyde, scopoletin, uracil, thymine, sitosterol, sitosteryl-D-glucoside, trigonelline, camptothecin, 9-*o*-methoxycamptothecin, 0-acetyl camtothecin[27,334]	Against cancer, arthritis, hernia, swelling; it has cytotoxic activity
Nymphaea tetragona Georgi	(Flower, leaf, root) Amino acids[6,16]	A cooling lotion to apply to eruptive fevers; treats colic, gonorrhea; lowers blood pressure
Nymphar shimadai Hayata	(Root, leaf)[3] No information is available from the literature	Kidney infection; lowers blood pressure; excessive sweating
Ocimum basilicum L.	(Root) Ocimene, α-pinene, 1,8-cineole, eucalyptole, linalool, geraniol, limonene, Δ^3-carene, methyl chavicol, eugenole, eugenol methyl ether, anethole, methyl cinnamate, 3-hexen-1-ol, 3-octanone, furfural, planteose, methyl eugenol[6]	Improves digestion, antitoxic; headache, irregular menses
Ocimum gratissimum L.	(Whole plant) Ocimene, bisabolene, citronenal, thymol, pentoses, hexoses, uronic acid, D-glucose, D-galactose, D-mannose, L-aralinose, D-galacturonic acid, D-mannuronic acid[28]	Improves stomach function, blood circulation; alleviates pain, flu, headache
Oenanthe javanica (Blume.) DC	(Whole plant) α-Pinene, myrcene, *n*-butyl-2-ethyl-butylphthalate, terpinolene, diethyl phthalate, *bis* (2-ethyl butyl) phthalate[6]	Diuretic, throat inflammation, hepatitis, high blood pressure

Oldenlandia hedyotidea (DC) Hand.-Mazz *O. diffusa* (Willd.) Roxb.	(Whole plant) Glycoprotein[49,335,336]	Anticancer, antipyretic, antitoxic; prevention of heatstrokes, gastroenteritis, bleeding hemorrhoids, furuncles, abscess, eczema, lumbago
Onychium japonicum (Thunb.) Kuntze	(Spores, aerial part) Kaempferol-rhamnoside[29,642]	Antihepatoxic actions, alleviates chest and abdominal pains, stops bleeding, diuretic, detoxicant, intestinal infections
Ophioglossum vulgatum L.	(Whole plant) 3-*O*-methylquercetin[2]	A hemostatic, abscesses; treats gangrene, externally for snakebite
Ophiopogon japonicus (L.) f. Ker-Gawl.	(Root) β-Sitosterol, stigmasterol, ophiopogenins, polysaccharides, kaempferol-3-glucosylgalactoside[2,8]	Antitussive, expectorant, emollient, anticancer; smooths lung functions, stops coughing
Opuntia dillenii (Ker.) Haw.	(Stem) α-Pyrone, 4-ethoxyl-6-hydroxymethyl-α-pyrone, 3-*o*-methyl isorhamnein, 1-hepstanecanol, vanillic acid, arabinogalactan[6,337,338]	Antitoxic, stomachache, sore throat, cough, diarrhea, hernia, bleeding
Oreocnide pedunculata (Shirai) Masamune	(Whole plant, root) Diterpenoids, flavonoids, Ionone-related compounds[3]	Stops bleeding, antitoxic, rheumatic bone pain, inflammation, sore throat
Orthosiphon aristatus (Blume) Miq. *O. stamineus* Benth.	(Whole plant) Orthosiphoni, ursolic acid, β-sitosterol, myoinositol, staminane- and isopimarane-type diterpenes[6,572]	Diuretic, antiinfection, kidney infection, arthritis, high blood pressure, bladder infection, hepatitis, urinary tract stone, kidney stone; it has nitric oxide inhibitory activity
Osbeckia chinensis L.	(Whole plant) Flavonoids, tannins[49]	Bacillary dysentery, enteritis, appendicitis, sore throat, asthmatic bronchitis, pulmonary tuberculosis, hemoptysis

TABLE 1 Major Constituents and Therapeutic Values of Taiwanese Native Medicinal Plants (continued)

Scientific Name	Major Constituents and Source	Claimed Therapeutical Values
Osmanthus fragrans Lour.	(Flower) β-Phellandrene, osmane, nerol, methyl-laurate, methylmyristate, methypalmintate, uvaol, r-decanolactone, α-ionone, β-ionone, trans-linalool oxide, cis-linalool, linalool, pelargonaldehyde, β-phellandrene, dihydro-β-ionone[2,49]	Reduces phlegm, removes blood stasis, cough, rheumatic stomachache
Oxalis corniculata L.	(Leaf) Niacin, vitamin C, β-carotene, oxalate, oxalic acid[10]	Neurasthenia, hypertension, hepatitis, anthelmintic, antiphlogistic, depurative, diuretic, emmenagogue, febrifuge, lithontriptic
Oxalis corymbosa DC	(Leaf) Ferritin, organoids[339] Oxalate, vitamin C, calcium, citric acid, malic acid, tartaric acid[8]	Antidote to arsenic and mercury; for bruises, clots, diarrhea, fever, influenza, snakebite, urinary tract infections
Paederia cavaleriei L.	(Whole plant, root) Paederoside, scandoside, asperuloside, iridoid[6,21]	Dissolves phlegm; for cough, malaria; improves digestion, antitoxic, antiinflammatory; for hepatitis, flu
Paederia foetida L.	(Whole plant) Corticosteroid (hydrocortisone)[341]	Antiinflammatory
Paederia scandens (Lour.) Merr.	(Whole plant, root) Paederoside, iridoid glucosides[10,340,342]	Treats gastrointestinal spasm, bronchitis, malabsorption, malnutrition, whooping cough, bronchitis enteritis, dysentery, pulmonary tuberculosis, iceteric hepatitis
Paliurus ramosissimus (Lour.) Poir.	(Stem, root) Zizyphine A-type cyclopeptide alkaloids, paliurines G-I, nummularine H, daechuine-S3, paliurines A-C, F[614,615]	Alleviates pain, antitoxic, antiinflammatory, throat pain, arthritis pain, vomiting blood, natural ejaculation

Plant	Constituents	Uses
Pandanus odoratissimus L. f. var. *sinensis* Kanehira	(Root) Phenolics, volatile oils, methyl phenylethyl ether, dipentene, D-linalool, phenylethyl acetate, ester of phthalic acid, citral, stearoptene[6]	Diuretic, flu, hepatitis, kidney infection, urinary tract infection, measles; stops diarrhea, thyroid gland inflammation; antitoxic, high blood pressure, hernia
Pandanus amaryllifolius Roxb. *P. pygmaeus* Thouars.	(Leaf) Pandanin, pyrrolidine, pandanamine, α-methyl, α, β-unsaturated gamma-lactone[343,344]	Hemagglutinating activity; antiviral against human herpes simplex virus type-1 and influenza virus
Parachampionella flexicaulis (Hayata) Hsieh et Huang *P. rankanensis* (Hayata) Bremek	(Whole plant)[27,154] No information is available in the literature	Antitoxic, antiinflammatory; alleviates pain, fever; treats flu, mouth cavity infection, hepatitis, throat inflammation, encephalitis, parotitis
Paracyclea gracillima (Diels) Yamamoto	(Root) Insularine, insulanoline, iso-chondrodendrine[27]	Gastritis, enteritis, stomachache, antitoxic; eliminates extravasated blood, alleviates pain, rheumatic arthritis, tonsil infection
Paracyclea ochiaiana Kudo et Yamamoto	(Root) Insularine, alkaloids[27,345]	Arthritis pain, headache, alleviates pain, eliminates extravasated blood, antiinflammatory, arthritis
Paris arisanensis Hayata *P. formosana* Hayata	(Root, stem) Formosanin-C[21,639]	Antiinflammatory, antitumor; stops coughing, lymphatic gland infection, stomachache
Paris lancifolia Hayata	(Root, stem) Pariphyllin, dioscin, diosgenin, steroidal saponins[6,21]	Anticancer, appendicitis, antitoxic, antiinflammatory; alleviates pain, encephalitis, parotitis, tonsil infection, lymphatic gland infection, windpipe infection
Paris polyphylla Smith	(Root) Polyphyllin D, α-paristyphnin, diosgenin, pariphyllin, dioscin, steroidal, saponins, furostanol, gracillin, picrolonic acid, benzo[a] pyrene, spirostanol[8,134,346-348] This herb is toxic[88]	Antitoxic, anticancer, antimutagenic activity; relieves coughing, alleviates pain, antiinflammatory, windpipe infection, treats carbuncle

TABLE 1 Major Constituents and Therapeutic Values of Taiwanese Native Medicinal Plants (continued)

Scientific Name	Major Constituents and Source	Claimed Therapeutic Values
Parthenocissus tricuspidata (Sieb. & Zucc.) Planch.	(Root, stem) Cyanidin, lysopine, octopinic acid, fatty acids[6,16]	Treats arthritis, stomachache, headache, stool with blood
Passiflora foetida L. var. *hispida* (DC ex Triana & Planch.)	(Whole plant, fruit) Passifloricins, polyketides, α-pyrones[20,349,350]	Diuretic, antitoxic, cough, swelling, improves lung function, alleviates pain
Passiflora suberosa L.	(Root, fruit) Anthocyanins, narcotic compounds, hydrocyanic acid, alkaloids[1,21]	Antitoxic, alleviates pain, stops bleeding, antiinfection, pain caused by inflammation
Pedilanthus tithymaloides (L.) Poit.	(Whole plant) Galactose-specific lectin[351] This herb is toxic[88]	Hemagglutination pattern in diabetes mellitus
Pemphis acidula J. R. & G. Forst.	(Leaf) Galloyl flavonol glycosides, quercetin[352]	Antioxidant activity
Pericampylus formosanus Diels	(Whole plant, rhizome) Narcotic alkaloid, mucilage, epifriedelinol, melissic acid, palmitic acid, stearic acid, bututic acid, daucosterol[8,154,353]	Antirheumatic, stomachache, hepatitis, diabetes
Pericampylus trinervatus Yamamoto *P. glaucus* (Lam.) Merr.	(Root, stem) Epifriedelinol, melissic acid, palmitic acid, stearic acid, bututic acid, daucosterol[27,354]	Antitoxic, stops bleeding, antiinflammatory, arthritis, sore throat, abdominal pain
Perilla frutescens (L.) Brit. *P. frutescens* (L.) Brit. var. *crispa* (Thunb.) Hand.-Mazz. *P. ocymoides* L.	(Leaf, stem, seed) Perilladehyde, 1-perilla, aldehyde, apigenin, luteolin, limonene, β-caryophyllene, α-bergamotene, linalool, 3-*p*-coumarylglycoside-5-glucoside of cyanidin, 7-caffeyl-glucosides of apigenin and luteolin, anthocyanins[2,8,11,135,136]	Productive cough, colds, headache, chest oppression, wheezing, nausea, vomiting, abdominal distention, diaphoretic, stomachache, dispels chills, antiasthmatic, antitussive, liquifies sputum

Plant	Constituents	Uses
Peristrophe japonica (Thunb.) Brem. P. *roxburghiana* (Schult.) Bremek	(Whole plant) Aliphatics[11,27,355,356]	Tuberculosis, eliminates phlegm, acute bronchitis, antitoxic, antiinflammatory, windpipe infection, hepatitis, diabetes
Petasites formosanus Kitamura	(Leaf, root) Phenylpropenolyl sulfonic acid, petasiformin A, petasiphyll A, sesquiterpene, s-petasin, s-isopetasin[27,357,358,612,625]	Antioxic, antiinflammatory, sore throat, cough, high blood pressure, antioxidant activity, inhibits vascular smooth muscle contraction
Petasites japonicus (Sieb. Et Zucc.) F. Schmidt.	(Flower, root, leaf) β-Sitosterol, β-carotene, thiamin, riboflavin, niacin, ascorbic acid[8]	For colds, asthma, cough, dyspnea, tuberculosis
Peucedanum formosanum Hayata	(Root) Anomalin, coumarin, penformosin[29]	Cooling function, alleviates pain, cough, treats colds, headache
Phellodendron wilsonii Hayata et Kanehira P. *amurense* Rupr. P. *chinensis* Schmeid.	(Bark, leaf) Berberine, palmatine, quaternary alkaloids, candicine, phellodendrine, obacunone[2,579,641,650]	Antibacterial, stimulates the phagocytic activity of leukocytes, against dysentery, hepatic protective
Phoenix dactylifera L.	(Seed) Heteroxylan, carbohydrates, vitamins, dietary fiber, high mineral ion content, fatty acids[359–361]	Anticancer, improves immune function
Phyla nodiflora (L.) Greene	(Whole plant)[20] Flavonoids, nodifloretin, β-sitosterol, nodifloridin A, B.	Antiinflammatory, antitoxic, adjusts irregular menses, sore throat
Phyllanthus multiflorus Willd. P. *emblica* L.	(Fruit, whole plant, root) Gallic acid, ellagic acid, 1-*o*-galloyl-β-D-glucose, 3,6-di-*o*-galloyl-D-glucose, chebulinic acid, quercetin, chebulagic acid, corilagin, isotrictiniin[553]	Treats flu, fever, cough, throat pain, diabetes, hypertension
Phyllanthus urinaria L.	(Whole plant, fruit) Phyllanthine, phyllantidine; in leaf, phyllanthin, hypophyllanthin, niranthin, nirtetralin, phylteralin, phenolics[2,49,642]	Dysentery, nephritic edema, urinary tract infection and stone, infantile malnutrition, enteritis, conjunctivitis, hepatitis; treats coughing, promotes digestion and secretion

TABLE 1 Major Constituents and Therapeutic Values of Taiwanese Native Medicinal Plants (continued)

Scientific Name	Major Constituents and Source	Claimed Therapeutical Values
Phyllodium pulchellum (L.) Desvaux	(Whole plant) *N,N*-Dimethyltryptamine, framine, befotenine, *N,N*-dimethyltryptamine oxide, physcim-1-gluco-rhamnodies[6]	Diuretic, alleviates swelling, treats colds, pain, regulates menses
Physalis angulata L.	(Whole plant) Physalin, hystonin[1,13]	Treats sore throat, gum pain, parotitis, acute hepatitis, antifebrile, laxative, diuretic, causes uterine contractions
Phytolacca acinosa Roxb. *P. americana* L. *P. japonica* Makino	(Root) Phytolacine, phytolaccatoxin, oxyristic acid, jaligonic acid, saponins[2,49] This herb is toxic[88]	Edema, ascites, cervical erosion, leukorrhea, furuncles, pyodermas
Pieris hieracloides L. *P. taiwanensis* Hayata *P. formosa* D. Don	(Leaf) Diterpenoids, pierisformosides G-I, diphenylamine derivative[554] This herb is toxic[88]	Fresh leaves treat tinea and scabies
Pilea microphylla (L.) Liebm. *P. rotundinucula* Hayata	(Whole plant)[29] No information is available in the literature	Antitoxic, stable uterus, lung diseases, hepatitis, sore throat, stroke, brain hemorrhage, lowers blood pressure
Pinellia pedatisecta Schott	(Rhizome) Hypnotic activity[362]	Treats flu, rheumatism, hypertension
Pinellia ternata (Thunb.) Breit.	(Tuber) 1-Ephedrine, choline, amino acids[2,137] This herb is toxic[88]	Antiemetic, antitussive, antidote for strychnine intoxication

Pinus massoniana Limb.	(Pine needles) Massonianoside E, monoepoxylignan, butanone, umbelliferon[363]	Pain caused by arthritis: flu, high blood pressure, diabetes
Pinus taiwanensis Hayata	(Root)[28] 1-α-Pinene, D-longifolene, D-candinene, comphene, sylvine, toluene, xylene, styrene, naphthalene, dipentene, bornyl acetate, vitamin C	Pain caused by arthritis: flu, high blood pressure, diabetes
Piper arboricola DC	(Fruit) Piperine, chavicine, piperanine, piperic acid, piperitone, piperonal, dihydrocarveol, caryophyllaneoxide, cryptone, 1-phellandrene[21]	Treats stomachache, abdominal pain, inflammation, malaria, arthritis pain
Piper betle L.	(Whole plant, fruit) Chavicol, chavibetol, allylpyrocatechol, carvacrol, eugenol, p-cymene, cineole, eugenol methyl ether, caryophyllene, cadinene, betelphenoal[13,20,566,567]	Asthmatic bronchitis, stomach discomfort
Piper kadsura (Choisy) Ohwi *P. kawakamii* Hayata	(Whole plant) Benzofurans, lignans, kadsurenone, pellitorine, piperlonguminine, piperanine, dihydropiperlonguminine, futoamide, guineensine, chingchengenamide, piperine, retrofractamides A, B, D, brachystamide B, piperaidine, sarmentine, pipataline, benzylbenzoate[364-366]	Antiinflammatory, inhibitory activity on prostaglandin and leukotriene biosynthesis
Piper nigrum L.	(Fruit) Piperine, chavicine, piperanine, piperonal, dihydrocarveol, cryptone, caryophyllene[2,138]	Anticonvulsive, sedative
Piper sarmentosum Roxb. *P. sanctum* (Miq.) Schltdl.	(Leaf, stem) Tetradecane, hexadecane, dodecane, tetrahydropyran, p-eugenol, methyleugenol, z-piperolide, piperolactam A, demethoxyyangonin, cepharanone B, cepharadione B, epoxipiperolid[555, 556]	Antimycobacteria

TABLE 1 Major Constituents and Therapeutic Values of Taiwanese Native Medicinal Plants (continued)

Scientific Name	Major Constituents and Source	Claimed Therapeutical Values
Plantago asiatica L. *P. major* L.	(Seed) D-Xylose, l-arabinose, D-galacturonic acid, l-rhamnose, plantasan, plantenolic acid, plantagin, homoplantagin, aucubin, ursolic acid, hentriacontane[16,584]	Antileukemic activity, diuretic, expectorant, intestinal infection, diarrhea caused by bacteria
Platycodon grandiflorum (Jacq.) A. DC	(Root) Platycodigenin, polygalacic acid, platycodigenic acid, platyconin, prosapogenin, betulin, 3-*O*-β-glucosylplatycodigenin, spinasterols, platycodonin[2,16,139]	An expectorant, antitussive, analgesic
Plectranthus amboinicus (Lour.) Spreng.	(Whole plant) Plectranthin[1,27]	Encephalitis, sore throat, antitoxic, antiinflammatory; alleviates pain, flu, mouth cavity infection, high blood pressure, tonsil infection
Pleione formosana Hayata	(Root) Glucomannan[29]	Antiinflammatory; eliminates phlegm, antitoxic, sore throat
Pluchea indica (L.) Less.	(Root, leaf) Lignan, sesquiterpenes, monoterpenes, triterpenes, steroid[20,367-369]	Antioxidant activity, antiulcer, pain caused by muscle twitch, antitoxic, arthritis pain, antiinflammatory
Plumbago zeylanica L.	(Root) Plumbagin, glucose, fructose, protease, invertase, naphthaquinone, siliptinone, 3-chloroplumbagin, 3,3'-biplumbagin[4,8] This herb is toxic[88]	Bactericidal, antifertility, antitoxic, rheumatic arthritis joint pain, scabies

Species	Part (compounds)	Uses
Plumeria rubra L. cv. *acutifolia* (Poir. ex Lam.) Bailey	(Leaf, stem bark, flower) Agoniadin, cerotinic acid, fulvoplumierin, lupeol, plumieric acid, plumieride, quercetin, pectins, cerotic acid, acetyl lupeol, essential oils, geraniol, citronellol, farnesol, phenylethyl alcohol, linalool, kaempferol, aldehydes, ketones[8]	Inhibits the tuberculosis bacterium, fungicidal, stimulant, emmenagogue, febrifuge, purgative; treats dropsy, herpes, and venereal infections
Podocarpus macrophyllus D. Don var. *nakaii* (Hayata) Li et Keng	(Stem bark, leaf, root, fruit) Pinene, camphene, cadinene, podocarpene, neocryuptomerin, kaurene, edysterone, ponasterone, makisterones, hinokiflavone, sciadopitysin, prodocarpus flavones, macrephyllic acid, podototarin, totarol[6]	For ringworms, blood disorders; tonic for heart, kidneys, lungs, stomach
Podocarpus nagi (Thunb.) Zoll. & Moritz.	(Bark) Diterpenoids, totarol, totaradiol, 19-hydroxytotarol, totaral, 4-β-carboxy-19-nortotarol, sugiol, nagilactosides C-E[154,370,371]	Antioxidant action
Pogonatherum crinitum (Thunb.) Kunth P. *paniceum* (Lam.) Hack.	(Whole plant)[21] No information is available in the literature	Treats windpipe infection, flu, heatstroke, kidney infection, hepatitis, liver infection, diabetes; refrigerant, antipyretic
Pogostemon amboinicus (Lour.) Spreng	(Whole plant) Patchouli oil, patchouli acid, pogostone, eugenol, chinamic aldehyde, benzaldehyde, pogostol, patchoulipyridine, β-elemene, epiguaipyridine, caryoptyllen, α-guriunene, alloaromadendrene, γ-patchoulene, β-guriunene, α-guailene, valencene[27]	Trypanocidal activity; antiseptic; for abdominal pain, colds, diarrhea
Pogostemon cablin (Blauco) Benth.	(Twig, leaf) Essential oils, sesquiterpene hydroperoxides, patchouli alcohol, β-patchoulene, α-patchoulene, aciphyllene, α-quaiene, pogostone[1,372,373]	Trypanocidal activity; antiseptic; for abdominal pain, colds, diarrhea

TABLE 1 Major Constituents and Therapeutic Values of Taiwanese Native Medicinal Plants (continued)

Scientific Name	Major Constituents and Source	Claimed Therapeutic Values
Pollia secundiflora (Blume) Bakh. *P. serzogonian* Blume.	(Rhizome)[374] No information is available in the literature	Anthelmintic action against human *Ascaris lubricates*
Polygala aureocauda Dunn.	(Whole plant) Xanthone, dimethoxyxanthone, methylendioxyxanthone, polysaccharide[375,376]	Immunity functions
Polygala glomerata Lour.	(Whole plant) Suchilactone, chisulactone, helioxanthis[13]	For coughs, hemoptysis, bronchitis, hepatitis
Polygonatum falcatum A. Gray *P. kingianum* Coll. Et Hemsl. *P. odoratum* (Mill.) Druce	(Stem, rhizome) Liquiritigenin, isollquiritigenin, salicylic acid, fructofuranoside, kinganone, indolizinone, isomucronulator, steroidal saponin, β-sitosterol, glucoside[154,379-381]	A tonic, alleviates hot temperature feeling, antibacterial, antifungal
Polygonum chinense L.	(Whole plant) 25R-Spirost-4-ene-3,12-dione, stigmas-4-ene-3,6-dione, stigmastane-3,6-dione, hecogenin, aurantiamide[6,377]	Antitoxic, diarrhea, dysentery, sore throat, irregular menses, antiinflammatory, antiallergic properties
Polygonum cuspidatum Sieb. et Zucc.	(Root) Emodin, emodin mono-methyl ether, chrysophanic acid, chrysophanol, anthraglycoside A, B, polydain, isoquercitrin, plastoguinoae[6]	Arthritis pain, hepatitis, irregular menses, burns
Polygonum multiflorum Thunb. ex Murray var. *hypoleucum* (Ohwi) Liu, Ying & Lai	(Root, stem) Chrysophanic acid, emodin, rhein, parietin, anthrone[27]	Eliminates phlegm, stops coughing, improves blood circulation; flu, tonsil infection, sore throat, rheumatic arthritis pain
Polygonum perfoliatum L.	(Whole plant) Indican, flavonoids, phenolics[6]	Antiinflammatory, improves blood circulation, antitoxic, alleviates dysentery, high blood pressure, sore throat, vomiting with blood, stool with blood

Plant	Constituents	Uses
Polygonum plebeium R. Br. *P. paleaceum* Wall.	(Whole plant, rhizome) Picryldydrazyl, 3,5-dihydroxyl hexanoic acid[6,378]	Diuretic, insecticide, aphrodisiac, treats scabies, silkworm diseases, treats chronic gastritis, duodenal ulcers, dysentery, wound pain, hemorrhage, irregular menstruation
Pometia pinnata forst.	(Root, bark) Polycyclic compounds[29]	Ulcer, abdominal inflammation, adjusts high body temperature, antiinfection, stops diarrhea, insecticide
Portulaca grandiflora Hook. *P. pilosa* L.	(Whole plant) Betanidin, portulal, betacyanin, betanin[27]	Antitoxic, sore throat, swelling
Portulaca oleracea L.	(Whole plant) Nordrenaline, potassium salt, dopamine, catecol, pyrocatechol, DOPA[27]	Gastritis, enteritis, improves liver function, antitoxic, antiinflammation, diabetes, hepatitis
Potentilla discolor Bunge	(Whole plant) Fumaric acid, gallic acid, *m*-phthalic acid, naringenin, protocatechuic acid, kaempferol, quercetin[28]	Antitoxic, stops bleeding, alleviates pain, antiinflammatory, windpipe infection, parotitis infection, pneumonia
Potentilla leuconta D. Don *P. multifida* L.	(Whole plant) 3β, 24-Dihydroxyl-urs-12-ene, ursolic acid, euscaphic acid, tormentic acid, epihedaragenin[28,382]	Rheumatic arthritis, antitoxic, stops bleeding, cough, abdominal pain, diarrhea, irregular menses, vaginal discharge
Potentilla tugitakensis Masamune	(Root, whole plant) Flavonoids, tannins, fumaric acid, gallic acid, protocatechuic acid[28]	Dysentery, improves blood circulation, antitoxic, antiinflammatory
Pothos chinensis (Raf.) Merr.	(Whole plant)[6] No information is available in the literature	Alleviates pain, antitoxic, improves digestion, stomach gas pain, hernia
Pouteria obovata (R) (Brown) Baehni.	(Root, fruit) Polyphenolic compounds, gallic acid, (+)-gallocatechin, (+)-catechin, (−)-epicatechin, dihydromyricetin, (+)-catechin-3-*O*-gallate, myricitrin[154,383]	Antioxidant activity

TABLE 1 Major Constituents and Therapeutic Values of Taiwanese Native Medicinal Plants (continued)

Scientific Name	Major Constituents and Source	Claimed Therapeutical Values
Pouzolzia elegans Wedd. P. pentandra Benn. P. zeylanica (L.) Benn.	(Leaf) Polyphenols[383,384]	Treats gastrointestinal ailments, chemopreventative activities
Pratia nummularia (Lam.) A. Br. & Asch.	(Callus, hairy root) Polyacetylene glycosides, lobetyol, lobetyolin, lobetyolinin, tryptophan[385]	Improves blood circulation, antiinfection, antitoxic
Premna obtusifolia R. Brown P. crassa Hand. P. serratifolia L. P. microphylla Turcz.	(Leaf, stem, root) Friedelin, epifriedelanol, stearic acid, β-sitosterol, glyceroglycolipid, ceramide[386,387]	Antitoxic, treats carbuncle, diarrhea, antiinflammatory, alleviates pain
Prinsepia scandens Hayata	(Root) Prinsepiol[29]	Irregular menses, rheumatic arthritis joint infection, antitoxic, improves blood circulation, antiinflammatory, stops coughing, lymphatic gland infection
Procris laevigata Blume	(Whole plant)[6] No information is available in the literature	Antiinflammatory, pneumonia, antitoxic, arthritis pain
Prunella vulgaris L.	(Whole plant) Oleanolic acid, ursolic acid, rutin, hyperoside, caffeic acid, vitamins B_1, C, K, delphindin, cyanidin, D-camphor, D-fenchone, D-camphor, ussonic acid, prunellin[6,348]	High blood pressure, hepatitis, antitoxic, antitumor, pneumonia, antimutagenic activity
Prunus persica (L.) Batsch	(Leaf, fruit) Aglycones, flavonoids, phenolics, chlorogenic acid[388]	Pain caused by menses, stops menses, bowel movement problem, treats hemorrhoids

Species	Constituents (part)	Uses
Pseudosasa usawai (Hayata) Makino et Nemoto *P. owatrii* Makino	(Leaf) Vitamins C, E, carotenoids[28,389]	Diuretic, stops vomiting, yellowish and reddish urine, antitoxic, flu, antitumor, antineoplastic agent
Psidium guajave L.	(Root) Arjunolic acid, luteic acid, crategolic, ellagic acid, amritoside, leucocyanidin, β-sitosterol, triterpenes, guaijaverin, psidiolic acid, maslinic acid, guajavolic acid, avicularin, quercetin, eugenol, methyl benzoate, phenyl ethyl acetate, methyl laurate, gallic acid[27]	Reduces sexual desire; stomachache, abdominal pain, diabetes, dysentery, headache, flu, encephalitis
Psophocarpus tetragonolobus (L.) DC	(Whole plant) Lectin[20,390]	Antiinfection, alleviates pain, sore throat, skin disease, urinary frequency, lectin II for capillary endothelium
Psychotria rubra (Lour.) Poir.	(Root, leaf) Sitosterol, phenolics[11]	Diphtheria, tonsillitis, pharyngitis, rheumatic pain, lumbago, dysentery, typhoid fever
Pteris ensiformis Burm *P. multifida* Poiret	(Whole plant) Phenolics, flavonoids, flavonoid glycoside[27,348]	Antitoxic, insecticide, flu, parotitis infection, throat infection, tonsil infection, liver disease, high blood pressure, antimutagenic activity
Pteris semipinnata L.	(Whole plant) Flavonoids[10]	Enteritis, dysentery, hepatitis, conjunctivitis
Pteris vittata L.	(Root, stem, leaf) cis-Dihydrodehydro-diconiferyl-9-o-β-D-glucoside, lariciresinol-9-o-β-D-glucoside[27]	Abdominal pain, dysentery, stops bleeding, diarrhea, antitoxic, insecticide, flu, prevents stroke
Pterocypsela indica (L.) C. Shih	(Root) Ginsenosides, saponins, dammarane, triterpenes, oleanolic acid[656]	Treats diabetes, hypertension, heart failure, hyperlipidemia

TABLE 1 Major Constituents and Therapeutic Values of Taiwanese Native Medicinal Plants (continued)

Scientific Name	Major Constituents and Source	Claimed Therapeutical Values
Pueraria lobata (Willd.) Ohwi P. montana (Loureiro) Merrill	(Root) Puerarin-xyloside, flavonoids, puerarin, daidzein, daidzin, β-sitosterol, arachidic acid[20]	Antitoxic, alleviates pain, typhoid, headache, dysentery, high blood pressure
Punica granatum L.	(Fruit skin, whole plant) Tannins[13] This herb is toxic[88]	Chronic diarrhea, dysentery, rectal bleeding, rectal prolapse, leukorrhea, intestinal parasitism, stomachache, massive uterine bleeding
Pyracantha fortuneana (Maxim.) Li	(Root) Fiber, proteins, carbohydrates, vitamins, β-carotene[20,391,680]	Febrifuge, poultice, alleviates pain, muscle pain, antiinflammatory, stops bleeding, improves blood circulation, stops vaginal discharge
Pyrola morrisonensis (Hayata) Hayata P. japonica Sieb.	(Whole plant, root) Glycosides, phenolic glycosides, androsin, (−)-syringaresinol glucoside, homoarbutin, pirolatin, hyperin, monotropein, chimaphilin, 5,8-dihydrochimaphilin[28,392–394]	Improves kidney function, blood circulation; treats abdominal pain, diarrhea, discharge; antisepsis, antiinflammatory, antipyretic
Pyrrosia adnascens (Sw.) Ching P. petiolosa (Christ et Baroni) Ching	(Whole plant) Chlorogenic acid, eriodictyol-7-O-β-D-glucuronide, glycosides, pyrropetioside, flavonoids[29,49,395,396]	Diuretic, arthritis, nerve pain, dysentery, parotitis infection, cervical diseases, lymphadenitis, urinary tract infection
Pyrrosia polydactylis (Hance) Ching	(Whole plant) Fumaric acid, caffeic acid, iso-mangiferin[27]	Stops coughing, bleeding; treats flu, fever, sore throat, kidney infection, swelling
Quisqualis indica L.	(Seed) Potassium quisqualate, trigonelline, fatty acid, cyanidin, monoglycoside[6] This herb is toxic[88]	Insecticide, improves spleen function, digestion; treats diarrhea, cough

Plant	Constituents	Uses
Rabdosia lasiocarpus (Hayata) Hara	(Whole plant) Plectranthin, enmein (isodorin)[21]; This herb is toxic[88]	Stomach tumor, gallbladder stone, urinary tract stone; improves digestion, antitoxic, reduces appetite; gastritis, enteritis, abdominal pain
Randia spinoa (Thunb.) Poir.	(Whole plant) Spinosic acid A, B, oleanolic acid, siaresinolic acid, stigmasterol, β-sitosterol, saponin[28]	Antiinflammatory, antitoxic, arthritis
Ranunculus japonicus Thunb.	(Whole plant) Protoanemonin, anemonin, volatile oil, ranunculin, anemonin, pyrogallol, tannin[6,154,397]	Treats hepatitis, migraine, stomachache, malaria, rheumatic arthritis; analgesic and antiinflammatory
Ronunculus sceleratus L.	(Whole plant) Protoanemonin, ranuncosislin, anemonin, pyrogallol, tannins[6,11]; This herb is toxic[88]	Treats tuberculosis, adenopathy, malaria, rheumatism, arthralgia; antitoxic, antiinflammatory
Rauvolfia verticillata (Lour.) Baillon	(Root) Reserpine, ajmalicine, raunescine, serpentinine, ajmaline, rauwelline, samatine, rauwolfia A, rellosimine, peraksine, δ-yahimbing, ursolic acid, aricine, robinin, teserpine[6]	Lowers blood pressure, reduces heartbeat, slows food movement in the intestine, and improves digestion
Rhamnus formosana Matsumura	(Root) Frangulin B, anthraquinone glycosides, flavonol triglycosides, anthraquinones, emodin, flavonoids[27,398-401]	Stomach disorder, antiinfection, diuretic, mouth cavity infection, hepatitis, kidney infection
Rhinacanthus nasutus (L.) Kurz.	(Whole plant) Rhinacanthin-M, N, Q, naphthoquinone esters, lignans[6,402-404]	Anticancer, against influenza, improves liver and lung function, stops coughing, antiinflammatory, antitoxic, high blood pressure, diabetes, liver disease, pneumonia, gastritis, enteritis

TABLE 1 Major Constituents and Therapeutic Values of Taiwanese Native Medicinal Plants (continued)

Scientific Name	Major Constituents and Source	Claimed Therapeutical Values
Rhodea japonica (Thunb.) Roth ex Kunth	(Leaf) Rhodexin A[154,405,406]	Cytotoxic, a cardiotonic agent, improves heart function, diuretic, antitoxic, alleviates fever
Rhododendron simsii Planch.	(Root) Matteucinol, matteucinin, ursolic acid, andromedotoxin, cyanidin 3-glucoside, cyanidin 3,5-diglucoside, azaleatin 3-rhamnosyl glucoside, myricetin 5-methyl ether[28] This herb is toxic[88]	Lowers blood pressure, abdominal pain, anticancer, stops coughing
Rhodomyrtus tomentose (Ait.) Hassk.	(Root) Phenolics, tannins, amino acids, flavonoid glycoside[29]	Stops bleeding, diarrhea; alleviates pain, treats hepatitis, arthritis, dysentery, hernia; antitoxic
Rhoeo spathacea (Sw.) Stearn.	(Leaf) Bretylium compounds[407] This herb is toxic[88]	Antiadrenergic action
Rhus chinensis Mill. *R. verniciflua* Stockes *R. typhina* L.	(Whole plant) Tannic acid, flavonoids, gallic acid[408,409,557,558] This herb is toxic[88]	Antiviral action against herpes simplex virus
Rhus javanica L.	(Fruit) Galloylglucoses, riccionidin A, semialactone, isofouquierone, peroxide, fouquierone, dammarane, triterpenes[410,411] This herb is toxic[88]	Antidiarrheal activity
Rhus semialata Murr. var. *roxburghiana* DC *R. microphylla* Sieb. & Zucc.	(Root) 6-Pentadecyl salicylic acid[154,412]	Antiinfection, antitoxic, antithrombin

Plant	Constituents	Uses
Rhus succedanea L.	(Fruit) Heptadecatrienyl, hydroquinone, biflavonoids, amentoflavone, agathisflavone, flavonoids[154,413-415]	Antiviral, anti-HIV, antioxidant; externally for carbuncle, scabies, hemorrhoids
Rhynchoglossum holglossum Hayata	(Whole plant)[21] No information is available in the literature	Relieves fever, diuretic, stops bleeding, antitoxic; acute hepatitis, thyroid gland infection
Rhynchosia volubilis Lour. *R. minima* (L.) DC	(Stem, leaf, seed) Gallic acid, methylester, 7-*o*-galloyl catechin, 1,6-di-*o*-galloylglucose, 1-*o*-galloylglucose, trigalloylgallic acid, ellagic acid, dilactone, dehydrodigallic acid, tergallic acid, flavogallonic acid[154,416]	Antitoxic, improves blood function, antiproliferative
Ribes formosanum Hayata *R. nigrum* L.	(Whole plant, root) Anthocyanins, phenolics, α-linolenic, stearidonic, and gamma-linolenic acids[21,417-419]	Pain caused by menses, intestinal infection, diarrhea, improves blood circulation, relaxes muscle, alleviates pain, hepatitis, arthritis
Ricinus communis L.	(Whole plant) Cytochrome C, lupensterol, 30-ner-lupan-3β-ol-20-one, ricinoleic acid, palmitic acid, linoleic acid, linolenic acid, dihydroxystearic acid, rutin, triricinolein, diricinolein, ricinine, nonricinolein, methyl-trans-2-decene-4,6,8-triynoate, 1-tridecene-3,5,7,9,11-pentyne, β-sitosterol, kaempferol-3-rutinoside, vitamin C, nicotiflorin, isoquercitrin, stearin, astragalin, reynoutrin, kaempferol, quercetin, β-eleostearic acid, oleic acid, fatty acid, gallic acid, shikimic acid, ricinolein, olein, stearic acid, stearodiricinolein, isoricinoleic acid, 9,10-dihydroxystearic acid[21] This herb is toxic[88]	Mammary gland infection, lymphatic gland infection, stops bleeding, insecticide, antitoxic, stops coughing, arthritis
Rollinia mucosa (Jacq.) Baill.	(Whole plant) Rollicosin[587]	Antitumor

TABLE 1 Major Constituents and Therapeutic Values of Taiwanese Native Medicinal Plants (continued)

Scientific Name	Major Constituents and Source	Claimed Therapeutical Values
Rosa taiwanensis Nakai *R. davurica* Pall.	(Root, fruit) Tetracyclic triterpene acids, flavonoids, ethyl β-fructopyranoside, methyl 3-*O*-β-glucopyranosyl-gallate[154,420,655]	Improves intestine and lung function, diuretic, stops diarrhea, prevents lipid peroxidation
Rotala rotundifolia (Roxb.) Koehne	(Whole plant) Phenolics, flavonoids, amino acids[6]	Diuretic, antiinflammatory, antitoxic, mammary gland inflammation, dysentery, menses pain, sore throat, high blood pressure
Rubia akane Nakai *R. lanceolata* Hayata *R. linii* Chao	(Root) Alizarin, purpurin, pseudopurpurin, munjistin[6,632]	Antiplatelet, stops coughing, eliminates phlegm, vomiting with blood, blood in urine and stool, stops bleeding, windpipe infection, diuretic, irregular menses
Rubus croceacanthus Levl. *R. lambertianus* Seringe	(Leaf) Tannins, dimeric, trimeric, tetrameric, ellagitannins, lambertianins A-D[3,635]	Adjusts heat from fever, improves blood circulation, resolves extravasated blood
Rubus formosensis Kize.	(Root, seed) Protein, oil, linoleic acid, linolenic acid, saturated fatty acid, phenolics, tocopherols, ellagitannins, ellagien acid[3,422]	Improves liver function, eyesight; treats hepatitis, pimples
Rubus hirsutus Thunb.	(Whole plant, flower petal) Vitamins C, E, superoxide dismutase[154,421]	Antitoxic; heatstroke, vomiting, headache
Rubus parvifolius Hayata	(Root, stem, leaf) Flavonoids, tannins[20,49]	Hepatosplenomegaly, urinary tract infection or stone, nephritis, contusion, hematoma, rheumatism; improves blood circulation, antiinflammatory; for flu, fever, sore throat, swelling, irregular menses, vaginal discharge, hepatitis

Plant	Parts / Constituents	Uses
Ruellia tuberosa L.	(Whole plant, root) Apigenin-7-D-glucuronide[27,423]	Diuretic, antitoxic, abdominal pain, flu, hepatitis, high blood pressure, diabetes
Rumex acetosa L.	(Root)[20] Chrysophanein, quercetin-3-galactoside, tannin, vitexin, violaxanthin, vitamin C This herb is toxic[88]	Diuretic, improves stomach function, antitoxic, insecticide, vomiting with blood, stool and urine difficulties
Rumex crispus L.	(Root) 1,8-Dihydroxy-3-methyl-9-anthrone, vitamin A, tannins, emodin, chrysophanic acid[20]	Stops bleeding, antitoxic, dissolves phlegm; for hepatitis, windpipe infection, vomiting with blood
Rumex japonicus Houtt.	(Root) Chrysuphanic acid, emodin, nepodin, vitamin C[20]	Influenza, diuretic, improves stool movement, stops bleeding, insecticide; for swelling tongue, vomiting with blood
Ruta graveolens L.	(Whole plant) Gravacridonodiol, rutamarin, gravacridonol chlorine, rhein, gravacridonediol, suberenon, gravelliferone, edulinine, xanthotoxin A, byak-angelicin, chalepensin, ribalinidin, 3-(1,1-dimethylallyl)-herniarin, gravacridonetriol, marmesin, marmesinin, savinin, nonan-2-one, 2-undecanone, nonan-2-ol, undecan-2-ol, cineole, α-pinene, β-pinene, linalool, camphene, camphorene, limonene, *p*-cymene, pangelin, graveoline, graveolinine, skimmianine, kokusaginine, 6-methoxy dictamnine, psoralen, arborinine, γ-fagarine, rutarin, rutacridone, *N*-methyl platydesmin, rutalinidin, bergapten, xanthotoxin, rutacultin, umbelliferone, scopoletin, isopimpinellin, *iso*-impertorin[28]	Antiinfection, treats ulcer, against cancer, lowers blood pressure, relieves convulsions, improves blood circulation; antiinflammatory, antitoxic, high blood pressure, pain caused by hernia, irregular menses

TABLE 1 Major Constituents and Therapeutic Values of Taiwanese Native Medicinal Plants (continued)

Scientific Name	Major Constituents and Source	Claimed Therapeutical Values
Saccharum officinarum L.	(Stem) 24-Ethylidene-lophenol, 24-methylene-lophenol[27]	Improves spleen, stomach, and intestine function; stops diarrhea; antitoxic; high blood pressure; inhibits vomiting; carbuncle
Salix warburgii O. Seemann	(Leaf) Salicylates[559]	Alleviates pain, antifungal
Salvia coccinea Juss. ex Murr.	(Whole plant) Saluianin[21,29]	Stops bleeding, cooling effect, stimulates sweating, alleviates swelling
Salvia hayatana Makino ex Hayata *S. japonica* Thunb. *S. roborowskii* Maxim	(Whole plant) Lupeol, β-hydroxy-lupeol, 3β-acetyl-11β-hydroxy-lupeol, ursolic acid, β-sitosterol, daucosterol[27,424]	Improves blood circulation; diuretic, antitoxic; an insecticide for parasites
Salvia plebeia R. Br.	(Aerial part) Flavonoids, homoplantaginin, hispidulin, eupafolin, essential oils[16]	Diuretic, vermifuge, astringent
Sambucus chinensis Lindl.	(Leaf) α-Amyrin, palmitate[3]	Antitoxic, antiinflammatory, diuretic; alleviates pain, heat from fever; externally, leaves applied to the head for headache
Sambucus formosana Nakai	(Leaf) α-Amyrin, palmitate[29]	Detoxicant, stops swelling, diuretic, alleviates pain
Sambucus javanica Reinw.	(Root, stem, leaf) Flavonoids[49]	Traumatic injury, rheumatism, nephritic edema
Sanguisorba formosana Hayata *S. officinalis* L. *S. minor* Scop.	(Whole plant) Carboxylic acid, gallic acid, ellagic acid, β-glucogallin, disaccharide[154,425,426]	Stops bleeding, antitoxic, alleviates fever, antiallergic activity

Plant	(Part) Constituents	Uses
Sanicula petagniodes Hayata *S. elata* D. Don	(Whole plant) Oleanane saponins[28,427]	Alleviates fever, pain; improves blood circulation and lung function, resolves phlegm
Sansevieria trifasciata Prain	(Leaf) Abamagenin, hemolytic sapogenin, organic acid[8]	Leaf sap for earache; treats itchiness
Sapindus mukorossi Gaertn.	(Flower, fruit, seed, root) Saponin, mukorosside[1,4]	For conjunctivitis, eye diseases, removes freckles and suntan
Sapium discolor Muell. *S. sebiferum* (L.) Merr.	(Root, bark) Epifriedelinol, flavonoids, friedelin, xanathoxylin, β-sitosterol, isoquercitrin[11,27] This herb is toxic[88]	Treats constipation, Liguria cirrhosis, ascites, and carbuncle
Sarcandra glabra (Thunb.) Nakai	(Whole plant) Glucosides, essential oils, fumaric acid, succinic acid, phenolics, tannins[2,49,140,141]	Treats malignant tumors, cancer of the pancreas, rectum, stomach, liver, and esophagus; epidemic influenza, encephalitis, pneumonitis, bacillary dysentery, appendicitis, furane colossus
Saurauia oldhamii Hemsl. *S. tristyla* DC var. *oldhamii* (Hemsi.) Finet & Gagncp.	(Root with stem) Steroidal[13,27]	Treats marrow infection, flu, abdominal pain, hepatitis; antitoxic; stops bleeding, coughing; treats fever, leukorrhea, urinary tract infection, schizophrenia, hepatitis; externally used in osteomyelitis, carbunculosis
Saururus chinensis (Lour.) Baill.	(Whole plant) Quercitrin, isoquercitrin, avicularin, hyperin, amino acids[20] This herb is toxic[88]	Cleans abscesses, antimalarial, diuretic, depurative, parasiticide
Saxifraga stolonifera (Lim.) Merrb.	(Whole plant) Arbutin, caffeic acid, esculitin[20]	Antitoxic, ear infection, cough, erysipelas, hemorrhoids, vomiting with blood, whooping cough

TABLE 1 Major Constituents and Therapeutic Values of Taiwanese Native Medicinal Plants (continued)		
Scientific Name	Major Constituents and Source	Claimed Therapeutical Values
Scaevola sericea Vahl.	(Whole plant) Scuevolin[21]	Diuretic, rheumatic pain, diarrhea
Scutellaria javanica Jungh. var. *playfairi* (Kudo) Huang & Cheng.	(Whole plant)[154] No information is available in the literature	Alleviates fever, pain, improves blood circulation, antitoxic
Schefflera octophylla (Lour.) Harms.	(Root)[11] No information is available in the literature	Improves blood circulation, hepatitis, rheumatic pain, antipyretic, antiinflammatory, antiswelling, stagnant blood dispelling
Schisandra arisanensis Hayata	(Stem) Schisantherin A, B, C, D, E[29,93]	For blood vomiting, pain caused by cold, overtiredness, wounds
Schizophragma integrifolium Oliv.	(Whole plant) D-Menthone, D-limonene[27]	Carminative, refrigerant
Scirpus ternatanus Reinw. ex Miq. *S. maritimus* L.	(Whole plant) Carotenoids, sterols, stilbens[3,428]	Diuretic, stops coughing, alleviates shortness of breath, cough
Scoparia dulcis L.	(Whole plant) Amellin, dulciol, hexacosanol, mannitol, β-sitosterol, tannins, salicyclic acid, scopanol, dulcilone, betulinic acid, iflaionic acid[8,142]	Cough remedy; induces labor, used as an opium substitute, therapeutic action in diabetes
Scrophularia yoshimurae Yamazaki	(Root) Harpagoside, iridoid glycoside[27,429]	Vomiting with blood, antitoxic, high blood pressure; sore throat, tonsil infection, lymphatic gland infection
Scurrula loniceriolius (Hayata) Danser *S. ritozonensis* (Hayata) Danser *S. liquidambaricolus* (Hayata) Danser *S. ferruginea* Danser	(Whole plant) Flavonols, quercetin, quercitrin, 4″-O-acetyl quercitrin[430]	Improves blood circulation, makes muscle and bone stronger, lowers blood pressure, antiinflammatory

Scutellaria barbata Don.	(Whole plant) Flavones, flavonoid congeners, scutellarin, E-1-(4'hydroxyphenyl)-but-1-en-3-one[431,432]	Antimicrobial, antiinflammatory, antitumor agents against human uterine leiomyoma, mammalian and ovarian cancer; treats liver, lung, and rectal tumors
Scutellaria formosana N. E. Brown	(Whole plant) Flavonoids, scutellarein, tannins, amino acids[21]	Alleviates fever, pain; treats coughing, hepatitis, hemorrhoids
Scutellaria indica L.	(Whole plant) Scutellarein, phenolics, amino acids[21]	Improves liver function, blood circulation; antitoxic, antiinflammatory, treats enteritis, dysentery, vomiting and cough with blood
Scutellaria rivularis Benth.	(Root) Baicalein, baicalin, wogonin, β-sitosterol, wognoside, 7-methoxy-baicalein, 7-methoxynorwogonin, skullcap falvones[2,140,144]	Antitumor, antibacterial, antiviral, antioxidant, antipyretic, antiinflammatory, antineoplastic
Securinega suffruticosa (Palas) Rehder	(Leaf, twig, flower) Securinine, allosecurinine, securinol, dihydrosecurinine, securitinine, phyllantidine[2] This herb is toxic[88]	Treats infantile paralysis, neurasthenia, and neuroparalysis
Securinega virosa (Roxb.) Pax & Hoffm.	(Leaf, root) Virosine, norsecurinine, fluggein, dihydrosecurinine, virosecuririn, viroallosecurinine[29]	Leaves used as a maturative, a detergent; has antibiotic activity; root to treat tooth and gum disease
Sedum formosanum N.E. Br.	(Whole plant) Triterpenes amyrenone, amyrenol[6]	Treats diabetes, alleviates swelling, pain, diarrhea, improves digestion
Sedum lineare Thunb. *S. sempervivoides* Ledebour *S. morrisonense* Hayata	(Whole plant) Sedoheptose, triterpenes, δ-amyrenone, δ-amyrenol[3,27,433] This herb is toxic[88]	Antioxidant properties; antiinflammatory, diabetes; externally for wounds

TABLE 1 Major Constituents and Therapeutic Values of Taiwanese Native Medicinal Plants (continued)

Scientific Name	Major Constituents and Source	Claimed Therapeutical Values
Selaginella delicatula (Desv.) Alston	(Whole plant) Biflavonoids[27,49,434]	Irregular menses, tonsil infection, hepatitis, liver disease, pulmonary disorders, tuberculosis, hemoptysis, cholecystitis, enteritis, dysentery, throat infection; improves blood circulation, antitoxic
Selaginella uncinata (Desv.) Spring	(Whole plant) Trehalose[28]	Hepatitis, pneumonia, antitoxic, improves blood circulation, gallstones, gastritis, enteritis, dysentery
Semnostachya longespicata (Hayata) Hsieh et Huang	(Whole plant)[27] No information is available in the literature	Throat pain, hepatitis, antitoxic, stops bleeding, antiinflammatory, flu, mouth cavity infection
Senecio nemorensis L.	(Whole plant) Cynorin, macrophylline, sarracine[29]	Antitoxic, treats dysentery, intestinal infection, hepatitis, inflammation caused by carbuncle
Senecio scandens Buch.-Ham. ex D. Don	(Aerial part) Lavoxanthin, macrophylline, cynarin, chlorogenic acid, chrysanthemaxanthin, sarracine[2,16]	Antibacterial, antiplasmodial; treats acute bacterial dysentery and bronchitis
Serissa foetida Comm.	(Whole plant) Tannins, glucosides[6]	Antitoxic, dysentery, swelling, sore throat, vaginal discharge, migraine
Serissa japonica (Thunb.) Thunb.	(Whole plant)[154,435] No information is available in the literature	Antitoxic, alleviates fever, anti–herpes simplex virus, antiadenovirus activities
Sesamum indicum L.	(Seed) Olein, linolein, palmitine, stearin, myristic acid, sesamin, sesamol, pentosan, phytin, lecithin, choline, calcium oxalate, chlorogenic acid, vitamins A, B[18]	A nutrient, laxative; hyperchlorhydria; a lenitive in seybalous constipation, as nutrient tonic in degenerative neuritis, neuroparalysis
Setaria italica (L.) Beauvois	(Fruit) Daphnin, β-alanine, β-carotene[21]	Antitoxic, treats vomiting and diarrhea, dysentery, improper digestion

Plant	Part and constituents	Uses
Setaria palmifolia (Koen.) Stapf. *S. viridis* (L.) Beauv.	(Root, aerial part) Tricin, *p*-hydroxycinnamic acid, vitexin 2''-oxyloside, orientin 2''-*o*-xyloside, tricin-7-*o*-β-D-glucoside, vitexin 2''-*o*-glucoside[20,436]	Improves stomach and intestinal function; improves digestion; eating disorder, arthritis, heart disease; antioxidant, free radical-scavenging activities on 1,1-diphenyl-2-picrylhydrazyl (DPPH)
Setcreasea purpurea Boom.	(Whole plant)[29] No information is available in the literature. This herb is toxic[88]	Improves blood circulation, antitoxic, antiinflammatory
Severinia buxifolia (Poir.) Tenore	(Root bark) Tetranortriterpenoids, 7-isovaleroylcycloepiatalantin, acridone alkaloids[154,437]	Resolves phlegm, diuretic; treats snakebite
Sida acuta Burm f.	(Whole plant, seed) Quindolinone, cryptolepinone, 11-methoxyquindoline, *N-trans*-feruloyltyramine, fatty acids, vomifoliol, loliolide, 4-ketopinoresinal, scopoletin, evofolin-A, evofolin-B, ferulic acid, sinapic acid, syringic acid, vanillic acid[20,438,440]	Antitoxic, antiinflammatory, alleviates pain, treats flu, hemorrhoids; it has chemotoxonomic value
Sida rhombifolia L.	(Root, seed) Phytoecdysteroids, cyclopropenoid fatty acids[20,439,440]	Enteritis, hepatitis, flu, pneumonia, improves blood circulation, resolves phlegm, alleviates pain
Siegesbeckia orientalis L.	(Whole plant) Darutin-bitter[6]	Treats high blood pressure, numb feeling in arms and legs, rheumatic arthritis, acute hepatitis
Silene morii Hayata *S. vulgaris* Garcke.	(Aerial part) Pectic polysaccharide, silenan[20,441]	Irregular menses, relieves fever; diuretic; treats kidney infection, swelling, blood in urine
Siphonostegia chinensis Benth.	(Whole plant) Caffeoylquinic acid, lignanoids, macranthoin, syringaresinol[20,442,443]	Acute kidney infection, improves blood circulation; antiinflammatory, antitoxic, dissolves phlegm, hepatitis, gall bladder infection
Smilacina formosana Hayata	(Root, stem) Convallarin[28]	Headache, migraine, irregular menses, weakness caused by tuberculosis: aphrodisiac

TABLE 1 Major Constituents and Therapeutic Values of Taiwanese Native Medicinal Plants (continued)

Scientific Name	Major Constituents and Source	Claimed Therapeutic Values
Smilax bracteata Presl.	(Leaf, rhizome) Phenolic compounds[3,444]	Improves blood circulation; antitoxic; rheumatic leg pain, scrofula
Smilax china L.	(Root) Phenolic compounds, amino acids[10,348]	Treats cancer, rheumatic arthralgia, enteritis, diabetes, dysentery, chyluria, psoriasis; antimutagenic activity
Solanum aculeatissimum Jacq.	(Fruit) Solanine, solasonine, β-solamargine, solasurine[20] This herb is toxic[88]	For cough, asthma; diuretic; for pain
Solanum biflorum Lour.	(Whole plant) Glycoside alkaloids, steroid[20,146,147]	A detoxicant, for cough, swelling, dog bites
Solanum capsicastrum Link ex Schauer *S. abutiloides* (Griseb.) Bitter & Lillo.	(Leaf) Solanocapsine, 3-β-acetoxysolavetivone[20,445,636]	Cytotoxic, cooling effect in the body, alleviates swelling, treats liver inflammation
Solanum incanum L.	(Root) β-Sitosterol, D-glucose, ursolic acid, alkaloids, solasodine, solamargine[20]	Treats liver inflammation, lymphatic gland infection; a detoxicant
Solanum indicum L.	(Root, leaf, fruit) Diosgenin, solanidine, solanine, solasodine, alkaloids, carbohydrates, maltase, saccharase, melibiase[8,636]	Antidote for poison, treats urinary disease, cytotoxic
Solanum lyratum Thunb.	(Root, leaf, flower, fruit) Trigonelline, stachydrine, choline, solanine, nasunin, shisonin, delphinidin-3-monoglucoside, adenine, imidazolylothylamine, solasodine, arginine glucoside[16] This herb is toxic[88]	For arthritis, respiratory disorder, swelling, cough, diarrhea, blood in the urine

Solanum nigraum L. *S. undatum* Lum.	(Whole plant) Solanigrines, saponin, riboflavin, nicotinic acid, vitamin C[2] This herb is toxic[88]	Antibacterial, diuretic, treats mastitis, cervicitis, chronic bronchitis, dysentery
Solanum verbascifolium L.	(Root, leaf) Solasonine[6,49]	Treats dysentery, intestinal pain, fever, stomachache, chronic granulocytic leukemia
Solidago altissima L.	(Whole plant) Allelopathic polyacetylene[154,446]	Antibacterial; for infection, stops bleeding
Solidago virgo-aurea L.	(Whole plant) Caffeic acid, chlorogenic acid, cyanidin-3-glucoside, flavonoids, astragalin, cyanidin-3-gentiobioside, kaempferol-rhamno glucoside, hydroxycinnamic acid, quinic acid, polygalactic acid[8,15]	Decoagulant, carminative; for bladder ailments, cholera, diarrhea, dysmenorrhea
Sonchus arvensis L. *S. oleraceus* L.	(Whole plant) Inositol, mannitol, taraxasterol, palmitic acid, stearic acid, tartaric acid, lactucerols[8]	Used as an insecticide; treats asthma, bronchitis, cough, ophthalmia, insomnia, pertussis, swellings, tumors
Sophora flavescens Aiton	(Root) D-Oxymatrine, D-sophoranol, cytisine, L-anagyrine, L-baptifoline, L-methylcytosine, trifolirhizin, D-matrine, norkurarinone, kuraridin[148]	Anthelmintic, antipruritic, treats irregular heartbeat, eczema, acute dysentery, trichomoniasis
Sophora tomentosa L.	(Seed, leaf, root) Cytisine (sophorine)	For diarrhea, cholera, colic, dysentery
Spilanthes acmella (L.) Murr. *S. acmella* var. *oleracea* Clarke	(Whole plant) α-Amyrenol, β-amyrenol, myricyl, stigmasterol, sitosteryl-*o*-β-D-glucoside, spilanthol[27]	Aphrodisiac, depurative, diuretic, ophthalmic; a tonic
Spinacia oleracea L.	(Stem, leaf) Polypeptides[154,447]	Improves blood function, stops bleeding and stool with blood; treats leukemia

TABLE 1 Major Constituents and Therapeutic Values of Taiwanese Native Medicinal Plants (continued)

Scientific Name	Major Constituents and Source	Claimed Therapeutical Values
Spiraea formosana Hayata	(Root) Spiradin A, B, C, D, F, G, spiraine[28]	Treats cough; alleviates pain, headache; antitoxic; marrow infection
Spiraea prunifolia Sieb. & Zucc. var. *pseudoprunifolia* (Hayata) Li	(Root with stem) Prunioside A[28,448]	Improves blood circulation; treats fever, sore throat
Spiranthes sinensis (Pers.) Ames.	(Aerial part, root) Homocyclotirucallane, sinetirucallol, dihytrophenanthirenes, sinensols G, H[10,449]	Tonsillitis, sore throat, debility, neurasthenia, cough, tuberculosis, hemoptysis
Stachys sieboldii Miq.	(Whole plant) Acteoside[28,450]	Treats pneumonia, improves blood circulation, flu, urinary tract infection
Stachytarpheta jamaicensis (L.) Vahl.	(Whole plant) Phenolics[6]	Antitoxic, antiinflammatory; alleviates pain, rheumatic arthritis pain
Stellaria media (L.) Cyr.	(Whole plant) r-Linolenic acid, octadecatetraenoic acid[16]	A postpartum depurative, emmenagogue, lactagogue; promotes circulation, treats mucus disorder; externally for rheumatic pains, ulcers, wounds
Stemona tuberosa Lour.	(Root) Stemonine, isotemonidine, stemondidine, protostemonine[2,8] This herb is toxic[88]	Suppresses excitation of the respiratory center and inhibits the cough reflex; antitubercular, antibacterial, antifungal
Stephania hispidula Yamamoto	(Stem) Tetrandrine, fangchinoline, menisin, demethyltetrandrine, menisidine, cyclanoline, flavonoid glycoside, phenals, stephanine, steponine, stemholine, protostephanine, prometaphanine, metaphanine, epistephanine, insularine, stemhanoline[29]	Pain caused by arthritis; antitoxic; alleviates pain, rheumatism

Species	Constituents	Uses
Stephania cephalantha Hayata	(Root) Cepharanthine, isotetrandrine, cycleanine, cepharanoline, berbamine, cepharamine, homoaromoline, tetrandrine, trilobine, hamoaromaline, papaverine, berberine, morphine, codeine, quine methyl-isochondodendrine, lycopene[2,29]	Diuretic, antiphlogistic, antirheumatic, analgesic, antiinflammatory
Stephania japonica (Thunb.) Miers	(Root) Stephanine, protostephanine, epistephanine, hypoepistephanine, homostephanoline, metaphanine, prometaphanine, hasubanonine, insularine, cyclanoline, steponine, stephanoline, stepinonine[2]	Treats nephritic edema, urinary tract infection, rheumatic arthritis, sciatic neuralgia
Stephania tetrandra S. Moore	(Whole plant) Longanine, stephanoline, tetrandrine, bisbenzylisoquinoline alkaloid[11,13,451]	Treats pyelonephritis, cystitis, chronic nephritis, enteritis, rheumatic arthritis, gastric duodenal ulcer
Sterculia nobilis (Salish.) R. Br. S. lychnophora Hance	(Fruit, seed) Fatty acids, 9,12 (Z,Z)-octadecadienoic acid, hexadecanoic acid, octadecanoic acid[6,452]	Insecticide; improves liver function, stops coughing, alleviates fever, treats abdominal pain, vomiting, hernia
Stevia rebaudiana Bertoni	(Stem, leaf) Stevioside, steviolbioside, rebaudiosides, austroinulin[21]	Treats diabetes, blood pressure; a tonic
Strychnos angustiflora Benth.	(Seed) Strychnine, brucine[11] This herb is toxic[88]	Treats rheumatic arthralgia, hemiplegia
Swertia randaiensis Hayata	(Whole plant) Swertianmarin, aleanonic acid, gentianine[29,646]	Antiinfection, antitoxic; treats gastritis, stomachache, hepatitis, tonsil infection
Symphytum officinale L.	(Leaf, root) Pyrrolizidine alkaloids, symphytine, sponbaneous, allantoin, anadoline, echimidine[6]	Improves blood function; stops diarrhea, high blood pressure; against cancer

TABLE 1 Major Constituents and Therapeutic Values of Taiwanese Native Medicinal Plants (continued)

Scientific Name	Major Constituents and Source	Claimed Therapeutical Values
Tabernaemontana amygdalifolia Jacq.	(Whole plant) Alkaloids, homocylindrocarpidine, 17-demethoxy-cylindrocarpidine, 10-oxo-cylindrocarpidin[453,454]	Antinflammatory, antipyretic, antinociceptive activities
Tabernaemontana pandacaqui Poir.	(Root, stem) Linoleic acid, alkaloids[455-457]	Anticancer, antitoxic; alleviates pain, lowers blood pressure; throat pain, arthritis pain, mammary gland infection
Tabernaemontana divaricata (L.) R. Br. ex Roem & Schult.	(Whole plant) Tabernaemontanin, coronaridin[21] This herb is toxic[88]	Improves blood vessels, lowers blood pressure; anticancer, headache, thyroid gland
Tagetes erecta L.	(Leaf, flower) α-Terthienyl, D-limonene, L-linalool, tagetone, *n*-nonyl aldehyde[8]	Treats sores and ulcers, colds, conjunctivitis, cough, mastitis, mumps
Taiwania cryptomeriodes Hayata	(Bark, heartwood) Sesquiterpenes, citerpene ferruginol, lignan helioxanthin, podocarpane-type trinorditerpenes[617,655]	Cytotoxic against human colon adenocarcinoma (HT-29) cell line; insecticide
Talinum paniculatum (Jacq.) Gaertn. *T. patens* (L.) Willd.	(Whole plant, root) Sitosterols, sugar alcohols[49,458]	Debilitation, weakness, sweating, cough, diarrhea, enuresis, irregular menses
Talinum triangulare Willd.	(Root, leaf) Cadmium, copper, iron, lead, manganese, zinc[1,459,663]	A tonic for general weakness; treats inflammation, swelling
Tamarix juniperina Bunge *T. chinensis* Lour.	(Young shoot, flower, gum) Quercetin-monomethylether[1,16,641]	Antihepatotoxic actions; treats colds, vomiting blood, respiratory infection
Taraxacum formosanum Kitamura	(Aerial part) Taraxasterol, choline, inulin, pectins[6]	Cure for swollen breasts, a diuretic; treats fever, trachealis, hepatitis, tonsillitis

Species	(Part) Constituents	Uses
Taraxacum mongolicum Hand-Mazz.	(Aerial part) Taraxasterol, taraxerol, taraxol, taraxacerin, taraxacin, cryptoxanthin, zeaxanthin, lutein, antheraxanthin, violaxanthin, neoxanthin, myristic acid, lauric acid, palmitic acid, stearic acid, β-sitosterol, β-amyrin, cysteic acid, cysteine, cystine, serine, glycine, asparagine, lysine, alanine[2,16]	Antibacterial, antispirochetic, antiviral; a choleretic agent
Taraxacum officinale Weber.	(Root) Inulin, essential oils, choline, hydroxycinnamic acids, carotenes, ether oils, monoterpene, oxalic acids, hydrocyanic acid, sesquiterpene glucosides, flavonoids, hydroxybenzoic acid, coumarins, anthocyanidines, anthraquinones, phytosterines, squalene, cerylic alcohol, arabinose, vitamins A, B, C[4,149,150]	Sudorific, stomachache; tonic; a remedy for sores, boils, ulcers, abscesses, snakebite
Taxillus matsudai (Hayata) Danser T. *levinei* (Merr.) H. S. Kiu	(Whole plant) Protocatechuic acid, isoquercitrin, quercetin-3-*O*-(6″-galloyl)-β-D-glucoside, quercetin-3-*O*-β-D-glucuronide[20,460]	Alleviates pain, arthritis pain, stroke; treats lymphatic disease, pimples
Taxus mairei (Lemee & Leveille) S. Y. Hu	(Bark, root) Taxoids, taxumairirols X–Z, taxane diterpenoids[590,628]	Anticancer
Terminalia catappa L.	(Leaf) Punicalagin, punicalin[624]	Antioxidant and hepatoprotective activity
Ternstroemia gymnanthera (Wight & Arn.) Sprague	(Bark, leaf) Tannins[6]	Alleviates pain, malaria, mammary gland infection; antitoxic
Tetrastigma dentatum (Hayata) L. T. *formosanum* (Hemsl.) Gagnep. T. *umbellatum* (Hemsl.) Nakai T. *hemsleyanum* Diels et Gilg.	(Whole plant) 6′-*O*-Benzoyldaucosterol, daucosterol, β-sitosterol[3,20,21,463,464]	Antitoxic; arthritis, inflammation, skin diseases; antiinflammatory: lymphatic gland infection, alleviates fever, rheumatic arthritis, sore throat
Tetrapanax papyriferus (Hook.) K. Koch.	(Stem) Fatty acids[20]	Improves digestion, swelling, lymphatic disease; diuretic
Teucrium viscidum Blume	(Whole plant) Phenolics, amino acids[49]	Hematemesis, epistaxis, melaena, dysmenorrhea, contusion, hematoma, furunculosis

TABLE 1 Major Constituents and Therapeutic Values of Taiwanese Native Medicinal Plants (continued)

Scientific Name	Major Constituents and Source	Claimed Therapeutical Values
Thalictrum fauriei Hayata	(Root) Flavonoids, fetidine, phetidine, thalfoetidine, thalpine, thalphinine, rhalidasine, hernandezine, thilic simidine, coptisine, oxypurpureine, berbamine, isotetrandrine, allocryptopine, oxycanthine, isothalidenzine, glaucine, berberine, palmatine, jatrorhizine, protopine, cryptopine, thalidezine[16,39]	Anticancer; treats fever, nausea, thirst, hemorrhages, conjunctivitis
Thevetia peruviana (Pers.) Schum.	(Seed, flower, leaf) Thevetin A, B, theveside, peruvosides, vertiaflavone, theviridoside[2] This herb is toxic[88]	Tranquilizing effect; treats congestive heart failure
Thladiantha nudiflora Hemsl. ex Forb. & Hemsl.	(Tuber) Dubiosides D-F, saponins, quillaic acid[28,662]	Diuretic, antitoxic, antiinflammatory; headache, enteritis, dysentery caused by fungus
Tinospora tuberculata Baumee	(Root, leaf) Tinosporin[154,467]	Treats stomachic, hepatitis, diabetes, rheumatic arthritis
Tithonia diversifolia (Hemsl.) Gray	(Stem, leaf, root) Tagitinin C, sesqiuterpene lactone[154,435,468-470]	Antitoxic, antinfection, antinflammatory; alleviates pain; antimalaria, analgesic properties; suppresses the replication of HSV-1 and HSV-2
Toddalia asiatica (L.) Lam.	(Root) Clelerythrine, toddaline, diosmin, dihydrocherythrine, toddalinine, skinmianine, berberine, eugenol toddaculine, citronella, β-sitosterol, pimpinellin, isopimpinellin, aculeatin, toddafolactone[6]	Alleviates pain, stops bleeding, improves stomach function; for arthritis, vomiting with blood, stomach diseases
Toona sinensis (Juss.) M. Roem.	(Root) Toosendanin, sterol, vitamins B, C, catechol, carotene	Insecticide; stops bleeding, alleviates pain; for pain caused by nerve, liver diseases; antitoxic

Plant	Components (Part)	Uses
Torenia concolor Lindley var. *formosana* Yamazaki	(Whole plant) Anthocyanin[20,471]	Antinfection, antitoxic; heatstroke, muscle pain, flu, dysentery
Tournefortia sarmentosa Lam.	(Root, stem) Supinin[20,593]	Antilipid peroxidative principles, antitoxic, antiinflammatory; treats ulcer
Trachelospermum jasminoides (Lindl.) Lem.	(Aerial part) Lignans, cyclitol, crystalline components[8,472-474] This herb is toxic[88]	Antiinflammatory
Trichosanthes cucumeroides (Seringe) Maxim. ex Fr. & Sav.	(Root) Kaempferitrin, choline, trichosanic acid, r-guanidinobutyric acid, α, β-diaminopropionic acid[3]	Resolves extravasated blood, treats thirst, hepatitis, stomach upset, irregular menses, throat infection
Trichosanthes homophylla Hayata *T. dioica* Roxb.	(Root, seed) Galactose-specific lectin, disaccharides, flavonoids, quercetin[3,475,476]	Antitoxic; alleviates pain, treats fever and thirst, hepatitis, throat inflammation and pain
Tricytis formosana Baker	(Whole plant) Anthocyanins, flavonoids[29,661]	Bladder infection, tonsil gland infection, pneumonia, diuretic; antitoxic, antiinflammatory; flu, sore throat
Tridax procumbens L.	(Whole plant) *bis*-Bithiophene, flavonoid, procumbenetin[154,477-479]	Diuretic; alleviates fever, liver diseases, high blood pressure
Tripterygium wilfordii Hook f.	(Root) Triptolide[29,480,569]	Pneumonia, itchy skin; an insecticide; antinfection, antitoxic; alleviates pain, arthritis
Tropaeolum majus L.	(Whole plant) Benzyl isothiocyanate, erucic acid, glucotropaolin, benzyl mustard oil, α-phenylcinnamic acid, nitrile, erucic acid, benzyl isothiocyanate, kaempferol glucoside, isoquercitroside[13]	For conjunctivitis; antitoxic; earache, sore eye
Tubocapsicum anomalum (Fr. & Sav.) Makino	(Whole plant) Ergostane derivatives[6,676]	Dysentery, carbuncle, kidney infection, inflammation

TABLE 1 Major Constituents and Therapeutic Values of Taiwanese Native Medicinal Plants (continued)

Scientific Name	Major Constituents and Source	Claimed Therapeutical Values
Turpinia formosana Nakai	(Root) Alkaloids[21,660]	Stomachache, spleen inflammation, antiinflammatory; alleviates pain; hepatitis, dysentery, pain caused by menses
Tylophora lanyuensis Liu & Lu *T. atrofolliculta* Liu & Lu	(Root) Phenanthrindolizidine alkaloids, tylophoridicines C–F, tylophorinine, tylophorinidine, R-(+)-deoxytylophorinidine[482]	Treats asthma, bronchitis
Tylophora ovata (Lindl.) Hook. ex Steud.	(Whole plant) Flavonoids, essential oil[6,570]	Dissolves phlegm, stops cough, cough with shortness of breath, arthritis
Typhonium divaricatum (L.) Decne.	(Leaf, tuber) Antineoplastic agents, phytogenic[8,14,483] This herb is toxic[88]	An expectorant, rubefacient; used for cough and pulmonary disorders
Uncaria hirsuta Haviland *U. rhynchophylla* Miq. *U. kawakamii* Hayata	(Stem) Rhynchophylline, corynoxeine, *iso*-rhynchophylline, isocorynoxeine, corynantheine, hirsutine, hirsuteine[2,630,648]	A sedative, anticonvulsive, lowers blood pressure; it has a triphasic effect; treats childhood epilepsy; oxygen-scavenging activity; dissolves artificial bladder calculi
Uraria crinita (L.) Desv. ex DC *U. lagopodioides* (L.) Desv.	(Leaf, root) Vitexin, vitexin-7-0-glucoside, orientin-7-0-glucoside, saponartin-4'-0-glucoside[29]	Treats hemorrhoids, dysentery, diarrhea, cough, pain, arthritis, irregular menses
Urena lobata L.	(Whole plant) Phenols, sterols[20]	Dysentery, vaginal discharge, antitoxic; alleviates pain, flu, fever, appendicitis, stomach diseases
Urena procumbens L.	(Leaf, twig) Phenols, flavonoid glucosides, amino acids[21]	Treats rheumatism, toothache

Plant	Part / Constituents	Medicinal uses
Urtica thunbergiana Sieb. & Zucc. *U. dioica* L.	(Leaf) Phenolic compounds, lectin, pyrocatechol equivalent, N-Acetylglucosamine[154,484-486,560] This herb is toxic[88]	Treats pain caused by hernia, snakebite; inhibits protease activity; inhibits adenosine deaminase activity; antimicrobial, antiulcer, and analgesic activities
Vaccinium japonium Miq. *V. myrtillus* L.	(Whole plant) Anthocyanin, sambubiosides[28,561]	Pain caused by rheumatic arthritis; antitoxic; alleviates pain; antiinflammatory; flu, fever, sore throat
Vaccinium emarginatum Hayata	(Root) Hentriacontane, friedelin, epipriedelinol, isoorientin, p-hydroxycinnamic acid, myoinositol, quercetol[11]	Urinary tract infection, antitoxic; bladder infection, acute arthritis
Vandellia crustacea (L.) Benth. *V. cordifolia* G. Don	(Whole plant) Phenoxy benzamine, isoproterenal[11,29,487]	Diuretic, hypertensive effect; enteritis, diarrhea, vaginal discharge, irregular menses, antitoxic, flu, hepatitis, kidney infection, swelling
Ventilago leiocarpa Benth.	(Root) Anthraquinones, ventilagolin, naphthoquinones[610,611]	Cytotoxic activity
Veratrum formosanum Loesener	(Root) Protoveratrine, jervine, alkaloids, veratramine[20] This herb is toxic[88]	Lowers blood pressure, stops vomiting; antifungal, a stimulant
Verbena officinalis L.	(Aerial part) Verbenalin, verbenalol, adenosine, tannins, essential oils[2]	Antiplasmodial, antibacterial, antitoxic, antiinflammatory
Vernonia cinerea (L.) Less.	(Leaf, root) Triterpinoid, alkaloids, saponin[153]	As restorative, febrifuge, antidiarrhetic; treats colic, stomachache
Vernonia gratiosa Hance	(Root with stem) Flavonoid glycoside, mannitol, 6-hydroxyluteolin[28]	Antitoxic; alleviates pain, flu, headache, parotitis, gland infection, sore throat

TABLE 1 Major Constituents and Therapeutical Values of Taiwanese Native Medicinal Plants (continued)

Scientific Name	Major Constituents and Source	Claimed Therapeutical Values
Veronicastrum simadai (Masamune) Yamazaki	(Whole plant) Catalpol, veronicoside, catalposide, amphicoside, verproside[29,488]	Diuretic, antiinflammatory, antitoxic; cough, swelling, lymphatic disease, hepatitis, irregular menses
Viburnum plicatum var. *formosanum* Y. C. Liu et C. H. Ou *V. odoratissimum* Ker Gowl. *V. awabuki* K. Koch. *V. luzonicum* Rolfe.	(Root, leaf, flower) Luzonoside A-D, luzonoid A-G, iridoid glucosides, *p*-coumaroyl iridoids, lupane triterpenes, diterpenoids, vibsane diterpenoids[21,489,573,577,592]	Antitoxic; improves digestion, flu, lymphatic gland infection, pain caused by rheumatic arthritis
Vigna radiata (L.) R. Wilez. *V. angularis* (Willd.) Ohwi & Ohashi *V. umbellata* (Thunb.) Ohwi & Ohashi.	(Seed) Phytic acid, tannic acid, phospholipids, phosphatidic acid, phosphatidylcholine, phosphatidylethanolamine[609,658,659]	Hepatoprotective and trypsin activities
Viola confusa Champ. ex Benth. *V. yedoensis* Makino.	(Leaf) Flavone c-glycosides[3,493,568]	Reduces hot feeling from fever; antitoxic, antibacterial activity; improves digestion, treats carbuncle, scrofula
Viola diffusa Ging. *V. tricolor* L. *V. betonicifolia* J. E. Smith	(Whole plant) Cyclotides[3]	Hepatitis, carbuncle; alleviates pain and inflammation, pertussis, whooping cough, acute conjunctivitis
Viola inconspicua Blume ssp. *nagasakiensis* (W. Becker) Wang & Huang *V. mandshurica* W. Becker	(Whole plant, fruit) Violutoside, violanin, violutin, rutin, rutinoside, violaxanthin, orientin, isoorientin, vitexin, saponaretin, oxycoccicyanin, myrtillin-a, auroxanthin, xeaxanthin, xanthophyll, phytofluene, flavoxanthin, myricetin, tocopherol, lycopene, saponin[3,28,154]	Antitoxic; carbuncle; antiinflammatory; treats poison caused by inflammation
Viola verecunda A. Gray *V. hondoensis* Becker et Boss. *V. philippica* Cav.	(Whole plant) Carboxylic acid[154,490,491]	Antibacterial; leaves used externally for inflammation and infection

Viscus alniformosanae Hayata *V. angulatum* Heyne.	(Whole plant) Oleanolic acid, β-sitosterol, β-amyrin, mesoinositol, lupeol, flavonoids, phenolic glycosides[20,594]	Improves liver, kidney function; treats arthritis, menses problems
Viscus multinerve (Hayata) Hayata	(Whole plant) Triterpenoids, β-amyrin, lupeol, campesterol, oleanolic acid, stigmasterol, β-sitosterol, stigmasterol, betulic acid, oleanolic acid, campesteryl, stigmasteryl, β-sitosteryl glucoside, flavonoids, homoeriodictyol, naringenin, rhamnazin-3-glucoside, homoeriodictyol-7-glucoside, *n*-heptacosane, *n*-octacosane, *n*-nonacosane, *n*-tehacosanol, *n*-octacotanol, palmitic acid, stearic acid, arachidic acid, *n*-tricosanoic acid, lignoceric acid, *n*-pentacosanoic acid, cerotic acid, *n*-octacosanoic acid, triterpenoids, rhamnazin, rhamnazin-3-glucoside, homoeriodidyol-7-glucoside, β-amyrin acetate, β-amyrenone[20]	Improves liver, kidney function; treats arthritis, menses problems
Vitex cannabifolia Sieb. et Zucc.	(Leaf, fruit, root) α-Pinene, linalool, terpinyl acetate, β-caryophyllene, caryophyllene oxide[49,494]	Influenza, malaria, enteritis, dysentery, genitourinary tract infection, eczema, dermatitis, asthma, epigastric pain, dyspepsia
Vitex negundo L.	(Leaf, fruit, root) Essential oil, phenolic, aucubin, cineol acid, pinene acid, dipentene, citronellol, geraniol, eugenol, camphene, delta-3-carene, tannic acid, nishindine, hydrocotylene, gluconomitol, hydroxybenzoic acid, iridoidglycoside-nishindaside, negundoside, agnuside, casticin, orientin, isoorientin, essential oil[4,8,495]	An astringent, sedative; for cholera, eczema, gravel, anxiety, convulsions, cough, headache, vertigo
Vitex rotundifolia L. f.	(Fruit, leaf, shoot) Camphene, pinene, vitricine, terpenylacetate, aucubin, agnuside, casticin, orientin, isoorientin, luteolin-7-glucoside, vitexicarpin, casticin, flavones[8,16]	For fever, analgesic, sedative, promotes beard growth, treats breast cancer

TABLE 1 Major Constituents and Therapeutic Values of Taiwanese Native Medicinal Plants (continued)

Scientific Name	Major Constituents and Source	Claimed Therapeutical Values
Vitis thunbergii Sieb. & Zucc.	(Root, stem) Tannins[3]	Dysentery, hepatitis; diuretic, antitoxic, antiinflammatory; stops bleeding, pain caused by rheumatic arthritis
Wedelia biflora (L.) DC *W. chinensis* (Osbeck) Merr.	(Root) Wedelolactone, caffeic acid derivatives, wedelosin, flavonoid glycosides[3,638,642,644,653]	Antihepatotoxic activity; dysentery; antitoxic; resolves phlegm; antiinflammatory; whooping cough, diphtheria, rheumatic arthritis infection; antihepatotoxic actions; externally used for skin ulcer
Wendlandia formosana Cowan	(Whole plant) Iridoid glucosides[496]	Antioxidant; against diphenyl picryl hydrazyl (DPPH) and hydroxyl radical and peroxynitrite
Wikstroemia indica (L.) C. A. Meyer	(Root) Wikstroemin, hydroxygenkwanin, daphnetin, acidic resin[2,20] This herb is toxic[88]	Treats pneumonia, parotitis, lymphatic gland infection; antitoxic, antiinflammatory; alleviates pain; diuretic
Xanthium sibiricum Patrin *X. strumarium* L.	(Fruit) Xanthinin, xanthumin, xanthanol, isoxanthanol, strumaroside, tetrahydro flavone, caffeic acid, dicaffeoxylquinic acid[2,16]	Antibacterial, antitussive, respiratory-stimulating effect; lowers blood pressure and blood sugar level
Youngia japonica (L.) DC	(Whole plant) Phenolic compounds, tannins[20,497]	Tonsil gland infection; diuretic, antitoxic, antiinflammatory; alleviates pain, flu, sore throat, hepatitis, mammary gland infection, vaginal discharge
Zanthoxylum ailanthoides Sieb. & Zucc.	(Aerial part) Essential oil, limonene, cumic alcohol, linalool, myrcene, benzene, sabinene, rerpinenol, piperitone, β-gurjunene, α-pinene, geraniol, estragole, cadinene, clovene[39]	Treats chills, influenza, sunstroke, indigestion

Species	Constituents	Uses
Zanthoxylum avicennae (Lam.) DC	(Whole plant) Avicine, avicinine, dictamnine, hesperidin, α-pinene, limonene, evicennin, furfuraldehyde, nitidine, oxynitidine, diosmin, skimmianine[28,498]	For epigastric pain, vomiting, diarrhea, abdominal pain due to intestine parasites, ascairasis; externally used for eczema
Zanthoxylum nitidum (Roxb.) DC; *Z. integrifoliolum* Merr.	(Root, leaf, fruit) Nitidine, oxynitidine, vitexin, 6-ethoxy-chelerythrin, diosmin, oxynitidine, oxychelerythrin, *N*-desmethylchelerythrine, skimmianine, bishordeninyl alkaloid, integramine, alfileramine, α-allocryptopine, pseudoprotopine, (+)-seasamin, zanthonitrile, diacetyltambulin, dimethylallyl ether, indolopyrido quinazoline[2,8,39,620,621,627]	Analgesic, anodyne, antitumor, against leukemia, carminative, detoxicant; increases blood circulation, antiplatelet aggregation activity
Zanthoxylum pistaciflorum Hayata; *Z. piperitum* (L.) DC; *Z. dimorphophylla* M. Kato	(Root, stem, fruit) Aliphatic acid amides, hydroxyl group, coumarins[3,499,500]	Relaxes muscles; antitoxic; alleviates pain
Zebrina pendula Schnizl.	(Stem, leaf) β-Ecdysone[3,657]	Antiarhythmic effect; treats pneumonia, vomiting with blood, sore throat, kidney infection, swelling, urinary tract infection
Zephyranthes candida (Lindl.) Herb.	(Aerial part) Lycorine, haemanthidien, nerinine, tazettin[8]	For convulsion, hepatitis, epilepsy; improves liver function; externally for snakebite
Zephyranthes carinata (Spreng.) Herb.	(Leaf, bulb) Alkaloids, lycorine[101,152]	Alleviates fever; used as a poultice for abscesses
Zingiber kawagoii Hayata; *Z. rhizoma* Recens.	(Root, stem) 6-Gingesulfonic acid, monoacyldigalactosyl glycerols, gingerglycolipids A-C[28,631,637]	Alleviates pain; antitoxic, antiinflammatory; abdominal pain, digestive disorders, stomachic, antiulcer

TABLE 1 Major Constituents and Therapeutic Values of Taiwanese Native Medicinal Plants (continued)		
Scientific Name	Major Constituents and Source	Claimed Therapeutical Values
Zingiber officinale Rosc.	(Root) Essential oils, zingiberol, zingiberene, phellandrene, camphene, citral, linalool, methylheptenone, nonylaldehyde, D-borneol, gingerol[39,151]	Antiinflammatory, antitumor; stimulates gastric secretion
Zornia diphylla (L.) Pers.	(Whole plant) Flavonoid glycoside, phenols, amino acids[29]	Antitoxic; treats carbuncle; antiinflammatory; flu, sore throat, gastritis, enteritis, diarrhea

References

1. Perry, L.M. and Metzger, J. *Medicinal Plants of East and Southeast Asia: Attributed Properties and Uses.* The MIT Press. London. 1980.
2. Huang, K.C. *The Pharmacology of Chinese Herbs.* 2nd ed. CRC Press. Boca Raton, FL. 1999.
3. Chang, Y.X. *Taiwan Native Medicinal Plants.* Committee on Chinese medicine and pharmacy. Dept. Health, Executive Yuan, Taipei, Taiwan. 2000 (in Chinese).
4. Chauhan, N.S. *Medicinal and Aromatic Plants of Himachal Pradesh.* Indus Publishing Co. New Delhi, India. 1999.
5. Brekhman, I.I. and Dardymov, I.V. Pharmacological investigation of glycosides from ginseng and *Eleutherococcus. Lloydia* 32, 46, 1969.
6. Chiu, N. and Chang, K. *The Illustrated Medicinal Plants in Taiwan.* vol. 1. SMC Publ. Inc. Taipei, Taiwan. 1995 (in Chinese).
7. Puffe, D. and Zerr, W. The content of various secondary plant products in widespread weeds in grassland with particular regard to higher lying areas. *Eichhof-Berichte* No. A. 11, 1989, 127.
8. Duke, J.A. and Ayensu, E.S. *Medicinal Plants of China.* vol. 1 and 2. Reference Publ. Inc. Algonac, Michigan. 1985.
9. Kanchanapoom, T. et al. Benzoxazinoid glucosides from *Acanthus ilicfolium, Phytochemistry* 58, 537, 2001.
10. Lin, K.H. *Chinese Medicinal Herbs of Taiwan.* vol. 2. How Xiong Di Publ. Inc. Taipei, Taiwan. 1998 (in Chinese).
11. Lin, K.H. *Chinese Medicinal Herbs of Taiwan.* vol. 3. How Xiong Di Publ. Inc. Taipei, Taiwan. 1998 (in Chinese).
12. Lin, K.H. *Chinese Medicinal Herbs of Taiwan.* vol. 5. How Xiong Di Publ. Inc. Taipei, Taiwan. 1998 (in Chinese).
13. Lin, K.H. *Chinese Medicinal Herbs of Taiwan.* vol. 7. How Xiong Di Publ. Inc. Taipei, Taiwan. 1998 (in Chinese).
14. Tomlinson, B. et al. Toxicity of complementary therapies: eastern perspective. *J. Chin. Pharmacol.* 40, 451, 2000.
15. Larry, D. Gas-chromatographic determination of β-asarome, a component of oil of calamus, in flavors and beverages. *J. Assn. Anal. Chem.* 56, 1281, 1973.
16. Zhu, Y.C. *Plantae Medicinales China Roreali-Orientalis.* Heilongjang Sci. Technology Publ. House. Ha-Er-Bing, Heilongjang, China. 1989 (in Chinese).
17. Fuentes-Granados, R.G., Widriechner, M.P., and Wilson, L.A. An overview of *Agastache* research. *J. Herbs Spices Med. Plants* 6, 69, 1998.
18. Keys, J.D. *Chinese Herbs — Their Chemistry and Pharmacodynamics.* Charles E. Tuttle. Rutland, Vermont. 1976.
19. Kasai, S. et al. Antimicrobial catechin derivatives of *Agrimonia pilosa. Phytochemistry* 31, 787, 1992.
20. Chiu, N. and Chang, K. *The Illustrated Medicinal Plants in Taiwan.* vol. 2. SMC Publ. Inc. Taipei, Taiwan. 1995 (in Chinese).
21. Chiu, N. and Chang, K. *The Illustrated Medicinal Plants in Taiwan.* vol. 4. SMC Publ. Inc. Taipei, Taiwan. 1995 (in Chinese).

22. Kawasaki, T., Yamauchi, T., and Itakura, N. Saponin of timo (*Anemarrhenae rhizoma*) I. *Yakugaku Zasshi* 83, 892, 1963.

23. Shibata, S., Mihashi, S., and Tanaka, O. The occurrence of (−) fimarane-type diterpene in *Aralia cordata* Thunb. *Tetrahedron Lett.* 51, 5241, 1967.

24. Kao, H.N. et al. Inhibition of tumor necrosis factor alpha–induced apoptosis by *Asparagus cochinensis* in Hep G2 cells. *J. Ethnopharmacol.* 73, 137, 2000.

25. Li, Z.X. et al. The chemical structure and antioxidative activity of polysaccharide from *Asparagus cochinensis*. *Yaoxue Xuebao* 35, 358, 2000.

26. Tsui, W.Y. and Brown, G.D. (+)-Nyasol from *Asparagus cochinensis*. *Phytochemistry* 43, 1413, 1996.

27. Chiu, N. and Chang, K. *The Illustrated Medicinal Plants in Taiwan.* vol. 5. SMC Publ. Inc. Taipei, Taiwan. 1995 (in Chinese).

28. Chiu, N. and Chang, K. *The Illustrated Medicinal Plants in Taiwan.* vol. 6. SMC Publ. Inc. Taipei, Taiwan. 1995 (in Chinese).

29. Chiu, N. and Chang, K. *The Illustrated Medicinal Plants in Taiwan.* vol. 3. SMC Publ. Inc. Taipei, Taiwan. 1995 (in Chinese).

30. Meyre, S.C. et al. Preliminary phytochemical and pharmacological studies of *Aleurites moluccana* leaves. *Phytomedicine* 5, 109, 1998.

31. Fujita, M., Itokawa, H., and Kumekawa, Y. The study on the constituents of Clematis and *Akebia* spp. 1. Distribution of triterpenes and other components. *Yakugaka Zasshi* 94, 189, 1974.

32. Koketsu, M., Kim, M., and Yamamoto, T. Antifugal activity against food-borne fungi of *Aspidistra elatior* Bulme. *J. Agr. Food Chem.* 44, 301, 1996.

33. Zheng, C., Feng, G., and Liang, H. *Bletilla striata* as a vascular embolizing agent in interventional treatment of primary hepatic carcinoma. *Chin. Med. J.* English ed. 111, 1060, 1998.

34. Bai, L. et al. Phenanthrene glucosides from *Bletilla striata*. *Phytochemistry* 33, 1481, 1993.

35. Yamaki, M., et al. Blespirol, a phenanthrene with a spirolactone ring from *Bletilla striata*. *Phytochemistry* 33, 1497, 1993.

36. Yamaki, M. et al. Bisphenanthrene ethers from *Blertilla striata*. *Phytochemistry* 34, 535, 1992.

37. Yamaki, M. et al. Blestrianol A, B, and C, bisphenanthrenes from *Bletilla striata*. *Phytochemistry* 30, 2733, 1991.

38. Yamaki, M. et al. Bisphenanthrene ethers from *Bletilla striata*. *Phytochemistry* 31, 3985, 1992.

39. Chen, F.C. *Active Ingredients and Identification in Common Chinese Herbs.* People Health Publ. Co. Beijing, China. 1997.

40. Kubota, T. and Hinoh, H. The constitution of saponins isolated from *Bupleurum falcatum* L. *Tetrahedron Lett.* 3, 303, 1968.

41. Takagi, S. et al. Minor basic constituents of evodia fruits. *Shoyakugaka Zasshi* 33, 30, 1979.

42. Izumi, S. et al. Wide range of molecular weight distribution of mitogenic substance(s) in the hot water extract of a Chinese herbal medicine, *Bupleurum chinense*. *Biol. Pharm. Bull.* 20, 759, 1997.

43. Sakurai, M.H. et al. Detection and tissue distribution of antiulcer pectic polysaccharides from *Bupleurum falcatum* by polyclonal antibody. *Planta Med.* 62, 341, 1996.

44. Yamade, H. et al. Characterization of anticomplementary neutral polysaccharides from the roots of *Bupleurum falcatum*. *Phytochemistry* 27, 3163, 1988.

45. Li, S.C. *Chinese Medicinal Herbs.* Georgetown Press, San Francisco. 1973.

46. Yadava, R.N. and Singh, R.K. A novel epoxyflavanone from *Atylosia scarabaeoides* roots. *Fitoterapia* 2, 122, 1998.

47. Arinathan, V., Mohan, V.R., and Britto, J.D.A. Chemical composition of certain tribal pulses in S. India, *Int. J. Food Sci. Nutr.* 54, 207, 2003.

48. Lin, K.H. *Chinese Medicinal Herbs of Taiwan*. vol. 6. How Xiong Di Publ. Inc. Taipei, Taiwan. 1998 (in Chinese).

49. Lin, K.H. *Chinese Medicinal Herbs of Taiwan*. vol. 1. How Xiong Di Publ. Inc. Taipei, Taiwan. 1998 (in Chinese).

50. Anonymous. Studies on the anti-cancer effect of *Broussonetia kazinkoki* extract. *Korean J. Soc. Food Sci.* 5(3), 1999. http://www.khuhomec.com.

51. Ryu, J.H., Ahn, H., and Lee, H.J. Inhibition of nitric oxide production on LPS-activated macrophages by kazinol B from *Broussonetia kazinoki*. *Fitoterapia* 74, 350, 2003.

52. Kwak, W.J. et al. Papyriflavonol A from *Broussonetia papyrifera* inhibits the passive cutaneous anaphylaxis reaction and has a secretory phospholipase A2-inhibitory activity. *Biol. Pharm. Bull.* 26, 299, 2003.

53. Tsukamoto, D., Shibano, M., and Kusano, G. Studies on the constituents of *Broussonetia* species X. Six new alkaloids from *Broussonetia kazinoki* Sieb. *Chem. Pharm. Bull.* (Tokyo) 49, 487, 2001.

54. Tsukamoto, D. et al. Studies on the constituents of *Broussonetia* species VIII. Four new pyrrolidine alkaloids, broussonetines R, S, T, and V, and a new pyrroline alkaloid, broussonetine U, from *Broussonetia kazinok* Sieb. *Chem. Pharm. Bull.* (Tokyo) 49, 492, 2001.

55. Morikawa, H. et al. Terpenic and phenolic glycosides from leaves of *Breynia officinalis* Hemsl. *Chem. Pharm. Bull.* (Tokyo) 52, 1086, 2004.

56. Lin, T.J. et al. Acute poisonings with *Breynia officinalis* — an outbreak of hepato-toxicity. *Toxicol. Clin. Toxicol.* 41, 591, 2003.

57. Bourdy, G. et al. Maternity and medicinal plants in Vanuatu. II. Pharmacological screening of five selected species. *J. Ethnopharmacol.* 52, 139, 1996.

58. Bader, G., Kulhanek, Y., and Ziegler-bohme, H. The antifungal action of polygalactic acid glycosides. *Pharmazie* 45, 618, 1990.

59. Siatka, T. and Kasparove, T. Seasonal changes in the hemolytic effects of the head of *Bellis perennis* L. *Ceska Slov. Farm.* 52, 39, 2003.

60. Nazaruk, J. and Gudej, J. Apigenin glycosides from the flowers of *Bellis perennis* L. *Acta Pol. Pharm.* 57, 129, 2000.

61. Gudej, J. and Nazaruk, J. Flavonol glycosides from the flowers of *Bellis perennis*. *Fitoterapia* 72, 839, 2001.

62. Knight V. et al. Anticancer effect of 9-nitrocamptothecin liposome aerosol on human cancer xenografts in nude mice. *Cancer Chemother. Pharmacol.* 44, 177, 1999.

63. Zhang, R. et al. Preclinical pharmacology of the natural product anticancer agent 10-hydroxycamptothecin, an inhibitor of topoisomerase. I. *Cancer Chemother. Pharmacol.* 41, 257, 1998.

64. Jaeger, E. et al. Irinotecan in second-line therapy of metastatic colorectal cancer. *Onkologie* 23, 15, 2000.

65. Zheng, S. et al. Initial study on naturally occurring products from traditional Chinese herbs and vegetables for chemoprevention. *J. Cell. Biochem. Suppl.* 27, 106, 1997.

66. Conney, A.H. Inhibitory effect of green and black tea on tumor growth. *Proc. Soc. Exp. Biol. Med.* 220, 229, 1999.

67. Chung, F.L. The prevention of lung cancer induced by a tobacco-specific carcinogen in rodents by green and black teas. *Proc. Soc. Exp. Biol. Med.* 220, 244, 1999.

68. Yang, G.Y. et al. Inhibition of growth and induction of apoptosis in human cancer cell lines by tea polyphenols. *Carcinogenesis* 19, 611, 1998.

69. Takido, M. et al. Torosachrysone, a new tetrahydroznthracene derivative from the seeds of *Cassia torosa*. *Lloydia* 40, 191, 1977.

70. Yen, G.C., Chen, H.W., and Duh, P.D. Extraction and identification of an antioxidative component from jue ming ze (*Cassia tora* L.). *J. Agric. Food Chem.* 46, 820, 1998.

71. Yen, G.C., Chen, H.W., and Duh, P.D. Extraction and identification of an antioxidative component from jue ming zi (*Cassia tora* L.). *J. Agric. Food Chem.,* 46, 820, 1998.

72. Jia, T. et al. Comparative research on the constituents of the volatile oil in the rhizome of *Cibotium barometz* (L.). J. Sm. and its processed products. *Zhongguo Zhongyao Zazhi* 21, 216, 1996.

73. Chu, C. and Yang, S. Thin-layer chromatographic differentiation of *Cinnamomum cassia* in Chinese traditional medicine. *Bull. Shin. Materia Med.* 11, 609, 1986.

74. Shan, B.E. et al. Stimulating activity of Chinese medicinal herbs on human lymphocytes *in vitro*. *Int. J. Immunopharmacol.* 21, 149, 1999.

75. Wei, M.J. et al. Study of chemical pattern recognition as applied to quality assessment of the traditional chinese medicine "Wei Ling Zian." *Acta Pharm. Sin.* 26, 772, 1991.

76. Chung, K.T. *A Chinese Native Medicinal Flora for Farmers.* Beijing University, Beijing, China, 1959.

77. Numata, M. et al. Antitumor components isolated from the Chinese herbal medicine *Coix lachryma-jobi*. *Plant Med.* 60, 356, 1994.

78. Li, B.S. et al. Antitumor activity of Kang Lai Te injection. *Zhongguo Yiyao Gongye Zuzhi* 29, 456, 1998.

79. Dong, Y.F. et al. Chemical analysis of fatty oil and polysaccharides in seeds from the genus *Coix* plants in China. *Plant Resour. Environ.* 9, 57, 2000.

80. Tian, R.H. et al. Study on fatty constituents in *Coicis* semen. *Nat. Med.* 51, 177, 1997.

81. Nagao, T. et al. Benzoxazinones from *Coix lachryma-jobi* var. *ma-yuen*. *Phytochemistry* 24, 2959, 1985.

82. Otsuka, H. et al. Phenolic compounds from *Coix lachryma-jobi* var. *ma-yuen*. *Phytochemistry* 28, 883, 1989.

83. Guo, P. et al. Determination of berberine hydrochloride in traditional Chinese medicine containing *Coptis chinensis* Franch. By reversed phase high performance liquid chromatography. *J. West China Univ. Med. Sci.* 22, 90, 1991.

84. Goda, Y. et al. Determination of digitoxigenin glycosides in "Moroheiya" (*Corchorus olitorium*) and its products by HPLC. *J. Food Hyg. Soc. Japan* 39, 415, 1998.

85. Goda, Y. et al. Identification and analyses of main cardiac glycoside in *Corchorus olitorius* seeds and their acute oral toxicity to mice. *J. Food Hyg. Soc. Japan* 39, 256, 1998.

86. Yoshikawa, M. et al. Medicinal food-stuffs. XIV. On the bioactive constituents of moroheiiya II: New fatty acids corchorifatty acids A, B, C, D, E, and F, from the leaves of *Carcharus olitorius* L. *Chem. Phar. Bull.* 46, 1008, 1998.

87. Bandara, B.M.R. et al. Antifungal activity of some medicinal plants of Sri Lanka. *J. Nat. Sci. Counc. Sri Lanka* 17, 1, 1989.

88. Zheng, Y.C. *Taiwan Toxic Plants.* Holiday Pub. Co. Ltd., Taipei, Taiwan. 2000 (in Chinese).

89. Qin, D. and Yu, L. A limonoid from the Chinese drug don-feng-jie (*Atalantia buxifolia*). *Yao Xue Xue Bao* 33, 34, 1998 (English abstract).

90. Qin, D.K. Studies on the chemical constituents of dong-feng-ju (*Atalantia buxifolia*). *Yao Xue Xue Bao* 21, 683, 1986 (English abstract).

91. Bates, S.H., Jones, R.B., and Bailey, C.J. Insulin-like effect of pinitol. *Br. J. Pharmacol.* 130, 1944, 2000.

92. Akihisa, T. et al. Sterols of the Cucurbitaceae. *Phytochemistry* 26, 1693, 1987.

93. Fang, S.D., Xu, R.S. and Gao, Y.S. Some recent advances in the chemical studies of Chinese herbal medicine. *Am. J. Bot.* 68, 300, 1981.

94. Vimala, S. Antitumour promoter activity in Malaysian ginger rhizobia used in traditional medicine. *Br. J. Cancer* 80, 110, 1999.

95. Surh, Y.J. Molecular mechanisms of chemopreventive effects of selected dietary and medicinal phenolic substances. *Mutat. Res. Fundam. Mol. Mech. Mutagen.* 428, 1–2, 305, 1999.

96. Mathes, H.W., Liu, D.B., and Gourisson, G. Cytotoxic components of *Zingiber zerumbet, Curcuma zedaria* and *C. domestica. Phytochemistry* 19, 2643, 1980.

97. Lin, K.H. *Chinese Medicinal Herbs of Taiwan.* vol. 4. How Xiong Di Publ. Inc. Taipei, Taiwan. 1998 (in Chinese).

98. Perez, C. et al. Hypotriglyceridaemic activity of *Ficus carica* leaves in experimental hypertriglyceridaemic rats. *Phytother. Res.* 13, 188, 191, 1999.

99. Chiu, C. *A New Manual of Chinese Materia Medica.* Fudan University. Shanghai, China. 1955.

100. Hall, I.H. et al. Antitumor agents XLI: effects of eupaformosnin on nucleic acid, protein and anerobic and aerobic glycolytic metabolism of *Ehrlich ascites* cells. *J. Pharm. Sci.* 69, 294, 1980.

101. Arthur, H.R. and Cheung, H.T. A phytochemical survey of the Hong Kong medicinal plants. *J. Pharm. Pharmacol.* 12, 567, 1960.

102. Li, C.S. *Appraisal of Chinese Medicine.* Shanghai Sci. Technology Publ. Co. Shanghai, China. 1996.

103. Kitamura, Y. et al. Different responses between anthocyanin-producing and nonproducing cell cultures of *Glehnia littoralis* to stress. *Proc. 9th Int. Congr. Int. Asso. Plant Tissue Culture and Biotechnology.* Jerusalem, Israel. 1998, 503.

104. Satoh, A. et al. Potent allelochemical falcalindiol from *Glehnia littoralis* F. Schm. *Biosci. Biotechnol. Biochem.* 60, 152, 1996.

105. Monmaney, T. A dose of caution. *The Los Angeles Times.* Sept. 1, 1998.

106. Eisenberg, D.M. et al. Unconventional medicine in the United States: prevalence, costs, and patterns of use. *N. Engl. J. Med.* 328, 246, 1993.

107. Shibata, S. The chemistry of Chinese drugs. *Am. J. Chin. Med.* 7, 103, 1979.

108. Takino, Y. et al. Quantitative determination of glycyrrhizic acid in liquorice roots and extracts by TLC-densitometry. *Hippokrates Veriag gimbH.* 36, 74, 1979.

109. Killacky, J., Ross, M.S.F., and Torner, R.D. The determination of β-glycyrrhetinic acid in liquorice by high pressure liquid chromatography. *Plant Med.* 30, 310, 1976.

110. Shi, G.Z. and Et, A.L. Blockage of *Glycyrrhiza uralensis* and *Chelidonium majus* in MNNG-induced cancer and mutagenesis. *Chin. J. Prevent. Med.* 26, 165, 1992.

111. Liao, D.F. et al. Effects of gypenosides on mouse splenic lymphocyte transformation and DNA polymerase II activity *in vitro. Acta Pharmacol. Sin.* 16, 322, 1995.

112. Hau, D.M. et al. Protective effects of *Gynostemma pentaphyllum* in gamma-irradiated mice. *Am. J. Chin. Med.* 24, 83, 1996.

113. Mei, K.F. Transformation of *Gynostemma pentaphyllum* by *Agrobacterium rhizogenes* and saponin production in hairy root cultures. *Acta Bot. Sin.* 95, 626, 1993.

114. Akihisa, T. et al. Sterols of the Cucurbitaceae. *Phytochemistry* 26, 1693, 1987.

115. Lu, C. et al. A new acylated flavonol glycoside and antioxidant effects of *Hedyotis diffusa. Plant Med.* 66, 374, 2000.

116. Nishihama, Y. et al. Three new iridoid glucosides from *Hedyotis diffusa*. *Plants Med.* 43, 28, 1981.

117. Ho, T.L. et al. An anthraquinone from *Hedyotis diffusa*. *Phytochemistry* 25, 1988, 1986.

118. Wu, H.M. et al. Iridoids from *Hedyotis diffusa*. *J. Nat. Prod.* 54, 254, 1991.

119. Wong, K.C. and Tan, G.I. Composition of the essential oil of *Hedyotis diffusa* Willd. *J. Essential Oil Res.* 7, 537, 1995.

120. Ishiguro, K. et al. A chromene from *Hypericum japonicum*. *Phytochemistry* 29, 1010, 1990.

121. Lin, C.C., Huang, P.C., and Lin, J.M. Antioxidant and hepatoprotective effects of *Anoectochilus formosanus* and *Gynostemma pentaphyllum*. *Am. J. Chin. Med.* 28, 87, 2000.

122. Lu, K.I. et al. The evaluation of the therapeutic effect of tao-shang-twao on alpha-naphthylisothiocyanate and carbon tetrachloride–induced acute liver damage in rats. *Am. J. Chin. Med.* 28, 361, 2000.

123. Lu, K.L. et al. Preventive effect of the Taiwan folk medicine *Ixeris laevigata* var. *oldhami* on alpha-naphthyl-isothiocyanate and carbon tetrachloride-induced acute liver injury in rats. *Phytother. Res.* 16, suppl. 1, S45, 2002.

124. Kuo, Y.H. et al. A new anti-HBeAg lignan, kadsumarin A from *Kadsura matsudai* and *Schizandra arisanensis*. *Chem. Phar. Bull.* 47, 1047, 1999.

125. Rittenhoue, J.R., Lui, P.D., and Lau, B.H.S. Chinese medicinal herbs reverse macrophage suppression induced by urological tumors. *J. Urol.* 146, 486, 1991.

126. Chung, Y.C. *New Chinese Material Medica. Vol. 1. Roots.* Beijing University, Beijing, China. 1959, 564 p.

127. Fujita, M. et al. Studies on the components of *Magnolia obovata* Thunb. II. On the components of the methanol extract of the bark. *Yakugaku Zasshi* 93, 422, 1973.

128. Hikino, H., Tamada, M., and Yen, K.Y. Mallorepine, eyano-gamma-pyridone from *Mallotus repandus*. *Plant Med.* 33, 385, 1978.

129. Saijo, R., Nonaka, G., and Nishioka, I. Tannins and related compounds LXXX-VII. Isolation and characterization of four new hydrolyzable tannins from the leaves of *Mallotus repandus*. *Chem. Pharm. Bull.* 37, 2624, 1989.

130. Streenis, C.G.S. Magic plants of the Dayak. *Sarawsk Mus. J.* n. s. 11, 430, 1958.

131. Lee-Huang, S. et al. Anti-HIV and anti-tumor activities of recombinant MAP30 from bitter melon. *Gene*, 161, 151, 1995.

132. Bourinbaiar, A.S. and Lee-Huang, S. Potentiation of anti-HIV activity of antiinflammatory drugs, dexamethasone and indomethacin, by MAP30, the antiviral agent from bitter melon. *Biochem. Biophys. Res. Commun.* 208, 779, 1995.

133. Hirazumi, A. et al. Anticancer activity of *Morinda citrifolia* (noni) on intraperitoneally implanted Lewis lung carcinoma in syngeneic mice. *Proc. West. Pharmacol. Soc.* 37, 145, 1994.

134. Youngken, H.W. *A Textbook of Pharmacognosy.* 4th ed. New York. 1936.

135. Singh, S.B. and Thakur, R.S. New furostanol and spirostanol saponins from tubers of *Paris polyphylla*. *Planta Med.* 40, 301, 1980.

136. Kasuya, S. et al. Lethal efficacy of leaf extract from *Perilla frutescens* (traditional Chinese medicine) or perillaldehyde on *Anisakis* larvae *in vitro*. *Jpn. J. Parasitol.* 39, 220, 1990.

137. Kang, R. et al. Antimicrobial activity of the volatile constituents of *Perilla frutescenes* and its synergistic effects with polygodial. *J. Agric. Food Chem.* 40, 2328, 1992.

138. Tomlinson, B. et al. Toxicity of complementary therapies: eastern perspective. *J. Chin. Pharmacol.* 40, 451, 2000.

139. Chang, H.M. *Advanced Chinese Medicinal Material Research.* World Scientific Publishing. Singapore. 1985.

140. Kubota, T., Kitani, H., and Hinoh, H. The structure of platycogenic acids A, B, and C, further triterpenoid constituents of *Platycodon grandiflorum* A. De Candolle. *Chem. Commun.* 1313, 1969.

141. You, Y. Determination of fumaric acid in *Sarcandra glabra* (Thunb.) Nakai by HPLC. *Zhongguo Zhongyao Zazhi* 22, 554,576, 1997.

142. Tsui, W.Y. and Brown, G.D. Cycloeudesmanolides from *Sarcandra glabra. Phytochemistry* 43, 819, 1996.

143. Petelot, A. Les plantes medicinales du Cambodge, du laos et du vietnam. I. *Arch. Rech. Agron. Pastor. Vietnam* 18, 1, 1952.

144. Yoshino, M. et al. Role of baicalein compounds as antioxidant in the traditional herbal medicine. *Biomed. Res.* 18, 349, 1997.

145. Goldberg, V.E. et al. Dry extract of *Scutellaria baicalensis* as a hemo-stimulant in antineoplastic chemotherapy in patients with lung cancer. *Eksp. Kim. Farmakol.* 60, 28, 1997.

146. Shinichi, I. et al. Antitumor effects of *Scutellariae radix* and its components baicalein, baicalin, and wogonin on bladder cancer cell lines. *Urology* 55, 951, 2000.

147. Bognar, R. Steroid alkaloid glycosides. VIII. Review of specific investigations on the occurrence of steroid alkaloid glycosides in genus *Solanum. Pharmzaie* 20, 40, 1965.

148. Sohreiber, K. On the appearance of glycoside alkaloid in tuber-bearing species of the genus *Solanum* L. Solanum alkaloids. XXVII. *Kulturpflanze* 11, 422, 1963.

149. Zhang, B.H. et al. Antiarrhythmic effects of matrine. *Acta Pharmacol. Sin.* 11, 253, 1990.

150. Roi, J. Traite des plantes medicinales chinoses. *Encyc. Biol.* 47, 1, 1955.

151. Puffe, D. and Zerr, W. The content of various secondary plant products in widespread weeds in grassland with particular regard to higher lying areas. *Eichhof-Berichte No. A.* 11, 127, 1989.

152. Vimala, S. Anti–tumour promoter activity in Malaysian ginger rhizobia used in traditional medicine. *Br. J. Cancer* 80, 110, 1999.

153. Gorter, K. Sur la distribution de la lycorine dans la famille des Amaryllidacees. *Bull. Jard. Bot. Buitenzong* III 1, 353, 1919.

154. Arthur, R. A phytochemical survey of some plants of North Borneo. *J. Pharm. Pharmacol.* 6, 66, 1954.

155. Lee, S.G. Development and utilization of Taiwan native nutraceutical and medicinal plants. In *Proc. Development and Utilization of Medicinal and Nutraceutical Plants Workshop*, Lee, S.G. Ed. Zhong-Shin University, Taichung, Taiwan. Mar. 23, 2002, 15 (in Chinese).

156. Wang, G.L. et al. Chemical constituents from *Alyxia sinensis* II. *Zhongguo Zhong Yao Za Shi.* 27, 199, 2002 (English abstract).

157. Zhang, S.X., Cau, S.Q., and Zhao, Y.Y. Studies on the chemical constituents of *Asarum longerhizomatosum* C.F. Liang and C.S. Yang. *Zhongguo Zhong Yao Za Zhi* 26, 762, 2001.

158. Ramesh, N. et al. Phytochemical and antimicrobial studies of *Begonia malabarica. J. Ethnopharmacol.* 79, 129, 2002.

159. Kikuchi, M. et al. Constituents of *Berchemia racemosa* Sieb. Et Zucc. *Yakugaku Zasshi* 110, 354, 1990.

160. Lin, W.C., Wu, S.C., and Kuo, S.C. Inhibitory effects of ethanolic extracts of *Boussingaultia gracilis* on the spasmogen-induced contraction of the rate isolated gastric fundus. *J. Ethnopharmacol.* 56, 89, 1997.

161. Morikawa, H. et al. Terpenic and phenolic glycosides from leaves of *Breynia officinalis* Hemsl. *Chem. Pharm. Bull.* 52, 1986, 2004.

162. Kim, Y.S. and Shin, D.H. Volatile constituents from the leaves of *Callicarpa japonica* Thunb. and their antibacterial activities. *J. Agric. Food Chem.* 52, 781, 2004.

163. Tripathi, S.M. et al. Enzyme inhibition by the molluscicidal agent *Punica granatum* L. and *Canna indica* L. root. *Phytother. Res.* 18, 501, 2004.

164. Kim, E.J. et al. Suppression by a sesquiterpene lactone from *Carpesium divaricatum* of inducible nitric oxide synthase by inhibiting nuclear factor kappa B activation. *Biochem. Pharmacol.* 61, 903, 2001.

165. Zee, O.P., Kim, D.K., and Lee, K.R. Thymol derivatives from *Carpesium divaricatum. Arch. Pharm. Res.* 21, 618, 1998.

166. Gholizadeh, A. and Kapoor, H.C. Modifications in the purification protocol of *Celosia cristata* antiviral proteins lead to protein that can be *N*-terminally sequenced. *Protein Pept. Lett.* 11, 551, 2004.

167. Priya, K.S. et al. *Celosia argentea* Linn. leaf extract improves wound healing in a rat burn wound model. *Wound Repair Regen.* 12, 618, 2004.

168. Shipochliev, T. Uterotonic action of extracts from a group of medicinal plants. *Vet. Med. Nauki.* 18, 94, 1981.

169. Enk, C.D. et al. Photoprotection by *Cichorum endivia* extracts: prevention of UVB-induced erythema, primdine dimer formation and IL-6 expression. *Skin Pharmacol. Physiol.* 17, 42, 2004.

170. Dupont, M.S. et al. Effect of variety processing and storage on the flavonoid glucoside content and composition of lettuce and endive. *J. Agric. Food Chem.* 48, 3967, 2000.

171. Seto, M. et al. Sesquiterpene lactones from *Cichorum endivia* L. and *C. intybus* L. and cytotoxic activity. *Chem. Pharm. Bull.* (Tokyo) 36, 2423, 1988.

172. Shen, Y.C. et al. Anti-inflammatory activity of the extracts from mycelia of *Antrodia camphorata* cultured with water-soluble fractions from five different *Cinnamomum* species. *FEMS Microbiol Lett.* 231, 137, 2004.

173. Whitman, S.C. et al. Nobiletin, a citrus flavonoid isolated from tangerines, selectively inhibits classa scavenger receptor-mediated metabolism of acetylated LDL by mouse macrophages. *Atherosclerosis* 178, 25, 2005.

174. Hillebrand, S., Schuaiz, M., and Winterhalter, P. Characterizations of anthocyanins and pyranoanthocyanins from blood orange [*Citrus sinensis* (L.) Osheck] juice. *Agric. Food Chem.* 52, 7331, 2004.

175. Chen, S.W. Pharmacognostic identification of *Claoxylon polot* and its fakery. *Zhong Yao Tong Bao* 8, 7, 1983 (English abstract).

176. Manosroi, A., Saraphanchotiwitthaya, A., and Manosroi, J. Immunomodulatory activities of fractions from hot aqueous extract of wood from *Clausena excavata. Fitoterapia* 75, 302, 2004.

177. Sunthitikawinsakul, A., Kongkathip, N., and Koogkathip, B. Anti-HIV-1 limonoid: first isolation from Clausena excavata. *Phytother. Res.* 17, 1101, 2003.

178. Ito, C. et al. Chemical constituents of *Clausena excavata*: isolation and structure elucidation of novel furanone-coumarins with inhibitory effects for tumor promotion. *J. Nat. Prod.* 63, 1218, 2000.

179. Mimak, Y. et al. Triterpene saponins from the roots of *Clematis chinensis. J. Nat. Prod.* 67, 1511, 2004.

180. He, M., Zhang, J.H., and Hu, C.Q. Studies on the chemical components of *Clematis chinensis. Yao Xue Xue Bao* 36, 278, 2001.

181. Thapliyal, R.P. and Bahuguna, R.P. An oleanolic acid based bisglycoside from *Clematis montana* roots. *Phytochemistry* 34, 861, 1993.

182. Thapliyal, R.P. and Bahuguna, R.P. Clemontanos-C, a saponin from *Clematis muntana. Phytochemistry* 33, 671, 1993.

183. Cheng, H.H. et al. Cytotoxic phenophorbide-related compounds from *Celerodendrum calamitosum* and *C. cyrtophyllum*. *J. Nat. Prod.* 64, 915, 2001.

184. Panthong, A. et al. Anti-inflammatory and anti-pyretic properties of *Clerodendrum petasites* S. Moore. *J. Ethnopharmacol.* 85, 151, 2003.

185. Hazekamp, A., Verpoorte, R., and Panthong, A. Isolation of a bronchodilator flavonoid from the Thai medicinal plant *Clerodendrum petasites*. *J. Ethnopharmacol.* 78, 45, 2001.

186. Mukhtar, N., Iqbal, K., and Malik, A. Novel sphingolipids from *Conyza canadensis*. *Chem. Pharm. Bull.* (Tokyo) 50, 1558, 2002.

187. Mukhtar, N., Iqbal, K., and Malik, A. Sphinolipids from *Conyza canadensis*. *Phytochemistry* 61, 1005, 2002.

188. Xu, L. et al. Chemical constituents of *Conyza blinii*. *Zhongguo Zhong Yao Za Zhi* 23, 552. 1998 (English abstract).

189. Kaneko, H. and Naruto, S. Constituents of *Corydalis* spp. VII. Studies on the constituents of *Corydalis pallida*. *Yakugaku Zasshi* 91, 101, 1971.

190. Kukai, T. et al. Anti fungal agents from the roots of *Cudrania cochinchinensis* against *Candida*, *Cryptococcus* and *Aspergillus* species. *J. Nat. Prod.* 66, 1118, 2003.

191. Hou, A. et al. Benzophenones and xanthones with isoprenoid groups from *Cudrania cochinchinensis*. *J. Nat. Prod.* 64, 65, 2001.

192. Arai, Y. et al. Fern constituents: dryocrassy formate, sitostanyl formate, and 12-hydroxyfern-9(11)-ene from *Cyathea podophylla*. *Chem. Pharm. Bull.* (Tokyo) 51, 1311, 2003.

193. Yang, Y., Shu, H.Y., and Min, Z.D. Anthraquinones isolated from *Morinda officinalis* and *Damnacanthus indicus*. *Yao Xue Xue Bao* 27, 358, 1992 (English abstract).

194. Kobayashi, J. et al. Daphniglaucins A and B, novel polycyclic quaternary alkaloids from *Daphniphyllum glaucescens*. *Org. Lett.* 5, 1733, 2003.

195. Akbar, E. and Malik, A. Antimicrobial triterpenes from *Debregeasia salicifolia*. *Nat. Prod. Lett.* 16, 339, 2002.

196. Zeng, X. et al. Modification of the analytical method for rotenoids in plants. *Se Pu* 20, 144, 2002 (English abstract).

197. Crombie, L. and Whiting, D.A. Review article number 135. Biosynthesis in the rotenoid group of national products: applications of isotope methodology. *Phytochemistry* 49, 1479, 1998.

198. Xu, L.R. et al. A new traterpenoid: taraxerol-3-β-0-tridecyl ether from *Derris triofoliata*. *Pharmazie* 59, 655, 2004.

199. Raj, R.K. Screening of indigenous plants for anthelmintic action against human *Ascaris lumbricoides*: part II. *Indian J. Physiol. Pharmacol.* 19, 1975.

200. Ghosal, S. et al. *Desmodium* alkaloids. IV. Chemical and pharmacological evaluation of *Desmodium triflorum*. *Planta Med.* 23, 321, 1973.

201. Rathi, A. et al. Antiinflammatory and antinociceptive activity of water decoction *Desmodium gangeticum*. *J. Ethnopharmacol.* 95, 259, 2004.

202. Govindarajan, R., Rastogi, S., and Vijayakumar, M. Studies on the antioxidant activities of *Desmodium gangetium*. *Biol. Pharm. Bull.* 26, 1424, 2003.

203. Malaviya, N., Pal, R., and Khanna, N.M. Saponins from *Deutzia corymbosa*. *Phytochemistry* 30, 2798, 1991.

204. Jadot, J. and Casimir, J. Separation and characterization of a hydroxyleucine from *Deutzia gracilis, Biochim. Biophys. Acta* 48, 400, 1961.

205. Blour, S.J. Deep blue anthocyanins from blue *Dianella* berries. *Phytochemistry* 58, 923, 2001.

206. Seple, S.J. et al. *In vitro* antiviral activity of the anthraquinone chrysophanic acid against polivirus. *Antiviral Res.* 49, 169, 2001.

207. Lin, L.C., Kuo, Y.C., and Chou, C.J. Immunomodulatory principles of *Dichrocephala bicolor. J. Nat. Prod.* 62, 405, 1999.

208. Luo, Y. et al. Glycosides from *Dicliptera riparia. Phytochemistry* 61, 449, 2002.

209. Wang, Y.Q. et al. Study of binding of REEs with water-soluble polysacchrides in fern. *Biol. Trace Elem. Res.* 1999 winter, 71, 103, 1999.

210. Pathak, A., Kulshreshtha, D.K., and Manya, R. Coumaroyl triterpene lactone, phenolic and naphthalene glycoside from stem bark of *Diospyros angustifolia. Phytochemistry* 65, 2153, 2004.

211. Orhan, I. and Sener, B. Fatty acid content of selected seed oil. *J. Herb Pharmacother.* 2, 29, 2003.

212. Ouyang, P. et al. Effects of flavone from leaves of *Diospyros kiki* on adventitial fibroblasts proliferation by advanced oxidation protein products *in vitro. Zhong Yao Cai* 27, 186, 2004.

213. Duan, J. et al. Structural analysis of a pectic polysaccharide from the leaves of *Diospyros kaki. Phytochemistry* 65, 609, 2004.

214. Inatomi, Y. et al. Constituents of a fern *Diplazium subsinuatum.* III. Four new hopane-triterpene lactone glycosides. *Chem. Pharm. Bull.* (Tokyo) 48, 1930, 2000.

215. Tervari, J.P. Fatty acids of the seed oil of *Dipteracanthus prostratus* Nees. *Arch. Pharm. Ber Dtsch Pharm Ges.* 304, 797, 1971.

216. Mukherjee, P.K. et al. Studies on antitussive activity of *Drymaria cordata* Willd. *J. Ethnopharmacol.* 56, 77, 1997.

217. Yuan, A.X. Chemical constituents of *Drymaria cordata.* Zhong Yao Tong Bao 12, 36, 1987 (English abstract).

218. Hsieh, P.W. et al. A new anti-HIV alkaloid, drymaritin and a new c-glycoside flavonoid, diandraflavone, from *Drymaria diandra. J. Nat. Prod.* 67, 1175, 2004.

219. Ding, Z. et al. Two new cyclic peptides from *Drymaria diandra. Planta Med.* 66, 386, 2000.

220. Kinjo, J., Suyama, K., and Nohara, T. Triterpenoidal saponins from *Dumasia truncata. Phytochemistry* 40, 1765, 1995.

221. Lin, L.C., Kuo, Y.C., and Chou, C.J. Immunodulatory proanthocyanidins from *Echysanthera utilis. J. Nat. Prod.* 65, 505, 2002.

222. Dong, M., Oda, Y., and Hirota, M. (10E, 12Z, 15Z)-9-hydroxy-10, 12, 15-octadecatrienoic acid methyl ester as an antiinflammatory compound from *Ehretic dicksonii. Biosci. Biotechnol. Biochem.* 64, 882, 2000.

223. Chow, S.Y., Chen, C.F., and Chen, S.M. Pharmacological studies of Chinese herbs (6) pharmacological effects of *Epiphyllum oxypetalum* Haw. Taiwan *Yi Xue Hui Za Zhi* 76, 916, 1977 (in Chinese).

224. Garcia, M.D. et al. Topical antiinflammatory activity of phytosterols isolated from *Eryngium foetidum* on chronic and acute inflammation models. *Phytother. Res.* 13, 78, 1999.

225. Simon, O.R. and Singh, N. Demonstration of anticonvulsant properties of an aqueous extract of spirit weed (*Eryngium foetidum* L.). *West Indian Med. J.* 35, 121, 1986.

226. Lo, W.L. et al. Antiplatelet and anti-HIV constituents from *Euchreta formosana. Nat. Prod. Res.* 17, 91, 2003.

227. Lo, W.L. et al. Coumaronochromones and flavanones from *Euchresta formosana* roots. *Phytochemistry* 60, 839, 2002.

228. Liu, Z.L. et al. Six new sesquiterpenes from *Euonymus nanoides* and their antitumor effects. *Planta Med.* 70, 353, 2004.

229. Kuo, Y.H. et al. A novel no-production-inhibiting triterpene and cytotoxicity of known alkaloids from *Euonymus laxiflorus. J. Nat. Prod.* 66, 554, 2003.

230. Huo, J. et al. Cytotoxic sesquiterpene lactones from *Eupatorium lindleyanum*. *J. Nat. Prod.* 67, 1470, 2004.

231. Yang, S.P. et al. Cytotoxic sesquiterpenoids from *Eupatorium chinensis*. *J. Nat. Prod.* 67, 638, 2004.

232. Norhanom, A.W. and Yadav, M. Tumor promoter activity in Malaysian teuphorbiaceae. *Br. J. Cancer* 71, 776, 1995.

233. Vamsidhar, I. et al. Antinociceptive activity of *Euphorbia heterophylla* roots. *Fitoterapia* 71, 562, 2000.

234. Nsimba-Lubaki, M., Peumans, W.J., and Carlier, A.R. Isolation and partial characterization of a lectin from *Euphorbia heterophylla* seeds. *Biochem. J.* 215, 141, 1983.

235. Cheng, H.Y. et al. Putranjivain A from *Euphorbia jolkini* inhibits both virus entry and late stage nephication of herpes simplex virus type 2 *in vitro*. *Antimicrob. Chemother.* 53, 577, 2004.

236. Dias-Baruffi, M., Sakamoto, M., and Rossetto, S. Neutrophil migration and aggregation induced by *Euphobia*, a lectin from the latex of *Euphorbia milii*, var. *milii*. *Inflamm. Res.* 49, 732, 2002.

237. Seshagirirao, K. and Prasad, M.N. Purification and partial characterization of a lectin from *Euphorbia neriifolia* latex. *Biochem. Mol. Biol. Int.* 35, 1199, 1995.

238. Yadav, R. et al. Larvicidal activity of latex and stem bark of *Euphobia tirucalli* plant on the mosquito *Culex quinquefasciatus*. *J. Commun. Dis.* 34, 264, 2002.

239. Lengye, E. and Gellert, M. Terpenoids from the root barks of *Evodia meliaefolia*. *Pharmazie* 33, 372, 1978.

240. Hirons, I. et al. Carcinogenic activity of *Farfugium japonicum* and *Senecio cannabifolius*. *Cancer Lett.* 20, 191, 1983.

241. Kurihara, T. and Suzuki, S. Studies on the constituents of *Farfugium japonicum* (L.) Kitam. III. On the components of the rhizome. *Yakugaku Zasshi* 100, 681, 1980.

242. Ito, K., Iida, T., and Funatani, T. Study on the ingredients of gigantic type of *Farfugium japonicum* (L.) Kitam. Isolation of 8–epi-eremophilenolides. *Yakugaku Zasshi* 99, 349, 1979.

243. Ito, K., Iida, T., and Takeichi, C. Study on the ingredients of *Farfugium japonicum* (L. f.) Kitamura var. *formosanum* (Hayata) Kitamura. Structure of new furanosesquiterpenes, farfuomolide A and B. *Yakugaku Zasshi* 98, 1592, 1978 (in Japanese).

244. Kitanaka, S. et al. Anti-tumor agents, 162. Cell-based assays for identifying novel DNA topoisomerase inhibitors: studies on the constituents of *Fatsia japonica*. *J. Nat. Prod.* 58, 1647, 1995.

245. Tomimon T. and Kizu, H. On the saponins from the leaves of *Fatsia japonica* Decne et Planch. *Yakugaku Zasshi* 99, 92, 1979.

246. Mousa, O. et al. Bioactivity of certain Egyptian *Ficus* species. *J. Ethnopharmacol.* 41, 71, 1994.

247. de Rosa, S.C. et al. Triterpene saponins and iridoid glucosides from *Galium rivale*. *Phytochemistry* 54, 751, 2000.

248. Wang, S.C. et al. *Gardenia* herbal active constituents: applicable separation procedures. *J. Chromatogr. B. Analyt Technol. Biomed. Life Sci.* 812, 193, 2004.

249. Urbain, A. et al. Xanthone from *Gentiana campestris* as new acetylcholinesterase inhibitors. *Planta Med.* 70, 1011, 2004.

250. Zhao, S. et al. Separation and determination of gentiopicroside and swertiamarin in Tibetan medicines by micellar electrokinetic electrophoresis. *Biomed. Chromatogr.* 18, 10, 2004.

251. Zhong, Y.Y., Li, S.H., and Tian, Z. Morphological and histological studies of the Chinese drug las-guan cao. *Yao Xue Xue Bao* 30, 46, 1995 (English abstract).

252. Otsuka, H. et al. Glochidiolide, isoglochidiolide, acuminaminoside, and glochida-cuminoside A-D from the leaves of *Glochidion acuminatum* Muell. *Chem. Pharm. Bull.* (Tokyo) 52, 591, 2004.

253. Otsuka, H. et al. Glochidionionosides A-D, megastigmane glucosides from leaves of *Glochidion zeylanicum* A. Juss. *Chem. Pharm. Bull.* (Tokyo) 51, 286, 2003.

254. Wu, M.J. et al. *Glossogyne tenuifolia* acts to inhibit inflammatory mediator production in a macrophage cell line by down regulating LPS-induced NF-Kappa B. *J. Biomed. Sci.* 11, 186, 2004.

255. Gao, X. et al. Antiproliferative activity of *Goldfussia psilostachys* ethanolic extraction on K562 leukemia cells. *Fitoterapia* 75, 639, 2004.

256. Du, X.M. Sedative and anticonvulsant activities of goodyerin, a flavonol glycoside from *Goodyera schlechtendaliana*. *Phytother. Res.* 16, 261, 2002.

257. Du, X. et al. Higher yielding isolation of kinsenoside in *Anoectochilus* and its anti-hyperlipiosis effect. *Biol. Pharm. Bull.* 24, 65, 2001.

258. Du, X.M., Sun, N.Y., and Shoyama, Y. Flavonoids from *Goodyera schlechtendaliana*. *Phytochemistry* 53, 997, 2000.

259. Qi, Y. et al. Chemical constituents of *Gossampinus malabarica* (L.) Merr. *Zhongguo Zhong Yao Za Zhi* 21, 234, 1996.

260. Hibasami, H. et al. 2-0-methylisohemigossylic acid lactone, a sesquiterpene, isolated from roots of Mokumen (*Gossampinus malabarica*) induces cell death and morpho-logical change indicative of apoptotic chromatin condensation in human promyelotic leukemia HL-60 cells. *Int. J. Mol. Med.* 14, 1029, 2004.

261. Lin, W.Y. et al. Antiplatelet aggregation and chemical constituents from the rhizome of *Gynura japonica*. *Plants Med.* 69, 757, 2003.

262. Zimmerman, D.C. and Fick, G.N. Fatty acid composition of sunflower (*Helianthus annus* L.) oil as influenced by seed position. *J. Am. Oil Chem.* 50, 273, 1973.

263. Deshpande, P.J., Pathak, S.N., and Shankaran, P.S. Healing of experimental wounds with *Helianthus annus*. *Indian J. Med. Res.* 53, 539, 1965.

264. Lin, Y.L. et al. Two new puriniums and three new pyrimidines from *Heterostemma brownii*. *J. Nat. Prod.* 60, 982, 1997.

265. Saiki, Y. et al. Gas-chromatographical studies on natural volatile oils. V. The gas-chromatography on the volatile oils of the plant belonging to *Heterotrope* genus. *Yakugaku Zasshi* 87, 1544, 1967.

266. Kwon, S.W. et al. Antioxidant properties of heat-treated *Hibiscus syriacus*. *Izv Akad Nauk. Ser. Biol.* 1, 20, 2003.

267. Lee, S.J. et al. An antioxidant lignan and other constituents from the root bark of *Hibiscus syriacus*. *Planta Med.* 65, 658, 1999.

268. Wu, P.L. et al. Cytotoxicity of phenylpropanoid esters from the stems of *Hibiscus taiwanensis*. *Bio. Org. Med. Chem.* 12, 2193, 2004.

269. Matsushita, A. et al. Hydrocotylosides I-VII, new oleanane saponins from *Hydro-cotyle sibthorpiodes*. *J. Nat. Prod.* 67, 384, 2004.

270. Mink, C.J. The glycosidic constituents of *Hydrocotyle vulgaris* L. *J. Pharm. Phar-macol.* 11, 244, 1959.

271. Masuda, T. et al. Simple detection method of powerful antiradical compounds in the raw extract of plants and its application for the identification of antiradical plant constituents. *J. Agric. Food Chem.* 51, 1831, 2003.

272. Chung, M.I. et al. Antiplatelet and antiinflammatory constituents and new oxygenated xanthones from *Hypericum geminiflorum*. *Planta Med.* 68, 25, 2002.

273. Chung, M.I. et al. A new chalcone, xanthones and a xanthonolignoid from *Hypericum geminiflorum*. *J. Nat. Prod.* 62, 1033, 1999.

274. Shen, C.C. et al. Furanolabdane diterpenes from *Hyposestes purpurea. J. Nat. Prod.* 67, 1947, 2004.

275. da Cruz Araujo, E.C., Sousa Lima, M.A., and Silveira, E.R. Spectral assignments of new diterpenes from *Hyptis martiusii* Benth. *Magn. Reson. Chem.* 42, 1049, 2004.

276. Costa-Lotufo, L.V. et al. Antiproliferative effects of abietane diterpenoids isolated from *Hyptis martiusii* Benth (Labiatae). *Pharmaxie* 59, 78, 2004.

277. Leite, S.P. et al. Embryotoxicity *in vitro* with extract of *Indigofera suffruticosa* leaves. *Reprod. Toxicol.* 18, 701, 2004.

278. Qiusheng, Z. et al. Protective effects of luteolin-7-glucoside against liver injury caused by carbon tetrachloride in rats. *Pharmaxie* 59, 286, 2004.

279. Lu, K.L. et al. Preventive effect of the Taiwan folk medicine *Ixeris laevigata* var. *oldhanii* on α-naphthyl-isothiocyanate and carbon tetrachloride–induced acute liver injury in rats. *Phytother. Res.* 16 suppl. 1, S45, 2002.

280. Huang, B. et al. Supercritical-CO_2 fluid extraction in extracting volatile constituents from *Juniperus formosana. Zhong Yao Cai* 20, 30, 1997 (in Chinese).

281. Kuo, Y.H. Studies on several naturally occurring lignans. *Gao Xiong Yi Xue Ke Yue Za Zhi* 5, 621, 1989 (English abstract).

282. Nguelefack, T.B. et al. Analgesic properties of the aqueous and ethanol extracts of the leaves of *Kalanchoe crenata. Phytother. Res.* 18, 385, 2004.

283. Yadav, N.P. and Dixit, V.K. Hepatoprotective activity of leaves of *Kalanchoe pinnata* Pers. *J. Ethnopharmacol.* 85, 197, 2003.

284. Supratman, U. et al. Anti–tumor promoting activity of bufadienolides from *Kalanchoe pinnata* and *K. daigremontiana × tubiflora. Biosci. Biotechnol. Biochem.* 65, 947, 2001.

285. Sho, Y. et al. An echinocystic acid saponin derivative from *Kalimeris shimadae. Phytochemistry* 43, 195, 1996.

286. Apers, S. et al. Characterisation of new oligoglycoside compounds in two Chinese medicinal herbs. *Phytochem. Anal.* 13, 202, 2002.

287. Hellion-Ibarrola, M.C. et al. Acute toxicity and general pharmacological effect on central nervous system of the crude rhizome extract of *Kyllinga brevifulia* Rottb. *J. Ethnopharmacol.* 66, 271, 1999.

288. Yadava, R.N. A new biologically active triterpenoid saponin from the leaves of *Lepidagathis hyalina Nees. Nat. Prod. Lett.* 15, 315, 2001.

289. Ravikanth, V. et al. An immunosuppressive tryptophan-derived alkaloid from *Lepidagathis cristata* Willd. *Phytochemistry* 58, 1263, 2001.

290. Sadhu, S.K. et al. Separation of *Leucas aspera,* a medicinal plant of Bangladesh, guided by prostaglandin inhibitory and antioxidant activities. *Chem. Pharm. Bull.* (Tokyo) 51, 595, 2003.

291. Ouyang, M.A., He, Z.D., and Wu, C.L. Antioxidative activity of glycosides from *Ligustrum sinense. Nat. Prod. Res.* 17, 381, 2003.

292. Nakamura, O. et al. Steriodal saponins from the bulbs of *Lilium speciosum × L. nobilissimum* "Star Gazer" and their antitumour-promoter activity. *Phytochemistry* 36, 463, 1994.

293. Mimaki, Y. and Sashida, Y. Steroidal and phenolic constituents of *Lilium speciosum. Phytochemistry* 30, 937, 1998.

294. Kuo, Y.C. et al. Samarangenir B from *Limonium sinense* suppresses herpes simplex virus type 1 replication in verocells by regulation of viral macromolecular synthesis. anti-microb. agents. *Chemotherapy.* 46, 2854, 2002.

295. Lin, L.C. and Chou, C.J. Flavonoids and phenolics from *Limonium sinense. Planta Med.* 66, 382, 2002.

296. Cai, L.H. et al. Determination of linderane in root tuber of *Lindera aggregata* by HPLC. *Zhongguo Zhong Yao Za Zhi* 29, 657, 2004 (English abstract).

297. Li, Y.M. et al. Extracts from the roots of *Lindera strychifolia* induce apoptosis in lung cancer cells and prolong survival of tumor-bearing mice. *Am. J. Chin. Med.* 31, 857, 2003.

298. Zhang, C.F. et al. Studies on constituents of the leaves of *Lindera aggregata* (Sims) Kosterm. *Zhongguo Zhong Yao Za Zhi* 26, 765, 2001 (English abstract).

299. Lindstrom, B. and Luning, B. Studies on orchidaceae alkaloids 23. Alkaloids from *Liparis loeselil* (L.) L. C. Rich and *Hammarbya paludosa* L. *Acta Chem. Scand.* 25, 895, 1971.

300. Choi, E.M. and Hwang, J.K. Effects of methanolic extract and fractions from *Litsea cubeba* bark on the production of inflammatory mediators in RAW 264.7 cells. *Fitoterapia* 75, 141, 2004.

301. Min, B.S. et al. Lactones from the leaves of *Litsea japonica* and their anticomplement activity. *J. Nat. Prod.* 66, 1388, 2003.

302. Shibano, M. et al. Two new pyrrolidine alkaloids, radicamines A and B, as inhibitors of α-glucosides from *Lobelia chinensis* Lour. *Chem. Pharm. Bull.* (Tokyo) 49, 1362, 2001.

303. Philipov, S. et al. Photochemical study and anti-inflammatory properties of *Lobelia laxiflora* L. *Z. Naturforsch.* 53, 311, 1998.

304. Ma, X. and Gang, D.R. The *Lycopodium* alkaloids. *Nat. Prod. Rep.* 21, 752, 2004.

305. Du, Z.Q. et al. Study on extraction process of *Zhanjin ruji*. *Shongguo Zhong Yao Za Zhi* 28, 32, 2003

306. Tian, J.K. et al. Two new triterpenoid saponins from *Lysimachia capillipes* Hemsl. *Yao Xue Xue Bao* 39, 722, 2004 (English abstract).

307. Tian, J.K. et al. Two new triterpenoid saponins from *Lysimachia davurica*. *Yao Xue Xue Bao* 39, 194, 2004 (English abstract).

308. Xie, C. et al. Flavonol glycosides from *Lysimachia capillipes*. *J. Asian Nat. Prod. Res.* 4, 17, 2002.

309. Lu, S.T. Studies on the alkaloids of Formosan louraceous plants II. Alkaloids of *Machilus kusanoi* Hayata (2). The isolation of di-coclaurine. *Jpn. J. Pharmacol.* 83, 19, 1963.

310. Koike, K. et al. New triterpenoid saponins from *Maesa japonica* Moritzi. *J. Nat. Prod.* 62, 228, 1999.

311. Sindambiwe, J.B. et al. Evaluation of biological activities of triterpenoid saponins from *Maesa lanceolata*. *J. Nat. Prod.* 61, 585, 1998.

312. Jiang, Z. et al. Six triterpenoid saponins from *Maesa laxiflora*. *J. Nat. Prod.* 62, 876, 1999.

313. Muhammad, I. et al. Cytotoxic and antioxidant activities of alkylated benzoquinones from *Maesa lanceolata*. *Phytotherapy.* 17, 887, 2003.

314. Picone, J.M., Matavish, H.S., and Clery, R.A. Emission of floral volatiles from *Mahonia japonica*. *Phytochemistry* 60, 611, 2002.

315. van Kiem, P. et al. Pentacyclic triterpenoids from *Mallotus apelta*. *Arch. Pharm. Res.* 27, 1109, 2004.

316. An, T. et al. Two new benzopyran derivatives from *Mallotus apelta*. *Nat. Prod. Res.* 17, 325, 2003.

317. Zhao, J. et al. The study on the antioxidation effect of root of *Mallotus apelta* in the rat model of liver fibrosis. *Zhong Yao Cai* 25, 185, 2002 (English abstract).

318. Reddy, Y.S., Venkatesh, S., and Suresh, B. Antinociceptive activity of *Malvastrum cormandelaiinum*. *Fitoterapia* 72, 278, 2001.

319. Rolston, D.D., Mathew, P., and Mathan, V.I. Food-based solutions are a viable alternative to glucose-electrolyte solutions for oral hydration in acute diarrhoea studies in a rat model of secretory diarrhoea. *Trans. R. Soc. Trop. Med. Hyg.* 84, 156, 1990.

320. Ito, K. and Lai, J. Studies on the constituents of *Marsdenia formosana* Masamune I. Isolation of triterpenoids and structure of marsformal. *Yakugaku Zasshi* 98, 249, 1978.

321. Lee, M.H. et al. Monoamine osidase Bard free radical scavenging activities of natural flavonoids in *Melastoma candidum* D. Don. *J. Agric. Food Chem.* 49, 5551, 2001.

322. Cheng, J.T., Hsu, F.L., and Chen, H. Antihypertensive principles from the leaves of *Melastoma candidum. Planta Med.* 59, 405, 1993.

323. Ivanova, D. et al. Polyphenols and antioxidant capacity of Bulgarian medicinal plants. *J. Ethnopharmacol.* 96, 145, 2005.

324. Marongiu, B. et al. Antioxidant activity of supercritical extract of *Melissa officinalis* ssp. officinalis and *Melissa officinalis* ssp. inodora. *Phytother. Res.* 18, 789, 2004.

325. de Sousa, A.C. et al. *Melissa officinalis* L. Essential oil: antitumoral and antioxidant activities. *J. Pharm. Pharmacol.* 56, 677, 2004.

326. Bundara, K.A. et al. Insecticidal piperidine alkaloid from *Microcos paniculata* stem bark. *Phytochemistry* 54, 29, 2002.

327. Ahmed, M. et al. Analgesic sesquiterpene dilactone from *Mikania cordata. Fitoterapia* 72, 919, 2001.

328. Panl, R.K., Jabbar, A., and Rashid, M.A. Antiulcer activity of *Mikania cordata. Fitoterapia* 71, 701, 2002.

329. Ito, C. et al. Chemical constituents of *Millettia taiwaniana*: structure elucidation of five new isoflavonoids and their cancer chemopreventive activity. *J. Nat. Prod.* 67, 1125, 2004.

330. Erazo, S. et al. Constituents and biological activities from *Muehlenbeckia hastulata. Z. Naturforsch.* 57, 801, 2002.

331. Intiyot, Y. et al. Antimutagenicity of *Murdannia loriformis* in the *Salmonella* mutation assay and its inhibitory effects on azoxymethane-induced DNA methylation and aberrant crypt focus formation in male F344 rats. *J. Med. Invest.* 49, 25, 2002.

332. Li, Y. et al. Antiviral activities of medicinal herbs traditionally used in southern mainland China. *Phytotherapy.* 18, 718, 2004.

333. Liao, Y. et al. Pharmacognostical studies on the stem and leaf of *Mussaenda pubescens. Zhong Yao Cai* 23, 195, 2000.

334. Zhao, W. et al. Triterpenes and triterpenoid saponins from *Mussaenda pubescens. Phytochemistry* 45, 1073, 1997.

335. Wu, T.S. et al. Constituents and cytotoxic principles of *Nothapodytes foetida. Phytochemistry* 39, 383, 1995.

336. Gupta, S. et al. Anti-cancer activities of *Oldenlandia diffusa. J. Herb. Pharmcother.* 4, 21, 2004.

337. Shan, B.E., Zhang, J.Y., and Du, X.N. Immunodulatory activity and antitumor activity of *Oldenlandia diffusa in vitro. Zhongguo Zhong Xi Yi Jie He Za Zhi* 21, 370, 2001.

338. Qiu, Y.K. et al. The isolation and identification of a new pyrone from *Opuntia dillenii. Yao Xue Xue Bao* 38, 523, 2003.

339. Srivastava, B.K. and Pande, C.S. Arabinogalactan from the pods of *Opuntia dillenii. Planta Med.* 25, 92, 1974.

340. Gori, P. Ferritin in the integumentary cells of *Oxalis coniculata. J. Ultrastruct. Res.* 60, 95, 1977.

341. Kim, Y. et al. Two new acylated iridoid glucosides from the aerial parts of *Paederia scandens. Chem. Pharm. Bull.* (Tokyo) 52, 1356, 2004.

342. Srivastava, M.C., Tewari, J.P., and Kant, V. Antiinflammatory activity of an indigenous plant — *Paederia foetida* (Gandhali). *Ind. J. Med. Sci.* 27, 231, 1973.

343. Quang, D.N. et al. Iridoid glucosides from roots of Vietnamese *Paederia scandens*. *Phytochemistry* 60, 505, 2002.

344. Ooi, L.S., Sun, S.S., and Ooi, V.E. Purification and characterization of a new antiviral protein from the leaves of *Pandanus amaryllifolius*. *Int. J. Biochem. Cell Biol.* 36, 1440, 2004.

345. Salim, A.A., Garson, M.J., and Craik, D.J. New alkaloids from *Pandanus amaryllifolius*. *J. Nat. Prod.* 67, 54, 2004.

346. Tomita, M. et al. Studies on the alkaloids of menispermaceous plants cc ILVI. Alkaloids of *Paracyclea ochiaiana* Kudo et Yammamoto. *Yakugaku Zasshi* 87, 1285, 1967.

347. Cheung, Y.N. et al. Polyphyllin D is a potent apoptosis inducer in drug-resistant HepG2 cells. *Cancer Lett.* 217, 203, 2005.

348. Wang, Q., Xu, G., and Jiang, Y. Analgesic and sedative effects of the Chinese drug *Rhizoma paridis*. *Zhongguo Zhong Yao Za Zhi* 15, 109, 128, 1990 (English abstract).

349. Lee, H. and Lin, J.Y. Antimutagenic activity of extracts from anticancer drugs in Chinese medicine. *Mutat. Res.* 204, 229, 1988.

350. Puricelli, L. et al. Preliminary evaluation of inhibition of matrix-metalloprotease MMP-2 and MMP-9 by *Passiflora edulis* and *P. foetida* aqueous extracts. *Fitoterapia* 74, 302, 2003.

351. Echeverri, F. et al. Passifloricins; polyketides pyrones from *Passiflora foetida* Vesin. *Phytochemistry* 56, 881. 2001.

352. Nagda, K. and Peshmukh, K. Hemagglutination pattern of galactose specific lectin from *Pedilanthus tithymaloides* in diabetes mellitus. *Ind. J. Exp. Biol.* 36, 426, 1998.

353. Masuda, T. et al. Isolation and antioxidant activity of gallogl flavonol glycosides from the seashore plant, *Pemphis acidula*. *Biosci. Biotechnol. Biochem.* 65, 1302, 2001.

354. Tomita, M., Kozuka, M., and Lu, S.T. Studies on the alkaloids of menispermaceous plants. Alkaloids of *Percicampylus formosanus* Diels. *Yakugaku Zasshi* 87, 315, 1967.

355. Liang, P., Zhou, Q., and Zhou, F. Chemical constituents of *Pericampylus glaucus* (Lam.) Merr. *Zhongguo Zhong Yao Zu Zhi* 23, 39, 60, 1998 (English abstract).

356. Yang, W. et al. Effect of the extract from *Peristrophe roxburghiana* on hemorheology in rats. *Zhong Yao Cai* 25, 727, 2002.

357. Zhuang, X. et al. Effects of *Peristrophe roxburghiaua* on blood pressure, NO, and ET in renal hypertensive rats. *Zhong Yao Cai* 26, 266, 2003.

358. Lin, C.H., Li, C.Y., and Wu, T.S. A novel phenylpropenoyl sulfonic acid and a new chlorophyll from leaves of *Petasites formosanus* Kitamura. *Chem. Pharm. Bull.* (Tokyo) 52, 1151, 2004.

359. Wang, G.T. et al. Calcium antagonizing activity of *S*-petasin, a hypotensive sesquiterpene from *Petasites formosanus*, on inotropic and chronotropic responses in isolated rat atria and cardiac myocytes. *Naunyn Schmiedebergs. Arch. Pharmacol.* 369, 322, 2004.

360. Ali-Mohamed, A.Y. and Khamis, A.S. Mineral ion content of the seeds of six cultivars of Bahraini date palm (*Phoenix dactylifera*). *J. Agric. Food Chem.* 52, 6522, 2004.

361. Ishurd, O. et al. An alkali-soluble heteroxylan from seeds of *Phoenix dactykufera* L. *Carbohydr. Res.* 338, 1609, 2003.

362. Al-Shahib, W. and Marshall, R.J. The fruit of the date palm: its possible use as the best food for the future? *Int. J. Food Sci. Nutr.* 54, 247, 2003.

363. He, Y., Wu, L., and Wang, X. Pharmacological study of the rhizome powder of *Pinellia pedatisecta* processed by different procedures. *Zhong Yao Cai* 20, 459, 1997 (English abstract).

364. Feng, W.S. et al. Isolation and structure identification of the chemical constituents from pine needles of *Pinus massoniana* Lamb. *Yao Xue Xue Bao* 39, 190, 2004 (English abstract).

365. Li, X. and Luo, H. Advances in pharmacological study of *Piper kadsura* (Choisy) Ohwi. *Zhong Yao Cai* 25, 214, 2002 (English abstract).

366. Li, R.W. et al. Antiinflammatory activity of Chinese medicinal vine plants. *J. Ethnopharmacol.* 85, 61, 2003.

367. Stohr, J.R., Xiao, P.G., and Rauer, R. Constituents of Chinese piper species and their inhibitory activity on prostaglandin and leukotriene biosynthesis *in vitro. J. Ethnopharmacol.* 75, 133, 2001.

368. Sen, T. et al. Antioxidant activity of the methanol fraction of *Pluchea indica* root extract. *Phytother. Res.* 16, 331, 2002.

369. Sen, T., Ghosh, T.K., and Chaudhuri, A.K. Studies on the mechanism of antiinflammatory and antiulcer activity of *Pluchea indica* — probable involvement of 5-lipoxygenase pathway. *Life Sci.* 52, 737, 1993.

370. Sen, T. and Nagchaudhur, A.K. Anti-inflammatory evaluation of a *Pluchea indica* root extract. *J. Ethnopharmacol.* 33, 135, 1991.

371. Haraguchi, H., Ishikawa, H., and Kubo, I. Antioxidative action of diterpenoids from *Podocarpus nagi. Planta Med.* 63, 213, 1997.

372. Xuan, L.J., Xu, Y.M., and Fang, S.D. Three diterpene dilactone glycosides from *Podocarpus nagi. Phytochemistry* 39, 1143, 1995.

373. Kiuchi, F. et al. New sesquiterpene hydroperoxides with trypanocidal activity from *Pogostemon cablin. Chem. Pharm. Bull.* (Tokyo) 52, 1495, 2004.

374. Luo, J., Guo, X., and Feng, Y. Constituents analysis on volatile oil of *Pogostemon cablin* from different collection time cultivated in Hainan. *Xhong Yao Cai* 25, 21, 2002.

375. www.hkherbarium.net/herbarium/ PDF/Eng%20part%203_Monocots.pdf

376. Huang, Z.H. et al. A new xanthone from *Polygala aureocauda* Dunn. *Yao Xue Xue Bao* 39, 752, 2004 (in Chinese).

377. Gin, H., Xia, X., and Li, Z. Effects of polysaccharide of *Polygala aureocauda* on the immunity functions of normal mouse. *Zhong Yao Cai* 21, 467, 1998.

378. Tsai, P.L. et al. Constituents and bioactive priniciples of *Polygonum chinensis. Phytochemistry* 49, 1663, 1998.

379. Wang, K.J., Zhang, Y.J., and Yang, C.R. Antioxidant phenolic compounds from rhizomes of *Polygonum paleaceum* Wall. *J. Ethnopharmacol.* 96, 483, 2005.

380. Wang, V.F. et al. Studies on chemical constituents from the root of *Polygonatum kingianum. Zhongguo Zhong Yao Za Zhi* 28, 524, 2003 (English abstract).

381. Wang, V.F. et al. A new indolizinone from *Polygonatum kingianum. Planta Med.* 69, 1066, 2003.

382. Lin, H.W., Han, G.Y., and Liao, S.X. Studies on the active constituents of the Chinese traditional medicine, *Polygonatum odoratum* (Mill.) Druce. *Yao Xue Xue Bao* 29, 215, 1994 (English abstract).

383. Xue, P.F. et al. Studies on the chemical consituents from *Potentilla multifida* L. *Beijing Da Xue Xud Bao* 36, 21, 2004 (English abstract).

384. Ma, J. et al. Analysis of polyphenolic antioxidants from the fruits of three *Pouteria* species by selected ion monitoring liquid chromatography–mass spectrometry. *J. Agric. Food Chem.* 52, 5873, 2004.

385. Bhamarapravati, S., Pendland, S.L., and Mahady, G.B. Extracts of spice and food plants from Thai traditional medicine inhibit the growth of the human carcinogen *Helicobacter pylori. In Vivo* 17, 541, 2003.

386. Ishimaru, K. et al. Polyacetylene glycosides from *Pratia nummularia* cultures. *Phyto-chemistry* 62, 643, 2003.

387. Wei, S., Si, X., and Xu, S. Chemical constituents in the stem of *Premna crassa* Hand. *Zhongguo Zhone Yao ZA Zhi* 15, 487, 1990 (English abstract).

388. Zhan, Z.J. and Yue, J.M. New glyceroglycolipid and ceramide from *Premna micro-phylla. Lipids* 38, 1299, 2003.

389. Chun, O.K., Kim, D.O., and Lee, C.Y. Superoxide radical scavenging activities of the major polyphenols in fresh plum. *J. Agric. Food Chem.* 51, 8067, 2003.

390. Kuroki, M. Studies on antitumor substance obtained from alkaline extract of sasa-leaves (leaves of *Pseudosasa owatarii* Makino). *Yokohama Med. Bull.* 19, 75, 1968.

391. Barkhordari, A. et al. Lectin histochemistry of normal lung. *J. Mol. Histol.* 35, 147, 2004.

392. Mei, X. et al. Effects of *Pyracantha fortuneana* extraction blood coagulation. *Zhong Yao Cai* 24, 874, 2001 (English abstract).

393. Kim, J.S. et al. Phenolic glycosides from *Pyrola japonica. Chem. Pharm. Bull.* (Tokyo) 52, 714, 2004.

394. Li, X. et al. Pharmacological actions of different species from genus *Pyrola. Zhong Yao Cai* 20, 402, 1997 (in Chinese).

395. Lee, S. et al. A new naphthoquinone from *Pyrola japonica. Arch. Pharm. Res.* 24, 522, 2001.

396. Ma, C., Zhou, Y., and Liu, A.R. Determination of chlorogenic acid and eriodictyol-7-0-β-D-glucuronide in *Pyrrosia* by RP-HPLC. *Yao Xue Xue Bao* 38, 286, 2003 (English abstract).

397. Yang, C. et al. Chemical constituents of *Pyrrosia petiolosa. J. Asian Nat. Prod. Res.* 5, 143, 2003.

398. Cao, B.J., Meng, Q.Y., and Ji, N. Analgesic and antiinflammatory effects of *Ranun-culus japonicus* extract. *Planta Med.* 58, 496, 1992.

399. Wei, B.L. et al. *In vitro* antiinflammatory effects of quercetin 3-0-methyl ether and other constituents from *Rhamnus* species. *Planta Med.* 67, 745, 2001.

400. Teng, C.M. et al. Frangulin B, an antagonist of collagen-induced platelet aggregation and adhesion, isolated from *Rhamus formosana* Thromb. *Haemost* 70, 1014, 1993.

401. Lin, C.N. et al. Flavonol and anthroquinone glycosides from *Rhamnus formosana. Phytochemistry* 30, 3103, 1991.

402. Lin, C.N. and Chung, M.I. Studies on the constituents of *Rhamus formosana. Gaoxing Yi Xye Ke Xue Za Zhi* 1, 684, 1985 (English abstract).

403. Kongkathip, N. et al. Synthesis of novel rhinacanthins and related anticancer naph-thoquinone esters. *J. Med. Chem.* 7, 4427, 2004.

404. Wu, T.S. et al. Rhinacanthin-Q, a naphthoquinone from *Rhinacanthus nascutus* and its biological activity. *Phytochemistry* 49, 2001, 1998.

405. Kernan, M.R. et al. Two new lignins with activity against influenza virus from the medicinal plant *Rhinacanthus nasutus. J. Nat. Prod.* 60, 635, 1997.

406. Masuda, T. et al. Cytotoxic screening of medicinal and edible plants in Okinawa, Japan and identification of the main toxic constituent of *Rhodea japonica* (Omoto). *Biosci. Biotechnol. Biochem.* 67, 1401, 2003.

407. Nawa, H. Studies on the components of *Rhodea japonica* Roth. XI. Structure of rhodeasapogenin. *Chem. Pharm. Bull.* (Tokyo) 6, 255, 1958.

408. Garcia, M. et al. Blockade of the antiadrenergic action of *Bretylium* by an aqueous extract of the leaves of *Rhoe spathacea. Can. J. Physiol. Pharmacol.* 49, 1106, 1971.

409. Fu, N. Protective effect of tannic acid against TPA- and cigarette smoke condensate-induced DNA strand breaks in human white cells and its antioxidant action. *Zhangguo Yi Xue Ke Xue Yuan Xue Bao* 13, 347, 1991.

410. Zheng, M. Experimental study of 472 herbs with antiviral action against the herpes simplex virus. *Zhong Xi Yi Jie He Za Zhi* 10, 39, 1990 (English abstract).

411. Tangpu, V. and Yadav, A.K. Antidiarrhoeal activity of *Rhus javanica* ripen fruit extract in albino mice. *Fitoterapia* 75, 39, 2004.

412. Lee, I.S. et al. Semialactone, isofouquierone peroxide, and fouquierone, three new dammarane triterpenes from *Rhus javanica. Chem. Pharm. Bull.* (Tokyo) 49, 1024, 2001.

413. Kuo, S.C. et al. 6-Pentadecyl salicylic acids, an antithrombin component isolated from the stem of *Rhus semialata* var. *ruxburghii. Planta Med.* 57, 247, 1991.

414. Wu, P.L. et al. Antioxidative and cytotoxic compounds extracted from sap of *Rhus succedanea. J. Nat. Prod.* 65, 1719, 2002.

415. Lin, Y.M. et al. Antiviral activities of bioflavonoids. *Planta Med.* 65, 120, 1999.

416. Wang, H.K. et al. Recent advances in the discovery and development of flavonoids and their analogues as antitumor and anti-HIV agents. *Adv. Exp. Med. Biol.* 439, 191, 1998.

417. Kinjo, J. et al. Antiproliferative constituents in the plant 8. Seeds of *Rhynchosia volubilis. Biol. Pharm. Bull.* 24, 1443, 2001.

418. Wu, X. et al. Characterization of anthocyanins and proanthocyanidins in some cultivars of riles, aronia, and sambacus and their antioxidant capacity. *J. Agric. Food Chem.* 52, 7846, 2004.

419. Del Castillo, M.L. et al. Fatty acid content and juice characteristics in black currant (*Ribes nigrum* L.) genotypes. *J. Agric. Food Chem.* 52, 948, 2004.

420. Suzutani, T. et al. Anti–herpes virus activity of an extract of *Ribes nigrum* L. *Phytother. Res.* 17, 609, 2003.

421. Jiao, S.P., Chen, B., and Du, P.G. Anti–lipid peroxidation effect of *Rosa davurica* Pall. Fruit. *Zhong Xi Yi Jie He Xue Bao* 2, 364, 2004 (English abstract).

422. Aritomi, M. On the components of the flower petals of *Rosa multiflora* Thunb. and *Rubus hirsutus* Thunb. *Japan J. Pharmacol.* 82, 771, 1962.

423. Bushman, B.S. et al. Chemical composition of caneberry (*Rubus* spp.) seeds and oils and their antioxidant potential. *J. Agric. Food Chem.* 52, 7982, 2004.

424. Wagner, H. et al. Synthesis of glucuronides in the flavonoid series 3. Isolation of apigenin-7-D-glucuronide from *Ruellia tuberosa* L. and its synthesis. *Chem. Ber.* 104, 2681, 1971 (in German with English abstract).

425. Liu, Y., Li, C., and Zhang, C. Studies on chemical constituents of *Salvia roburowskii* Maxim. *Zhong Yao Cai* 25, 792, 2002 (English abstract).

426. Park, K.H. et al. Antiallergic activity of a disaccharide isolated from *Sanguisorba officinalis. Phytother. Res.* 18, 658, 2004.

427. Ayoub, N.A. Unique phenolic carboxylic acids from *Sanguisorba minor. Phytochemistry* 63, 433, 2003.

428. Matsushita, A. et al. *Oleanane saponins* from *Sanicula elata* var. *Chinesis. J. Nat. Prod.* 67, 377, 2004.

429. Powell, R.G., Bajaj, R., and McLaughlin, J.L. Bioactive stilbenes of *Scirpus maritimus. J. Nat. Prod.* 50, 293, 1987.

430. Sagare, A.P. et al. *De novo* regeneration of *Scrophularia yoshimrae* yamazaki (Scrophulariaceae) and quantitative analysis of harpagoside, an iridoid glucoside formed in aerial and underground parts of *in vitro* propagated and wild plants by HPLC. *Biol. Pharm. Bull.* 24, 1311, 2001.

431. Lohezic-Le Devehat, F. et al. Flavonols from *Scurrula ferruginea* Darser. *Z. Naturforsch.* 57, 1092, 2002.

432. Yu, J. et al. Chemical composition and antimicrobial activity of the essential oil of *Scutellaria barbata. Phytochemistry* 65, 881, 2004.

433. Sato, Y. et al. Phytochemical flavones isolated from *Scutellaria barbata* and antibacterial activity against methicillin-resistant *Staphylococcus aureus*. *J. Ethnopharmacol.* 72, 483, 2000.

434. Mavi, A. et al. Antioxidant properties of some medicinal plants: *Prangos ferulacea, Sedum sempervivoides, Malva negelcta, Cruciata taurica, Rosa pimpinellifolia, Galium verum* subsp. *verum, Urtica dioica. Biol. Pharm. Bull.* 27, 702, 2004.

435. Lin, L.C., Kuo, Y.C., and Chou, C.J. Cytotoxic biflavonoids from *Selaginella delicatula. J. Nat. Prod.* 63, 627, 2000.

436. Chiang, L.C. et al. *In vitro* anti–herpes simplex viruses and antiadenoviruses activity of 12 traditionally used medicinal plants in Taiwan. *Biol. Pharm. Bull.* 26, 1600, 2003.

437. Kwon, Y.S. et al. Antioxidant constituents from *Setaria viridis. Arch. Pharm. Res.* 25, 300, 2002.

438. Wu, T.S., Chen, C.M., and Lin, F.W. Constituents of the root bark of *Severinia buxifolia* collected from Hainan. *J. Nat. Prod.* 64, 1040, 2001.

439. Jang, D.S. et al. Compounds obtained from *Sida acuta* with the potential to induce quinone reductase and to inhibit 7, 12-dimethylbenz9a0 anthracene-induced preneoplastic lesions in a mouse mammary organ culture model. *Arch. Pharm. Res.* 26, 585, 2003.

440. Dinan, L., Bourne, P., and Whiting, P. Phytoecdysteroid profiles in seeds of *Sida* spp. *Phytochem. Anal.* 12, 110, 2001.

441. Ahmad, M.U. et al. Cyclopropenoid fatty acids in seed oils of *Sida acuta* and *Sida rhombifolia. J. Am. Oil Chem. Soc.* 53, 698, 1976.

442. Bushneva, O.A. et al. Structure of silenan, a pectic polysaccharide from *Silene vulgaris* Garcke. *Biochemistry* 68, 1360, 2003.

443. Jiang, H.L. et al. Quinic acid esters from herba of *Siphonostegia chinensis. Zhongguo Zhong Yao Za Zhi* 27, 923, 2002 (English abstract).

444. Zhang, H. et al. Constituents of lignonoids in *Siphonostegia chinensis* Benth. *Zhongguo Zhong Yao Za Zhi* 1995 April 20(4), 230, 253, 1995.

445. Li, S.Y. et al. New phenolic consituents from *Smilax practeata. J. Nat. Prod.* 65, 262, 2002.

446. Yokose, T. et al. Antifungal sesquiterpenoid from the root exudate of *Solanum abutiloides. Biosci. Biotechnol. Biochem.* 68, 2640, 2004.

447. Inoguchi, M. et al. Production of an allelopathic polyacetylene in hairy root cultures of goldenrod (*Solidago altissima* L.). *Biosci. Biotechnol. Biochem.* 67, 168, 2003.

448. van Berkel, J.M. et al. Polypeptides of the chloroplast envelope membranes as visualized by immunochemical techniques. *J. Histochem. Cytochem.* 34, 577, 1986.

449. Oh, H. et al. The absolute configuration of prunioside A from *Spiraea prunifolia* and biological activities of related compounds. *Phytochemistry* 64, 1113, 2003.

450. Lin, Y.L. et al. Homocyclotirucallance and two dihydrophenanthrenes from *Spiranthes sinensis. Chem. Pharm. Bull.* (Tokyo) 49, 1098, 2001.

451. Hayashi, K. et al. Acteoside, a component of *Stachys sieboldii* MIQ, may be a promising antinephritic agent (3): effect of aceteoside on expression of intercellular adhesion molecule-1-in experimental nephritic glomeruli in rats and cultured endothelial cells. *Jpn. J. Pharmacol.* 70, 157, 1996.

452. Zhao, X. et al. Tetrandrine, a bisbenzylisoquinoline alkaloid from Chinese herb Radix, augmented the hypnotic effect of pentobarbital through the serotonergic system. *Eur. J. Pharmcol.* 506, 101, 2004.

453. Naik, D.G. et al. Analysis of fatty acids in the seeds of *Steroulia lychnophora* by GC-MS. *Zhongguo Zhong Yao Za Zhi* 28, 533, 2003.

454. Achenbach, H. Homocylindrocarpidine and 17-demethoxy-cylindrocarpidine, two new alkaloids from *Tabernaemontana amygdalifolial. Z. Naturforsch. B.* 22, 955, 1967.

455. Achenbach, H. 10-Oxo-cylindrocarpidin, a new alkaloid from *Tabernaemontana amygdalifolia. Tetrahedron Lett.* 19, 1793, 1967.

456. Taesolikul, T. et al. Antiinflammatory, antipyretic and antinociceptive activities of *Tabernaemontana pandacaqui* Poil. *J. Ethnopharmacol.* 84, 31, 2003.

457. Ingkaninan, K. et al. Interference of linoleic acid fraction in some receptor-binding assays. *J. Nat. Prod.* 62, 912, 1999.

458. Lathui Uiere, P. et al. On the alkaloids of *Tabernaemontana pandacaqui* Poir. Apocynaceae. *Ann. Phar. Fr.* 24, 547, 1966.

459. Komatsu, M. et al. Studies on the constituents of *Talinum paniculatum* Gaertner. I. *Yakugaku Zasshi* 102, 499, 1982 (in Japanese with English abstract).

460. Udo, U.F. et al. *In vivo* evaluation of iron bioavailability in some Nigerian peasant meals by haemoglobin regeneration technique. *Cent. Afr. J. Med.* 39, 142, 1993.

461. Li, L., Li, M., and Zhu, A. Chemical constituents of *Taxillus levinei* (Merr.) H. S. Kiu. *Zhongguo Zhong Yao Sa Shi* 21, 34, 63, 1996.

462. Fukunaga, T. et al. Studies on the constituents of Japanese mistletoes from different host trees, and their antimicrobial and hypotensive properties. *Chem. Pharm. Bull.* (Tokyo) 37, 1543, 1989.

463. Ikuta, A. et al. Ursane- and oleanane-type triterpenes from *Ternstroemia gymnanthera* callus tissues. *J. Nat. Prod.* 66, 1051, 2003.

464. Liu, D. and Yang, J. A study on chemical components of *Tetrastigma hemsleyanum* Diels et Gilg native to China. *Zhongguo Zhong Yao Za Zhi* 24, 611, 1999 (English abstract).

465. Yang, D. et al. Chemical constituents of *Tetrastigma hemsleyanum* Diels. et Gilg. *Zhongguo Zhong Yao Za Zhi* 23, 419, 1998 (English abstract).

466. Jakinovich, W., Jr. Evaluation of plant extracts for sweetness using the Mongolian gerbil. *J. Nat. Prod.* 53, 190, 1990.

467. Hussain, R.A. et al. Plant-derived sweetening agents: saccharide and polyol constituents of some sweet-tasting plants. *J. Ethnopharmacol.* 28, 103, 1990.

468. Yokozawa, T., Tanaka, T., and Limura, T. Examination of the nitric oxide production-suppressing component in *Tinospora tuberculata. Biol. Pharm Bull.* 24, 1153, 2001.

469. Elufioye, T.O. and Agbedahunsi, J.M. Antimalarial activities of *Tithonia diversifolia* and *Crossoptery × febrifuga* on mice *in vivo. J. Ethnopharmacol.* 93, 167, 2004.

470. Owoyele, V.B. et al. Studies on the antiinflammatory and analgesic properties of *Tithonia diversifolia* leaf extract. *J. Ethnopharmacol.* 90, 317, 2004.

471. Goffin, E. et al. Quantification of tagitinin C in *Tithonia diversifolia* by reversed-phase high performance liquid chromatography. *Phytochem. Anal.* 14, 378, 2003.

472. Nagira, Y. and Ozeki, Y. A system in which anthocyanin synthesis is induced in regenerated torenia shoots. *J. Plant Res.* 117, 377, 2004.

473. Li, R.W. et al. Antiinflammatory activity of Chinese medicinal vine plants. *J. Ethnopharm.* 85, 61, 2003.

474. Rao, M.A. and Rao, E.V. Crystalline components from the leaves and twigs of *Trachelospermum fragrans. Planta Med.* 32, 46, 1977.

475. Nishibe, S., Hisada, S., and Inagaki, I. Cyclitol of several *Trachelospermum* species. *Yakuga Ku Zasshi* 93, 539, 1973.

476. Sultan, N.A., Kenoth, R., and Swamy, M.J. Purification, physicochemical characterization, saccharide specificity, and chemical modification of a gal/gal NAC specific lectin from the seeds of *Trichosathes dioica. Arch. Biochem. Biophys.* 432, 212, 2004.

477. He, X.J. et al. Research on active constituents research of gualou xieba; baijiutang (III). The active flavonoids. *Zhongguo Zhong Yao Za Zhi* 28, 420, 2003 (English abstract).

478. Tiwari, U. et al. Immunomodulator effects of aqueous extract of *Tridax procumbens* in experimental animals. *J. Ethnopharmacol.* 92, 113, 2004.

479. Ali, M.S. and Jahangir, M. A *bis*-bithiophene from *Tridax procumbens* L. *Nat. Prod. Lett.* 16, 217, 2002.

480. Ali, M., Ravinder, E., and Ramachandram, R. A new flavonoid from the aerial parts of *Tridax procumbens*. *Fitoterapia* 72, 313, 2001.

481. Wang, B. et al. Triptolide, an active component of the Chinese herbal remedy *Tripterygium wilfordii* Hook f., inhibits production of nitric oxide by decreasing inducible nitric oxide synthase gene transcription. *Arthritis Rheum.* 50, 2995, 2004.

482. Yu, Q. et al. Turpinionosides A-E: megastigmane glucoside from leaves of *Turpinia ternata* Nakai. *Chem. Pharm. Bull.* (Tokyo) 50, 640, 2002.

483. Huang, X. et al. Cytotoxic alkaloids from the roots of *Tylophora atrofolliculta*. *Planta Med.* 70, 441, 2004.

484. Neoh, C.K. *Typhonium divaricatum* (rodent tuber): a promising local plant in the fight against cancer. *Med. J. Malaysia* 47, 86, 1992.

485. Gul, N., Ahmed, S.A., and Smith, L.A. Inhibition of the protease activity of the light chain of type A botulinum neurotoxin by aqueous extract from stinging nettle (*Urtica dioica*) leaf. *Basic Clin. Pharmcol. Toxicol.* 95, 215, 2004.

486. Durak, I. et al. Aqueous extract of *Urtica dioica* makes significant inhibition on adenosine deaminase activity in prostate tissue from patients with prostate cancer. *Cancer Biol. Ther.* 3, 9, 2004.

487. Gulcin, I. et al. Antioxidant, antimicrobial, antiulcer and analgesic activities of nettle (*Urtica dioica* L.). *J. Ethnopharmacol.* 90, 205, 2004.

488. Tsai, H.Y. et al. The effects of *Vandellia cordifolia* on renal functions and arterial blood pressure. *Am. J. Chin. Med.* 17, 203, 1989.

489. Albach, D.C., Gotfredsen, C.H., and Jensen, S.R. Iridoid glucosides of *Paederota lutea* and the relationships between *Paederota* and *Veronica*. *Phytochemistry* 65, 2129, 2004.

490. Fukuyame, Y., Minoshima, V., and Kishimoto, Y. Iridoid glucosides and *p*-coumaroyl iridoids from *Viburnum luzonicum* and their cytotoxicity. *J. Nat. Prod.* 67, 1833, 2004.

491. Moon, H.I. et al. Triterpenoid saponin from *Viola hondoensis* W. Becker et H. Boss and their effect on MMP-1 and type I procollagen expression. *Arch. Pharm. Res.* 27, 730, 2004.

492. Xie, C. et al. Antibacterial activity of the Chinese traditional medicine Zi Hua Di Ding. *Phytotherapy* 18, 497, 2004.

493. Svangard, E. et al. Cytotoxic cyclotides from *Viola tricolor. J. Nat. Prod.* 67, 144, 2004.

494. Xie, C. et al. Flavone c-glycosides from *Viola yedoensis* Makino. *Chem. Pharm. Bull.* (Tokyo) 51, 1204, 2003.

495. Pan, J.G., Xu, Z.L., and Fan, J.F. GC-MS analysis of essential oils from four *Vitex* species. *Zhongguo Zhong Yao Za Zhi* 14, 357, 1989 (English abstract).

496. Fang, H.J., Chen, L.S., and Xhou, T.H. Studies on the components of essential oils III. Studies of chemical constituents of essential oil from *Rhododendron racemosum* Franch, comparison of the constituents of *Vitex negundo* var. *Cannabifolia* (Sieb et Zucc.) Hand-Mazz. *Yao Xue Xud Bao* 15, 284, 1980 (English abstract).

497. Raju, B.L. et al. Antioxidant iridoid glucosides from *Wendlandia formosana*. *Nat. Prod. Res.* 18, 357, 2004.

498. Ooi, L.S. et al. Anticancer and antiviral activities of *Youngia japonica* (L.) DC. *J. Ethnopharmacol.* 94, 117, 2004.

499. Xiong, Q.B. and Shi, D.W. Morphological and histological studies of Chinese traditional drug "hua jiao" (*Pericarpium, Zanthoxyli*) and its allied drugs. *Yao Xue Xue Bao* 26, 938, 1991 (English abstract).

500. Hatano, T. et al. Alophatic acid amides of the fruits of *Zanthoxylum piperitum*. *Phytochemistry* 65, 2599, 2004.

501. Tao, Z.Y. et al. Studies on the coumarins in the root of *Zanthoxylum dimorphophyllum*. *Zhongguo Zhong Yao Za Zhi* 28, 344, 2003 (English abstract).

502. Tien, N.Q. et al. New ceramide from *Alocasia macrorhiza. Arch. Pharm. Res.* 27, 1020, 2004.

503. Wang, H.X. and Ng, T.B. Alocasin, an antifungal protein from rhizomes of the giant taro *Alocasia macrorrhiza. Protein Expr. Purif.* 28, 9, 2003.

504. Kooiman, D. Structures of the galactomanans from seeds of *Annona muricata, Arenga saccharifera, Cocos nucifera, Convolvulus tricolor,* and *Sophora japonica. Carbohydr. Res.* 20, 329, 1972.

505. Takemoto, T. et al. Isolation of insect moulting substances from *Blechnum amabile* and *Blechnum niponicum. Chem. Pharm. Bull.* (Tokyo) 17, 1973, 1969.

506. Luo, D.Q. et al. Antifungal properties of pristimerin and celastrol isolated from *Celastrues hypoleucus. Pest Manag. Sci.* 61, 85, 2005.

507. Quo, Y.Q. et al. Sesquiterpene esters from the fruits of *Celastrus orbiculatus. Chem. Pharm. Bull.* (Tokyo) 52, 1134, 2004.

508. Borrelli, F. et al. New sesquiterpenes with intestinal relaxant effect from *Celastrus paniculatus. Plants Med.* 70, 652, 2004.

509. Schedlbauer, M.D., Cave, C.F., and Bell, P.R. The incorporation of DL-(3-14C) cysteine during spermatogenesis in *Ceratopteris thalictroides. J. Cell Sci.* 12, 765, 1973.

510. Quilez, A.M. et al. Phytochemical analysis and antiallergic study of *Agave intermixta* Trel. and *Cissus sioyoides* L. *J. Pharm. Pharmacol.* 56, 1185, 2004.

511. Beltrame, F.L., Ferreira, A.G., and Cortex, D.A. Coumarin glucoside from *Cissis scyoides. Nat. Prod. Lett.* 16, 213, 2002.

512. Tewtrakul, S.H. et al. HIV-1 integrase inhibitory substances from *Coleus parvifulius. Phytother. Res.* 17, 232, 2003.

513. Atta, A.H. and El-Sooud, K.A. The antinociceptive effect of some Egyptian medicinal plant extracts. *J. Ethnopharmacol.* 95, 235, 2004.

514. Su, Y. et al. Conyzasaponins I-Q, nine new triterpenoid daponins from *Conyza blinii. J. Nat. Prod.* 66, 1593, 2003.

515. Kongsaeres, P. et al. Antimalaria dihydroisocoumarins produced by *Geotrichum* sp., an endophytic fungus of *Crassocephalum crepidioides. J. Nat. Prod.* 66, 709, 2003.

516. Chagnon, M. et al. Antiinflammatory activity of extracts of *Crassocephalum multicorymbosum. Planta Med.* 49, 255, 1983.

517. Lakshmi, V. and Chaulan, J.S. A new pentacyclic triterpene from the root bark of *Crateva nurvala. Planta Med.* 32, 214, 1977.

518. Yoo, H.S. et al. Flavonoides of *Crotalaria sessiliflora. Arch. Pharm. Res.* 27, 544, 2004.

519. Zhang, W. et al. A new flavan from *Daphne odora* var. *atrocaulis. Fitoterapia* 75, 799, 2004.

520. Rathi, A. et al. Antiinflammatory and antinociceptive activity of the water decoction *Desmodium gangeticum. J. Ethnopharmacol.* 95, 259, 2004.

521. Govindarajan, R., Rastogi, S., and Vijayakumar, M. Studies on the antioxidant activities of *Desmodium gangeticum*. *Biol. Pharm. Bull.* 26, 1424, 2003.

522. Jabbar, S.M., Khan, T., and Choudhuri, M.S. The effects of aqueous extracts of *Desmodium gangeticum* DC on the central nervous system. *Pharmazie* 56, 506, 2001.

523. Malviya, N., Pal, R., and Khanna, N.M. Saponins from *Deutzia corymbosa*. *Phytochemistry* 30, 2798, 1991.

524. Jadot, J. and Casimir, J. Separation and characterization of a hydroxyleucine from *Deutzia gracilis*. *Biochim. Biophys. Acta* 48, 400, 1961.

525. Watanabe, M. Antioxidative phenolic compounds from Japanese barnyard millet (*Echinochloa utilis*) grains. *J. Agric. Food Chem.* 47, 4500, 1999.

526. Cao, S.G. et al. Flavonol glycosides from *Elaeagnus lanceollata* Warb. *Nat. Prod. Lett.* 15, 211, 2001.

527. Cao, S.G. et al. Flavonol glycosides from *Elaeagnus backii. Nat. Prod. Lett.* 15, 1, 2001.

528. Ramezani, M., Hosseinzadeh, H., and Daneshmand, N. Antinociceptive effect of *Elaeagnus angustifolia* fruit seeds in mice. *Fitoterapia* 72, 255, 2001.

529. Chow, S.Y., Chen, D.F., and Chen, S.M. Pharmacological studies of Chinese herbs (6). Pharmacological effects of *Epiphyllum oxypetalum* Haw. *Taiwan Yi Xue Hui Za Zhi* 76, 916, 1977.

530. Aritomi, M., Shimoioe, M., and Mazaki, T. Chemical consitituents in flowers of *Gnaphalium affine* D. Don. *Yakugaku Zasshi* 84, 895, 1964.

531. Aoshima, Y., Hasegawa, Y., and Hasegawa, S. Isolation of Gnafc, a polysaccharide constituent of *Gnaphalium affine*, and synergistic effects of Gnafc and asorbate on the phenotypic expression of osteroblastic MC3T3-E1 cells. *Biosci. Biotechnol. Biochem.* 67, 2068, 2003.

532. Shui, G. and Peng, L.L. An improved method for the analysis of major antioxidants of *Hibiscus osculentus* L. *J. Chromatgr. A.* 1048, 17, 2004.

533. Holser, R.A., Bost, G., and van Boven, M. Phytosterol composition of hybrid *Hibiscus* seed oils. *J. Agric. Food Chem.* 52, 2546, 2004.

534. Wu, P.L. et al. Cytotoxicity of phenylpropanoid esters from the stem of *Hibiscus taiwanensis. Bioorg. Med. Chem.* 12, 2193, 2004.

535. Pragada, R.R. et al. Cardioprotective activity of *Hydrocotyle asiatica* L. in ischemia-reperfusion induced myocardial infection in rats. *J. Ethnopharmacol.* 93, 105, 2004.

536. Matsushita, A. et al. Hydrocotylosides I-VII, new oleanane saponins from *Hydrocotyle sibthorpioides. J. Nat. Prod.* 67, 384, 2004.

537. Potter, D.M. and Barid, M.S. Carcinogenic effects of ptaquiloside in bracken fern and related compounds. *Br. J. Cancer* 83, 914, 2000.

538. Thangadurai, D., Viswanathan, M.B., and Ramesh, N. Indigoferabietone, a novel abietane diterpenoid from *Indigofera longeracemosa* with potential antituberculous and antibacterial activity. *Pharmazie* 57, 714, 2002.

539. Zhang, S. and Zhu, W. Pharmacognostical studies on cortex *Indigoferae pseudotinctorial. Zhong Yao Cai* 24, 30, 2001 (in Chinese).

540. Terahara, N. et al. Characterization of acylated anthocyanins in callus induced from storage root of purple-fleshed sweet potato, *Ipomoea hatatas* L. *J. Biomed. Biotechnol.* 279, 2004.

541. Leon, I. et al. Isolation and characterization of five new tetrasaccharide glycosides from the roots of *Ipomoea stans* and their cytotoxic activity. *J. Nat. Prod.* 67, 1552, 2004.

542. Robertson, P.A. and Macfarlane, W.V. Pain-producing substances from the stinging bush *Laportea moroides. Aust. J. Exp. Biol. Med. Sci.* 35, 381, 1957.

543. Lindikett, R. and Jung, F. The substances of the irritant of *Laportea*. *Pharmazie* 8, 78, 1953.

544. Sadhu, S.K. et al. Separation of *Leucas aspera*, a medicinal plant of Bangladesh, guided by prostaglandin inhibitory and antioxidant activities. *Chem. Pharm. Bull.* (Tokyo) 51, 595, 2003.

545. Mukherjee, K., Saha, B.P., and Mukherjee, P.K. Psychopharmacological profiles of *Leucas lavandulaefolia* Rees. *Phytother. Res.* 16, 696, 2002.

546. Kang, O.H. et al. Inhibition of trypsin-induced mast cell activation by water fraction of *Lonicera japonica*. *Arch. Pharm. Res.* 27, 1141, 2004.

547. Chai, X.Y., Li, P., and Tang, L.Y. Studies on chemical constituents in dried buds *of Lonicera confusa*. *Zhongguo Zhong Yao Za Zhi* 29, 865, 2004.

548. van Kiem, P. et al. Pentacyclic triterpenoids from *Mallotus apelta*. *Arch. Pharm. Res.* 27, 1109, 2004.

549. Watanabe, C. et al. Inhibitory mechanisms of glycoprotein fraction derived from *Miscanthus sinensis* for the immediate phase response of an IgE-mediated cutaneous reaction. *Biol. Pharm. Bull.* (Tokyo) 22, 26, 1999.

550. Misra, L. and Wagner, H. Alkaloidal constituents of *Mucuna puriens* seeds. *Phytochemistry* 65, 2565, 2004.

551. Singh, D.D. et al. Purification, crystallization and preliminary X-ray structure analysis of the banana lectin from *Musa paradisiace*. *Acta Crystallogr D. Biol. Crystallogr.* 60, 2104, 2004.

552. Rabbani, G.H. et al. Green banana and pectin improve small intestinal permeability and reduce fluid loss in Bangladeshi children with persistent diarrhea. *Dig. Dis. Sci.* 49, 475, 2004.

553. Kuo, P.L. et al. Prodelphinidin B-2, 3, 3'-di-0-gallata from *Myrica rubra* inhibits proliferation of A 549 carcinoma cells via blocking cell cycle progression and inducing apoptosis. *Eur. J. Pharmacol.* 501, 41, 2004.

554. Zhang, L.Z. et al. Studies on chemical constituents in fruits of Tibetan medicine *Phyllanthus emblica*. *Zhongguo Zhong Yao Za Zhi* 28, 940, 2003 (in Chinese).

555. Wang, L.Q. et al. Three diterpene glucoside and a diphenylamine derivative from *Pieris formosa*. *Fitoterapia* 72, 779, 2001.

556. Mata, R. et al. Antimycobacterial compounds from *Piper sanctum*. *J. Nat. Prod.* 67, 1961, 2004.

557. Pelter, A. and Hansel, R. Epoxipiperolid from *Piper sanctum*. *Z. Naturforsch. [B]* 27, 1186, 1972.

558. Son, Y.O. et al. Selective antiproliferative and apoptotic effects of flavonoids purified from *Rhus verniciflua* stokes on normal versus transformed hepatic cell lines. *Toxicol. Lett.* 15, 115, 2005.

559. Werner, R.A. et al. Biosynthesis of gallic acid in *Rhus typhina*: discrimination between alternative pathways from natural oxygen isotope abundance. *Phytochemistry* 65, 2809, 2004.

560. Reisner, L. Biologic poisons for pain. *Curr. Pain Headache Rep.* 8, 427, 2004.

561. Balzarini, J. et al. The mannose-specific plant lectins from *Cymbidium* hybrid and *Epipactis helleborine* and the (*N*-acetylglucosamine) *n*-specific plant lectin from *Urtica dioica* are potent and selective inhibitors of human immunodeficiency virus and cytomegalovirus replication *in vitro*. *Antiviral Res.* 18, 191, 1992.

562. Du, Q., Jerz, G., and Winterhalter, P. Isolation of two anthocyanin sambubiosides from bilberry (*Vaccinium myrtillus*) by high-speed counter current chromatography. *J. Chromatogr. A.* 1045, 59, 2004.

563. Nakane, T. et al. Fern constituents: triterpenoids from *Adiantum capillus-veneris*. *Chem. Pharm. Bull.* (Tokyo) 50, 1273, 2002.
564. Okuyama, T. et al. Effects of Chinese drugs "xiebai" and "dasuan" on human platelet aggregation (*Allium bareri, A. sativum*). *Planta Med.* 55, 242, 1989.
565. Seidlova-Wuttke, D. et al. *Belamcanda chinensis* and their purified tectorigenin have selective estrogen receptor modulator activities. *Phytochemistry* 11, 392, 2004.
566. Gilani, A.H. et al. Presence of cholinominetic and acetylcholinesterase inhibitory constituents in betel nut. *Life Sci.* 15, 2377, 2004.
567. Wu, M.T. et al. Constituents of areca chewing related to esophageal cancer risk in Taiwanese men. *Dis. Esophagus* 17, 257, 2004.
568. Gilani, A.H. et al. The presence of cholinomimetic and calcium channel antagonist constituents in piper betle L. *Phytother. Res.* 14, 436, 2000.
569. Xie, C. et al. Antibacterial activity of Chinese traditional medicine "Zi Hua Di Dong." *Phytother. Res.* 18, 497, 2004.
570. Lue, Y., Sinha Hikim, A.P., and Wang, C. Triptolide, a potential male contraceptive. *J. Androl.* 19, 479, 1998.
571. Anonymous. Herbal pharmacology in the People's Republic of China. A trip of the American Herbal Pharmacology Delegation. National Academy of Sciences, Washington DC, 1975. (www.swsbm.com/Ephemera/China_herbs.pdf)
572. Wang, L.W. et al. New alkaloids and a tetraflavonoid from *Cephalotaxus wilsenians*. *J. Nat. Prod.* 67, 1182, 2004.
573. Nguyen, M. et al. Staminane- and isopimarane-type diterpenes from *Orthosiphon stamineus* of Taiwan and their nitric oxide inhibitory activity. *J. Nat. Prod.* 67, 654, 2004.
574. El-Gamal, A.A. et al. New diterpenoids from *Viburnum awabuki*. *J. Nat. Prod.* 67, 333, 2004.
575. Shen, Y.C. et al. New cytotoxic clenodane diterpenoids from the leaves and twigs of *Casearia membranacea*. *J. Nat. Prod.* 67, 316, 2004.
576. Negi, N. et al. Two new dimeric acridone alkaloids from *Glycosmis citrifolia*. *Chem. Pharm. Bull.* (Tokyo) 52, 362, 2004.
577. Chang, C.I. et al. Three new oleanane-type triterpenes from *Ludwigia octovalvis* with cytotoxic activity against two human cancer cell lines. *J. Nat. Prod.* 67, 91, 2004.
578. Shen, Y.C. et al. Vibsane diterpenoids from the leaves and flowers of *Viburnum odoratissimum*. *J. Nat. Prod.* 67, 74, 2004.
579. Itoh, A. et al. Six secoiridoid glucosides from *Adina racemose*. *J. Nat. Prod.* 66, 1212, 2003.
580. Wu, T.S. et al. Constituents from the leaves of *Phellodendron amurense* var. *Wilsonii* and their bioactivity. *J. Nat. Prod.* 66, 1207, 2003.
581. Wu, P.L., Su, G.C., and Wu, T.S. Constituents from the stems of *Aristolochia manshuriensis*. *J. Nat. Prod.* 66, 996, 2003.
582. Su, C.R. et al. Acetophenone derivatives from *Acronychia pedunculata*. *J. Nat. Prod.* 66, 990, 2003.
583. Hou, C.C. et al. Antidiabetic dimeric guianolides and a lignan glucoside from *Lactuca indica*. *J. Nat. Prod.* 66, 625, 2003.
584. Hou, W.C. et al. Free radical–scavenging activity of Taiwanese native plants. *Phytomedicine* 10, 170, 2003.
585. Chiang, L.C. et al. Antileukemic activity of selected natural products in Taiwan. *Am. J. Chin. Med.* 31, 37, 2003.
586. Lan, Y.H. et al. Cytotoxic styrylpyrones from *Goniothalamus amuyon*. *J. Nat. Prod.* 66, 554, 2003.

587. Weng, J.R. et al. Antiinflammatory constituents and new pterocarpanoid of *Crotalaria pallica. J. Nat. Prod.* 66, 404, 2003.
588. Liaw, C.C. et al. A novel constituent from *Rollinia mucosa*, rollicosin, and a new approach to develop annonaceous acetogenins as potential antitumor agents. *J. Nat. Prod.* 66, 279, 2003.
589. Hou, C.C. et al. Bacopaside III, bacopasaponin G, and bacopasides A-C from *Bacopa monniera. J. Nat. Prod.* 65, 1759, 2002.
590. Stough, C. et al. The chronic effects of an extract of *Bacopa monniera* (Brahmi) on cognitive function in healthy human subjects. *J. Psychopharmacol.* (Berlin.) 156, 481, 2001.
591. Shen, Y.C. et al. Taxumairols A-Z: new taxoides from Taiwanese *Taxus mairei* (Lemee & Leveille) S. Y. Hu. *Chem. Pharm. Bull.* (Tokyo) 50, 1561, 2002.
592. Kuo, Y.H., Lee, P.H., and Wein, Y.S. Four new compounds from the seeds of *Cassia fistula. J. Nat. Prod.* 65, 1165, 2002.
593. Shen, Y.C. et al. New vibsane diterpenes and lupane triterpenes from *Viburnum odoratissimum. J. Nat. Prod.* 64, 1052, 2002.
594. Lin, Y.L. et al. Anti–lipid-peroxidatve principles from *Tournefortia sarmentosa. J. Nat. Prod.* 65, 745, 2002.
595. Lin, J.H., Chiou, Y.N., and Lin, Y.L. Phenolic glycosides from *Viscum angulatum. J. Nat. Prod.* 65, 638, 2002.
596. Day, S.H. et al. Potent cytotoxic lignans from *Justicia procumbens* and their effects on nitric oxide and tumor necrosis factor-α production in mouse macrophages. *J. Nat. Prod.* 65, 379, 2002.
597. Chang, F.R. et al. Cytotoxic constituents of the stem bark of *Neolitsea acuminatissina. J. Nat. Prod.* 65, 255, 2002.
598. Chang, F.R. et al. Cytotoxic constituents of the stem bark of *Neolitsea acaminatissima. J. Nat. Prod.* 65, 255, 2002.
599. Su, H.J. et al. Lanostanoids of *Amentotaxus formosana. J. Nat. Prod.* 65, 79, 2002.
600. Wu, T.S. and Lin, F.W. Alkaloids of the wood of *Cryptocarya chinensis. J. Nat. Prod.* 64, 1404, 2001.
601. Bentley, K.W. β-Phenylethylamines and the isoquinoline alkaloids. *Nat. Prod. Rep.* 21, 395, 2004.
602. Kuo, Y.H. et al. Five new cadinane-type sesquiterpenes from the heartwood of *Chamaecyparis obtusa* var. *formosana. J. Nat. Prod.* 64, 1502, 2002.
603. Chen, H.I. et al. New cytotoxic butanolides from *Litsea acutivena. J. Nat. Prod.* 64, 1502, 2001.
604. Chang, Y.S. et al. Analysis of three lupane type triterpenoids in *Helicteres angustifolia* by high-performance liquid chromatography. *J. Pharm. Biomed. Anal.* 26, 849, 2001.
605. Hsieh, T.J. et al. The alkaloids of *Artabotrys uncinatus. J. Nat. Prod.* 64, 1157, 2001.
606. Chen, I.S. et al. Quinoline alkaloids and other constituents of *Melicope semecarpifolia* with anti–platelet aggregation activity. *J. Nat. Prod.* 64, 1143, 2001.
607. Lin, T.H. et al. Two phenanthraquinones from *Dendrobium moniliforme. J. Nat. Prod.* 64, 1084, 2001.
608. Chang, F.R. and Wu, Y.C. Novel cytotoxic annonaceous acetogenins from *Annona muricata. J. Nat. Prod.* 64, 925, 2001.
609. Lin, M.T. et al. Allelopathic prenyl flavanones from the fallen leaves of *Macaranga tanarius. J. Nat. Prod.* 64, 827, 2001.
610. Wu, S.J. et al. Evaluation of hepatoprotective activity of legumes. *Phytomedicine* 8, 213, 2001.

611. Lin, L.C., Chou, C.J., and Kuo, Y.C. Cytotoxic principles from *Ventilago leiocarpa* Benth. *J. Nat. Prod.* 64, 674, 2001.

612. Lin, L.C., Chou, C.J., and Kuo, Y.C. Cytotoxic principles from *Ventilago leiocarpa* Benth. *J. Nat. Prod.* 64, 674, 2001.

613. Ko, W.C. et al. Mechanisms of relaxant action of s-petasin and s-isopetasin, sesqui-terpenes of *Petasites formosanus,* in isolated guinea pig trachea. *Planta Med.* 67, 224, 2001.

614. Chen, C.Y. et al. Four alkaloids from *Annona cherimola. Phytochemistry* 56, 753, 2001.

615. Lin, H.Y. et al. Cyclopeptide alkaloids from *Paliurus ramossisimus. J. Nat. Prod.* 63, 1338, 2000.

616. Lee, S.S., Su, W.C., and Lin, K.C. Cyclopeptide alkaloids from stems of *Paliurus ramossisimus. Phytochemistry* 58, 1271, 2001.

617. Wu, T.S., Leu, Y.L., and Chan, Y.Y. Constituents from the stem and root of *Aristolo-chia kaempferi. Biol. Pharm. Bull.* 23, 1216, 2000.

618. Kuo, Y.H. and Chang, C.I. Pedocarpane-type trinorditerpenes from the bark of *Tai-wania cryptomenoides. J. Nat. Prod.* 63, 650, 2002.

619. Lin, W.Y. et al. Anti–platelet aggregation constituents from *Gynura elliptica. Phyto-chemistry* 53, 833, 2000.

620. Kuo, Y.H. et al. A novel cytotoxic methylated biflavone from the stem of *Cephalotaxus wilsoniana. Chem. Pharm. Bull.* (Tokyo) 48, 440, 2000.

621. Chen, I.S. et al. Chemical constituents and biological activities of the fruit of *Zan-thoxylum integrifoliolum. J. Nat. Prod.* 62, 833, 1999.

622. Lin, S.L. et al. Bishordeninyl terpene alkaloids from *Zanthoxylum integrifoliolum. J. Chin. Chem. Soc.* 47, 571, 2000.

623. Tsai, C.C. and Lin, C.C. Antiinflammatory effects of Taiwan folk medicine "Teng-Khia-U" on carrageenan and adjuvant-induced paw edema in rats. *J. Ethnopharmacol.* 64, 85, 1999.

624. Li, Y.C. and Kuo, Y.H. Five diterpenoids from the wood of *Cunninghamia konishii. J. Nat. Prod.* 61, 997, 1998.

625. Lin, C.C. et al. Antioxidant and hepatoprotective activity of punicalagin and punicalin on carbon tetrachloride–induced liver damage in rats. *Phytother. Res.* 15, 206, 2001.

626. Lin, Y.L. et al. Four new sesquiterpenes from *Petasites formosanus. J. Nat. Prod.* 61, 887, 1998.

627. Chan, S.C. et al. Three new flavonoids and antiallergic, antiinflammatory contituents from the heart wood of *Dalbergia odoriferer. Planta Med.* 64, 153, 1998.

628. Sheen, W.S., Tsai, T.L., and Teng, C.M. Indolopyrido quinazoline, a alkaloids with antiplatelet aggregation activity from *Zanthoxylum integrifoliolum. Planta Med.* 62, 175, 1996.

629. Shen, Y.C., Tai, H.R., and Chen, C.Y. New taxane diterpenoids from the roots of *Taxus mairei. J. Nat. Prod.* 59, 173, 1996.

630. Lin, C.C., Tsai, C.C., and Yen, M.H. The evaluation of hepatoprotective effects of Taiwan folk medicine "Teng-Khia-U." *J. Ethnopharmacol.* 45, 113, 1995.

631. Lin, J.M. et al. Studies on Taiwan folk medicine, Thang-Kau-Tin (II): measurement of active oxygen scavenging activity using an ESR technique. *Am. J. Chin. Med.* 23, 43, 1995.

632. Yoshikawa, M., Yamaguchi, S., and Kunimi, K. Stomachic principles in ginger III. An antiulcer principle, 6-gingesulfonic acid, and three monoacyldigalactosylglycer-ols, gingerglycolipids A-C from *Zingiberis rhizoma* originating in Taiwan. *Chem. Pharm. Bull.* (Tokyo) 42, 1226, 1994.

633. Chung, M.I. et al. Antiplatelet constituents of Formosan *Rubia akane. J. Nat. Prod.* 57, 513, 1994.

634. Lin, S.C. et al. Hepatoprotective effects of Taiwan folk medicine *Ixeris chinensis* (Thunb.) Nak. on experimental liver injuries. *Am. J. Chin. Med.* 22, 243, 1994.

635. Chang, F.R. et al. Studies on the acetogenins of Formosan annonaceous plants. II. Cytotoxic acetogenins from *Annona reticulata* L. *J. Nat. Prod.* 56, 1688, 1993.

636. Tanaka, T., Tachibana, H., and Nonaka, G. Tannins and related compounds. CXXII. New dimeric, trimeric, and tetrameric ellagitannins, lambertianins A-D from *Rubus lambertianus* Seringe. *Chem. Pharm. Bull.* (Tokyo) 41, 1214, 1993.

637. Gan, R.H., Lin, C.N., and Won, S.J. Cytotoxic principles and their derivatives of Formosan solanum plants. *J. Nat. Prod.* 56, 15, 1993.

638. Yoshikawa, M., Hatakeyama, S., and Taniguchi, K. 6-Gingesulfonic acid, a new antiulcer principle and gingerglycolipids A-C, three new monoacyldigalacta acylglycerols from *Zingiberis rhizoma* originating from Taiwan. *Chem. Pharm. Bull.* (Tokyo) 40, 2239, 1992.

639. Chen, J.Y., Lin, C.C., and Namba, T. Development of natural crude drug resources from Taiwan (X). Pharmacognostical studies on the Chinese crude drug "Han-Lian-Cao." *Am. J. Chin. Med.* 20, 51, 1992.

640. Wu, R.J. et al. Formosanin-C, an immunomodulator with antitumor activity. *Int. J. Immunopharmacol.* 12, 777, 1990.

641. Chen, C.P., Lin, C.C., and Namba, T. Screening of Taiwanese crude drugs for antibacterial activity against *Streptococcus mutans. J. Ethnopharmacol.* 27, 285, 1989.

642. Chiu, H.F. et al. The pharmacological and pathological studies on several hepatic protective crude drugs from Taiwan (I). *Am. J. Chin. Med.* 16, 127, 1988.

643. Yang, L.L. et al. Antihepatotoxic actions of Formosan plant drugs. *J. Ethnopharmacol.* 19, 103, 1987.

644. Hsu, S.Y. Effects of the constituents of *Alpinia spociosa* rhizome on experimental ulcers. *Taiwan Yi Xue Hui Za Zhi* 86, 58, 1987.

645. Yang, L.L. et al. Antihepatotoxic principles of *Wedelia chinese* herbs. *Plants Med.* 6, 499, 1986.

646. Wang, J.P., Hsu, M.F., and Teng, C.M. Antiplatelet effect of "hsien-ho-t'sao" (*Agrimonia pilosa*). *Am. J. Chin. Med.* 13, 109, 1985.

647. Chung, M.I. et al. Studies on the constituents of Formosan gentianaceous plants. Part VII. Constituents of *Swertia randaiensis* Hayata and pharmacological activity of norswertianolin. *Gaoxiong Yi Xue Ke Xue Za Zhi* 2, 131, 1986.

648. Iinuma, M., Tanaka, T., and Matsurra, S. Synthesis of new flavones of *Bauhinia championii* in Formosa. *Yaku Gaku Zasshi* 103, 994, 1983.

649. Wu, C.C., Chang, C.C., and Chen, R.L. The antihypertensive effect of *Uncaria rhynchophylla* in essential hypertension. *Taiwan Yi Xue Hui Za Zhi* 79, 749, 1980.

650. Liao, L.L., Chen, C.H., and Chen, G.C. Formosan medicinal herb, *Hedyotis diffusa* Willd. as an antitumor agent. *Taiwan Yi Xue Hui Za Zhi* 78, 658, 1979.

651. Wu, W.N., Mitscher, L.A., and Beal, J.L. A note on the isolation and identification of quaternary alkaloids of *Phellodendron wilsonii. Lloydia* 39, 249, 1976.

652. Kuramoto, M. et al. Proceedings: effect of the extract of *Uncaria kawakamii* of Formosa origin in dissolving artificial bladder calculi. *Jpn. J. Pharmcol.* 24, 137, 1974.

653. Takemo, T. et al. Studies on the constituents of *Achyranthes radix.* IV. Isolation of the insect moulting hormones from Formosan *Achyranthes* spp. *Yakugaku Zasshi* 87, 1478, 1967.

654. Apers, S. et al. Characterisation of new oligoglycosidic compounds in two Chinese medicinal herbs. *Phytochem. Anal.* 13, 202, 2002.
655. Asongalem, E.A. et al. Analgesic and antiinflammatory activities of *Erigeron floribundus* (Syn. *Conyza sumatrensis*). *J. Ethnopharmacol.* 91, 301, 2004.
656. Kuang, H.X. et al. Chemical constituents of pericarps of *Rosa davurica* Pall., a traditional Chinese medicine. *Chem. Pharm. Bull.* (Tokyo) 37, 2232, 1989.
657. Dharmananda, S. The nature of ginseng: from traditional use to modern research and question of dosage. *Herbalgram* 54, 34, 2002.
658. He, K. et al. Bioactive compounds from *Taiwania cryptomerioides*. *J. Nat. Prod.* 60, 38, 1997.
659. Tanaka, R. et al. Synthesis of digalactosyl diacylglycerols and their structure-inhibitory activity on human lanosterol synthase. *Bioorg. Med. Chem. Lett.* 15, 159, 2005.
660. Yang, C. et al. Studies on the isolation and identification of β-ecdysone from *Zebrian pendula* Schnizl. and its antianhythmic effect. *Nat. Prod. Res. Dev.* 8(3), 17, 1996 (English abstract).
661. Philip, J. and Prema, L. Variability in the antinutritional constituents in greengram *Vigna radiata*. *Plant Foods Hum. Nutr.* 53, 99, 1999.
662. Takeda, Y. and Kasamo, K. Transmembrane topography of plasma membrane constituents of mung bean (*Vigna radiata* L.) Hypocotyl cells. I. Transmembrane distribution of phospholipids. *Biochim. Biophys. Acta* 1513, 38, 2001.
663. Cordell, G.A., Quinn-Beattie, M.L., and Farnsworth, N.R. The potential of alkaloids in drug discovery. *Phytother. Res.* 15, 183, 2001.
664. Williams, C.A. and Grayer, R.J. Anthocyanins and other flavonoids. *Nat. Prod. Rep.* 21, 539, 2004.
665. Nagao, T. and Tanak, R. Studies on the constituents of *Thladiantha dubia* Bunge: II. Structures of dubiosides D, E, and F, neutral saponins of quillaic acid isolated from the tuber. *Chem. Pharm. Bull.* 38, 378, 1990.
666. Adeniyi, A.A. Determination of cadmium, copper, iron, lead, manganese, zinc in water leaf (*Talinum triangulare*) in dumpsites. *Env. Int.* 22, 259, 1996.
667. Kohler, I. et al. Herbal remedies traditionally used against malaria in Ghana: bioassay-guided fractionation of *Microglossa pyrifolia* (Asteraceae). *Z. Naturforsch.* 57, 1022, 2002.
668. Ogihara, K. et al. Pyrrolizidine alkaloids from *Messerschmidia argentea*. *Phytochemistry* 44, 545, 1997.
669. Tian, J.K. et al. Studies on chemical constituents in herba of *Lysimachia dauvurica*. *Zhongguo Zhong Yao Za Zhi* 27, 283, 2002 (in Chinese).
670. Op de Back, P. et al. Volatile constituents from leaves and wood of *Leea guineensis* G. Don (Leeaceae) from Cameroon. *Flavour Fragrance J.* 15, 182, 2000.
671. http://www.sahealthinfo.org/traditionalmeds/monographs/hypoxis.htm
672. http://info.kib.ac.cn/kibinfoEN/soft/2603.htm
673. Wilson, D.M. et al. Habenariol, a freshwater feeding deterrent from the aquatic orchid *Habenaria repens*. *Phytochemistry* 50, 1333, 1999.
674. http://ift.confex.com/ift/2004/techprogram/paper_22911.htm.
675. Kakuda, R. et al. Studies on the constituents of *Gentiana* species. II. A new triterpenoid and (S)-(+) and (R)-(−)-gentiolactones from *Gentiana lutea*. *Chem. Pharm. Bull.* (Tokyo) 51, 885, 2003.
676. Kulate, J.R. et al. Composition of the essential oil from leaves and flowers of *Dichrocephala integrifolia* (L.) O. Kuntze Chev. from Cameroon. *Flavour Fragrance J.* 14, 419, 2000.

677. Koike, K. et al. Triterpenoid saponins from *Dianthus chinese*. *Tetrahedron* 50, 12, 811, 1994.

678. Lanksy, P.S. Plants that lower cholesterol. *Acta-Horticulturae* 332, 131, 1993.

679. Shingu, K. et al. Two new ergostane derivatives from tubocapsicum anomalum solanaceae. *Chem. Pharm. Bull.* (Tokyo) 38, 1107, 1990.

680. Hui, W.H., Li, M.M., and Lee, Y.C. Triterpenoids from two Hong Kong Euphorbiaceae species (*Claoxylon polot, Fluggea virosa*). *Phytochemistry,* 16, 607, 1977.

681. Gowrikumar, G. et al. *Diplocyclos palmatus*. A new seed source of punicic acid. *Lipids* 16, 558, 1981.

682. Kanchanapoom, T. et al. Chemical constituents of Thai medicinal plant *Diplocyclos palmatus*. *Nat. Med.* 56, 274, 2002.

683. Yuan, J. et al. Nutritional components of fruits from *Prinsepia utilis* Royle and *Pyracantha fortuneana* (Maxim.) Li. *J. Plant Res. Environ.* 11, 63, 2002.

Chemical Components
and Their Sources

Component	Source
(+)-catechin	*Pouteria obovata*
(+)-catechin-3-*O*-gallate	*Pouteria obovata*
(+)-gallocatechin	*Pouteria obovata*
(+)-gynunone	*Gynura formosana, G. elliptica*
(+)-nyasol	*Asparagus cochinchinensis*
(+)-seasamin	*Zanthoxylum nitidum, Z. integrifoliolum*
(−) pimaradene	*Aralia chinensis*
(−) kaurene derivatives	*Aralia chinensis*
(−)-catechin	*Berchemia formosana, B. lineata*
(−)-epicatechin	*Pouteria obovata*
(−)-syringaresinol glucoside	*Pyrola morrisonensis, P. japonica*
(10E,12Z,15Z)-9-hydroxy-10,12,15-octadecatrienoic acid	*Clerodendrum calamitosum, C. cyrtophyllum*
(10S)-hydroxypheophytin	
(2′S)-7-hydroxy-2-(2′-hydroxypropyl)-5-methylchromone	*Clerodendrum calamitosum, C. cyrtophyllum*
(2′S)-7-hydroxy-5- hydroxymethyl-2-(2′-hydroxypropyl)chromone	*Cassia fistula*
(24S)-ethylcholesta-5,22,25,3 β-o1 apigenin-7-0-glucurondes	*Cassia fistula*
	Clerodendrum inerme
(S)-(+)-and (R)-(−)gentiolactones	*Gentiana scabrida, G. scabrida var. horaimontana, G. lutea*
α-agarpfiram	*Aquilaria sinensis, A. sibebsus*
α-agarofuran hydroagarofuran	*Aquilaria sinensis, A. sibebsus*
α-allocryptopine	*Thalictrum fauriei, Zanthoxylum nitidum, Z. integrifiolium*
α-amyrenol	*Spilanthes acmella, S. acmella var. oleracea*
α-amyrin	*Cirsium japonicum, C. japonicum var. australe, Jatropha curcas, Myrica adenophora, Sambucus formosana, S. chinensis*
α-bergamotene	*Perilla frutescens, P. frutescens var. crispa, P. ocymoides*
α-butenolide alkaloid	*Artabotrys uncinatus*

α-camphorene	*Cinnamomum camphora*
α-cholesterol	*Eryngium foetidum*
α-crocetin	*Gardenia angusta* var. *kosyunensis, G. oblongifolia*
α-curcumene	*Curcuma zedoaria*
α-diaminopropionic acid	*Trichosanthes cucumeroides*
α-elacostearic acid	*Aleurites fordii, A. moluccana, A. montana*
α-epimer stigmasterol	*Clerodendrum paniculatum*
α-guaiene	*Pogostemon amboinicus*
α-ionone	*Osmanthus fragrans*
α-lemonene	*Citrus medica* var. *gaoganensis*
α-linolenic	*Ribes formosanum, R. nigrum*
α-methyl	*Pandanus amaryllifolius, P. pygmaeus*
α-methyl ether	*Morinda citrifolia*
α-naphthyl-isothiocyanate	*Ixeris laevigata* var. *oldhamii*
α-obscunine	*Lycopodium salvinioides*
α-paristyphnin	*Paris polyphylla*
α-patchoulene	*Pogostemon cablin*
α-phenylcinnamic acid	*Tropaeolum majus*
α-pinene	*Acorus calamus, A. gramineus, Chrysanthemum indicum, Dendranthema indicum, Dendropanax pellucidopunctata,Glechoma hederacea* var. *grandis, Lantana camara, Ocimum basilicum, Oenanthe javanica, Ruta graveolens, Vitex cannabifolia, Zanthoxylum ailanthoides, Z. avicennae*
α-pyrones	*Opuntia dillenii, Passiflora foetida* var. *hispida*
α-quaiene	*Pogostemon cablin*
α-sitosterol	*Eupatorium tashiroi*
α-spinasterol	*Clinopodium laxiflorum*
α-taralin	*Aralia chinensis, A. taiwaniana*
α-terpineol	*Chamaecyparis formosensis, C. obtusa* var. *filicoides, C. obtusa* var. *formosana, Cymbopogon citratus, Glechoma hederacea* var. *grandis*
α-terthienyl	*Tagetes erecta*
α-toxicarol	*Derris trifoliata*

Component	Source
α-tropolone	*Jasminum hemsleyi*
α-unsaturated gamma-lactone	*Pandanus amaryllifolius, P. pygmaeus*
β-epimer poriferasterol	*Clerodendrum paniculatum*
β-agarofuran	*Aquilaria sinensis, A. sibebsus*
β-alanine	*Setaria italica*
β-amyrenl	*Lactuca indica*
β-amyrenol	*Spilanthes acmella, S. acmella* var. *oleracea*
β-amyrenone	*Viscus multinerve*
β-amyrin	*Aspidixia articulata, A. liquidambaricala, Balanophora spicata, Bischofia javanica, Coleus scutellarioide* var. *crispipilus, C. parvifolius, Cirsium japonicum, C. japonicum* var. *australe, Diospyros angusifolia, Jatropha curcas, Kalanchoe spathulata, K. pinnata, K. gracillis, K. crenata, K. tubiflora, Myrica adenophora, Lysimachia ardisloides, L. capillipes, L. davurica, Viscus multinerve, V. alniformosanae, V. angulatum, Taraxacum mongolicum*
β-amyrin acetate	*Artocarpus altitis, Aspidixia articulata, A. liquidambaricala, Ficus pumila* var. *awkeotsang, Viscus multinerve*
β-asarone	*Acorus calamus, A. gramineus*
β-butenolide alkaloid	*Artabotrys uncinatus*
β-caritebem deguelin	*Derris trifoliata*
β-carotene	*Achyranthes aspera* var. *indica, A. aspera* var. *rubro-fusca, Basella alba, B. rubra, Boehmeria densiflora, Cycas revoluta, Duranta repens, Oxalis corniculata, Petasites japonicus, Pyracantha fortuneana, Setaria italica*
β-caryophyllene	*Ageratum conyzoides, A. houstonianum, Murraya paniculata, Lantana camara, Perilla frutescens, P. frutescens* var. *crispa, P. ocymoides, Vitex cannabifolia*
β-cyamines	*Gomphrena globosa*
β-D-glucoside	*Allium bakeri, A. scorodoprasum*
β-diaminopropionic acid	*Trichosanthes cucumeroides*
β-dihydropseudoionone	*Cymbopogon citratus*
β-ecdysone	*Zebrina pendula*
β-elaterin	*Momordica charantia*
β-elemene	*Pogostemon amboinicus*
β-eleostearic acid	*Ricinus communis*
β-eudesmol	*Magnolia liliflora*

β-glucogallin	Sanguisorba formosana, S. officinalis, S. minor
β-glycyrrhetinic acid	Glycyrrhiza uralensis
β-guaienen	Artemisia lactiflora, A. princeps
β-guriunene	Pogostemon amboinicus, Zanthoxylum ailanthoides
β-hydroxy-lupeol	Salvia hayatana, S. japonica, S. roborowskii
β-ionone	Osmanthus fragrans
β-p-glucophyranoside	Bauhinia championi
β-patchoulene	Pogostemon cablin
β-phellandrene	Osmanthus fragrans
β-phenylethylamines	Cryptocarya chinensis
β-pinene	Agastache rugosa, Glechoma hederacea var. grandis, Ruta graveolens
β-sitosterol	Acanthopanax senticosus, Adina pilulifera, A. racemose, Allium bakeri, A. scorodoprasum, Akebia longeracemosa, A. quinata, Aletris formosana, Angelica acutiloba, A. citriodor, Anisomeles indica, Asarum hypogynum, A. macranthum, A. hongkongense, A. longerhizomatosum, Asparagus cochinchinensis, Bauhinia championi, Bischofia javanica, Blumea lanceolaria, B. aromatica, B. lacera, Cirsium japonicum, C. japonicum var. australe, Crotalaria pallida, Dicliptera chinensis, D. riparia, Dodonaea viscosa, Dumasia pleiantha, Eryngium foetidum, Eupatorium tashiroi, Evolvulus alsinoides, Gardenia jasminoides, Glehnia littoralis, Glochidion lanceolarium, G. rubrum, Hedyotis uncinella, H. pinifolia, H. diffusa, Hemerocallis fulva, H. longituba, Hydrangea chinensis, Hylocereus undatus, Hypoxis aurea, Ilex asprella, Jasminum hemsleyi, Jatropha curcas, Kalanchoe spathulata, K. pinnata, K. gracillis, K. crenata, K. tubiflora, Lespedeza cuneata, Lindera aggregata, L. okoensis, Lonicera kawakamii, L. confusa, Morinda umbellata, Mussaenda parviflora, Ophiopogon japonicus, Orthosiphon aristatus, O. stamineus, Petasites japonicus, Phyla nodiflora, Polygonatum falcatum, P. kingianum, P. odoratum, Prenna obtusifolia, P. crassa, P. serratifolia, P. microphylla, Psidium guajave, Randia spinosa, Pueraria lobata, P. montana, Ricinus communis, Salvia hayatana, S. japonica, S. roborowskii, Sapium discolor, S. sebiferum, Scoparia dulcis, Scutellaria rivularis, Solanum incanum, Taraxacum mongolicum, Tetrastigma dentatum, T. formosanum, T. umbellatum, T. hemsleyanum, Toddalia asiatica, Viscus multinerve, V. alniformosanae, V. angulatum
β-sitosterol, latex	Ficus pumila var. awkeotsang
β-sitosterol-β-D-glucoside	Momordica charantia
β-sitosterol-3-β-D-glucopyranoside	Begonica fenicis, B. laciniata, B. malabarica
β-sitosterol-3-0-β-D-glucoside	Jatropha curcas

Component	Source
β-sitosterol-D-glucoside	*Ficus pedunculosa var. mearnsii, F. religiesa, Hedyotis uncinella, Viscus multinerve*
β-solamargine	*Solanum aculeatissimum*
β-sotpsterols	*Dioscorea bulbifera*
β-taralin	*Aralia chinensis, A. taiwaniana*
β-tropolone	*Jasminum hemsleyi*
β-unsaturated gamma-lactone	*Pandanus amaryllifolius, P. pygmaeus*
γ-cadinene	*Cymbopogon nardus*
γ-fagarine	*Ruta graveolens*
γ-patchoulene	*Pogostemon amboinicus*
γ-sitosterol	*Clerodendrum japonicum, C. kaempferi*
γ-terpinene	*Lantana camara*
δ-amyrenol	*Sedum lineare, S. sempervivoides, S. morrisonense*
δ-amyrenone	*Sedum lineare, S. sempervivoides, S. morrisonense*
δ-cardinol	*Jasminum hemsleyi*
δ-yohimbine	*Rauvolfia verticillata*
Δ³-carene	*Ocimum basilicum*
0-acetyl camtothecin	*Nothapodytes foetida, N. nimmoniana*
1,8-*p*-menthadien-5-ol	*Cymbopogon citratus*
1-pinene	*Chamaecyparis formosensis, C. obtusa var. filicoides, C. obtusa var. formosana*
1-α-pinene	*Piper arboricola*
1-0-galloyl-β-D-glucose	*Phyllanthus multiflorus, P. emblica*
1-0-galloylglucose	*Rhynchosia volubilis, R. minima*
1-6-hydroxy-7-methoxynaphthalene	*Bombax malabarica*
1-8 lacone	*Bombax malabarica*
1-acetyl-6-E-geranyl gerandol-19-oic acidandsinaphy diangelate	*Microglossa pyrifolia*
1-arabinose	*Plantago asiatica, P. major*
1-borneol	*Cymbopogon citratus, Liquidambar formosana*
1-camphor	*Hedyotis corymbosa*

1-curine	*Cyclea insularis, C. barbata*
1-cyano-2-hydroxy methylprop-1-ene-3-ol	*Cardiospermum halicacabum*
1-cyano-2-hydroxy methylprop-2-ene-1-ol	*Cardiospermum halicacabum*
1-ephedrine	*Pinellia ternata*
1-heptanecanol	*Opuntia dillenii*
1-hydroxy-2-methylanthraquinone	*Damnacanthus indicus*
1-hydroxyanthraquinone	*Damnacanthus indicus*
1-menthone	*Glechoma hederacea* var. *grandis*
1-methoxy-2-methylanthraquinone	*Morinda umbellata*
1-octen-3-ol, 2-hexenal	*Callicarpa formosana, C. japonica*
1-perilla	*Perilla frutescens, P. frutescens* var. *crispa, P. ocymoides*
1-phellandrene	*Piper arboricola*
1-pinocamphone	*Glechoma hederacea* var. *grandis*
1-pulegone	*Glechoma hederacea* var. *grandis*
1-rhamnose	*Plantago asiatica, P. major*
1-tridecene-3,5,7,9,11-pentyne	*Ricinus communis*
10-hydroxycamptothecin	*Camptotheca acuminata*
10-oxo-cylindrocarpidin	*Tabernaemontana amygdalifolia*
11-hydroxylated kauranic acids	*Adenostemma lavenia*
11-methoxyquindoline	*Sida acuta*
12α-hydroxyfern-9(11)-ene	*Cyathea lepifera, C. podophylla*
12a-hydroxy- and 6a,12a-dehydro-analogs	*Derris elliptica*
1,1-Diphenyl-2-picrylhydrazyl	*Hypericum chinese, H. patulum*
1,4-dihydroxy-2-methylanthraquinone	*Damnacanthus indicus*
15α-acetyl-dehydrosulfurenic acid	*Cinnamomum insulari-montanum, C. kotoense, C. micranthum*
1,6-di-0-galloylglucose	*Rhynchosia volubilis, R. minima*
1,6-dihydroxy-2,4-dimethoxyanthraquinone	*Rhynchosia volubilis, R. minima*
1,6-hentriacontanol	*Euphoria longana*
1,7-demethoxy-cylindrocarpidine	*Tabernaemontana amygdalifolia*
1,8-cineol, isopinocamphone	*Glechoma hederacea* var. *grandis*
1,8-cineole	*Ocimum basilicum*

Component	Source
1,8-dihydroxy-3-methyl-9-anthrone	*Rumex crispus*
1,9-hydroxytotarol	*Podocarpus nagi*
2′,4,4′-trihydroxy-3′-[(E)3-methyl-6-oxo-2-hexenyl]	*Angelica keiskei*
2-amino-4-hydroxyhept-6-ynoic acid	*Euphoria longana*
2-amino-4-hydroxymethylhex-5-ynoic acid	*Euphoria longana*
2-hydroxyanthraquinone	*Morinda umbellata*
2-hydroxyphenylacetic acid	*Astilbe longicarpa*
2-O-methylisohemigossylic acid lactone	*Gossampinus malabarica*
2-undecanone	*Ruta graveolens*
24-ethylidene-lophenol	*Saccharum officinarum*
24-methylene-lophenol	*Saccharum officinarum*
2,4,4′,t-tetrahydroxybenzophenone	*Morus alba*
2,4,5-trimethoxybenzaldehyde	*Asarum hypogynum, A. macranthum, A. hongkongense, A. longerhizomatosum*
2,5-dimethoxy-p-benzoquinone	*Dicliptera chinensis, D. riparia*
25R-spirost-4-ene-3,12-dione	*Polygonum chinense*
3 β-hydroxystigmast-5-en-7-one	*Asarum hypogynum, A. macranthum, A. hongkongense, A. longerhizomatosum*
3,3′-biplumbagin	*Plumbago zeylanica*
3′,4′,5′-trihydroxyisoflaone	*Euchresta formosana*
3-(1,1-dimethylallyl)-herniarin	*Ruta graveolens*
3-(4-hydroxyphenyl)-2 (E)-propenoate	*Costus speciosus*
3-β-acetoxysolavetivone	*Solanum capsicatstrum, S. abutiloides*
3-o-methyl isorhamnein	*Opuntia dillenii*
3-chloroplumbagin	*Plumbago zeylanica*
3-epicaryoptin	*Clerodendrum inerme*
3-hexen-lo1	*Ocimum basilicum*
3-indolylmethylgluco-sinolate	*Clerodendrum cyrtophyllum*
3-methoxypyridine	*Equisetum ramosissimum*
3-methy-1-5-isopropyl	*Bombax malabarica*

Component	Source
3-O-β-glucosylplatycodigenin	*Platycodon grandiflorum*
3-O-methylquercetin	*Ophioglossum vulgatum*
3-octanone	*Ocimum basilicum*
3-oxykojie acid	*Maytenus diversifolia*
3-p-coumarylglycoside-5-glucoside of cyanidin	*Perilla frutescens*
3α-hydroxyhop-22(29)-ene	*Mallotus apelta*
3β, 24-dihydroxyl-urs-12-ene	*Potentilla leuconta, P. multifida*
3β-acetyl-11β-hydroxy-lupeo	*Salvia hayatana, S. japonica, S. roborowskii*
30-ner-lupan-3β-ol-20-one	*Ricinus communis*
3,3'-tdi-O-gallate	*Myrica rubra*
3,5-dicaffeuyl quinic acid	*Dichrocephala bicolor*
3,5-dihydroxyl hexanoic acid	*Polygonum plebeium, P. paleaceum*
3,6-di-o-galloyl-D-glucose	*Phyllanthus multiflorus, P. emblica*
4"-O-acetyl quercitrin	*Scurrula loniceriolius, S. ritozonensis, S. liquidambaricolus, S. ferruginea*
4-(2,4,5-trimethoxyphenyl)-3-en-butylone	*Ananas comosus*
4-β-carboxy-19-nortotarol	*Podocarpus nagi*
4-butyrolactone	*Farfugium japonicum*
4-ethoxyl-6-hydroxymethyl-α-prone	*Opuntia dillenii*
4-hydroxy-1,2-dimethoxysanthone	*Hypericum geminiflorum*
4-hydroxybenzaldehyde	*Clinopodium laxiflorum*
4-hydroxycinnamic acid	*Ananas comosus*
4-hydroxyderricin	*Angelica keiskei*
4-hydroxyrottlerine	*Mallotus tiliaefolius*
4-ketopinoresinal	*Sida acuta*
4-quinazolone	*Dichroa febrifuga*
4,5-dicaffeoyl quinic acid	*Dichrocephala bicolor*
5,25-stigmastardien-3β-ol-β-D-glucoside	*Momordica charantia*
5-(2-hydroxyphenoxymethyl)furfural	*Cassia fistula*
5-β-cholanic	*Abrus precatorius*
5-7-dimethoxy-8-(2,3-dihydroxyisopenty) coumarin	*Murraya paniculata*

Component	Source
5-avenasterol	*Hibiscus tillaceus, H. esculentus*
5-caffeoyl quinic acid	*Dichrocephala bicolor*
5-gualzulene osthol	*Murraya paniculata*
5-hydroxymethylfurfural	*Cassia fistula*
5-methoxy-8-hydroxypsoralen	*Angelica hirsutiflora*
5-stigmautena-3β-7d-diol	*Ananas comosus*
5,7-dihydroxy flavonone 7-glucoside	*Hyphea kaoi*
5,7,4'-trihydroxy-8-ethoxycarbonyl flavan	*Daphne arisanensis, D. odora*
5,8-dihydrochimophilin	*Pyrola morrisonensis, P. japonica*
6'-O-benzoyldaucosterol	*Tetrastigma dentatum, T. formosanum, T. hemsleyanum, T. umbellatum*
6-α, 12α-12a-hydroxyelliptone	*Derris trifoliata*
6-acetyl-2,2-dimethylchroman-4-one	*Gynura formosana, G. elliptica*
6-ethoxy-chelerythrin	*Zanthoxylum nitidum, Z. integrifoliolum*
6-gingesulfonic acid	*Zingiber kawagoii, Z. rhizoma*
6-hydroxy-luteolin	*Vernonia gratiosa*
6-methoxy dictamnine	*Ruta graveolens*
6-pentadecyl salicylic acid	*Rhus semialata var. roxburghiana, R. microphylla*
7-α-methoxy-deoxyptojaponol	*Jasminum hemsleyi*
7-β-methoxy-deoxyptojaponol	*Jasminum hemsleyi*
7-o-galloyl catechin	*Rhynchosia volubilis, R. minima*
7-caffeyl-glucosides	*Perilla frutescens, P. frutescens var. crispa, P. ocymoides*
7-deto-β-sitosterol	*Jatropha curcas*
7-hydroxyhinokinin	*Hypoestes purpurea*
7-hydroxylathyrol	*Euphorbia lathyris, E. lectin, E. milli, E. neriifolia, E. thymifolia*
7-isovaleroylcycloepiatalantin	*Severinia buxifolia*
7-methoxy-2,2-dimethylchromene	*Ageratum conyzoides, A. houstonianum*
7-methoxy-baicalein	*Scutellaria rivularis*
7-methoxynorwogonin	*Scutellaria rivularis*
7α-diol, stigmast-5-ene-3β	*Jatropha curcas*

7α-L-rhamnosyl-6-methoxyluteolin	Alternanthera philoxeroides, A. sessilis
7β-diol	Jatropha curcas
8-isopentyl-limettin	Murraya paniculata
9-o-methoxycamptothecin	Nothapodytes foetida, N. nimmoniana
9-methoxycamptothecine	Nothapodytes foetida, N. nimmoniana
9,10-dihydroxystearic acid	Ricinus communis
9,12 (Z,Z)-octadecadienoic acid	Sterculia nobilis, S. lychnophora
Abamagenin	Sansevieria trifasciata
Abietane diterpenoids	Hyptis martiusii, H. suaveolens, Indigofera zollingeriana, I. longeracemosa
Abrine	Abrus precatorius, A. cantoniensis
Acacetin	Leucas aspera, L. chinensis
Acacetin-7-glucoside	Chrysanthemum morifolium
Acacetin-7-glucurono-(1,2)-glucuronide	Chrysanthemum morifolium
Acacetin-7-rhamnoglucoside	Chrysanthemum morifolium, Cirsium albescens
Acacetin-7-rhamnosidoglucoside	Dendropanax pellucidopunctata
Acalyphine	Acalypha australis, A. indica
Acetic acid	Luffa cylindrica, Jasminum sambac, Kalanchoe spathulata, K. pinnata, K. gracillis, K. crenata, K. tubiflora, Michelia alba
Acetogenins	Annona muricata, A. cherimola, A. reticulata
Acetone	Coriandrum sativum
Acetophenone derivatives	Acronychia pedunculata
Acetycophalotaxine	Cephalotaxus wilsonianer
Acetyl lupeol	Plumeria rubra cv. acutifolia
Acetylcamptothecin	Nothapodytes foetida, N. nimmoniana
Acetylcholine	Diospyros kaki
Acetylcholinerase	Canna flaccida, C. indica
Acetyldaidzin	Hibiscus syriacus
Acetylgenistin	Hibiscus syriacus
Acetyllindenanolide B-1	Lindera aggregata, L. okoensis

Component	Source
Acetyllindenanolide B-2	*Lindera aggregata, L. okoensis*
Acetylsalicylic acid	*Malvastrum coromandelianum*
Achillin	*Achillea millefolium*
Acidic resin	*Wikstroemia indica*
Aciphyllene	*Pogostemon cablin*
Aconitine	*Aconitum bartletii, A. fukutomel, A. formosanum, A. kojimae, A. kojimae var. lassiocarpium, A. kojimae var. ramosum, A. yamamotoanum*
Aconitum	*Aconitum bartletii, A. fukutomel, A. formosanum, A. kojimae, A. kojimae var. lassiocarpium, A. kojimae var. ramosum, A. yamamotoanum*
Acoric acid	*Acorus calamus, A. gramineus*
Acridone alkaloids	*Severinia buxifolia*
Acrylamide	*Murdannia keisak, M. loriformis*
Acteoside	*Stachys sieboldii*
Aculeatin	*Toddalia asiatica*
Acuminaminoside	*Glochidion eriocarpum, G. acuminatum, G. zeylanicum*
Acyclic diterpene glycosides	*Lycium chinense*
Acyl flavonol	*Hedyotis diffusa*
Adenine	*Chryanthemum segetum, C. morifolium, Morus australis, Solanum lyratum*
Adenosine	*Coix lacryma-jobi, Verbena officinalis*
Adenosine guanosine	*Allium bakeri, A. scorodoprasum*
Adenosinetriphoshatase	*Canna flaccida, C. indica*
Adiantone	*Adiantum capillus-veneris, A. flabellulatum*
Adipedatol	*Adiantum capillus-veneris, A. flabellulatum*
Adynerin	*Nerium indicum*
Aesculitannin C	*Ecdysanthera rosea, E. utilis*
Aflatoxins	*Coriandrum sativum*
Afzelin	*Dicranopteris linearis*
Agarospirol	*Aquilaria sinensis, A. sibebsus*
Agathisflavone	*Rhus succedanea*

Agerato-chromene	*Ageratum conyzoides, A. houstonianum*
Aglycones	*Corchorus capsularis, C. olitorius, Glycyrrhiza uralensis, Prunus persica*
Agnuside	*Vitex negundo, V. rotundifolia*
Agoniadin	*Plumeria rubra* cv. *acutifolia*
Agrimols	*Agrimonia pilosa*
Agrimonine	*Agrimonia pilosa*
Agrimonolide	*Agrimonia pilosa*
Agrimophol	*Agrimonia pilosa*
Ajmalicine	*Rauvolfia verticillata*
Ajmaline	*Rauvolfia verticillata*
Ajugalactone	*Akebia longeracemosa, A. quinata*
Ajugoside	*Eucommia ulmoides*
Akebigenin	*Akebia longeracemosa, A. quinata*
Akebin	*Akebia longeracemosa, A. quinata*
Akolactone B	*Litsea japonica. L. hypophaea*
Alanine	*Taraxacum mongolicum*
Alcorhic acid	*Ipomoea batata, I. obscura, I. stans*
Aldehydes	*Perilla frutescens, P. frutescens* var. *crispa, P. ocymoides, Plumeria rubra* cv. *acutifolia*
Aleanolic acid	*Chenopodium album, Swertia randaiensis*
Alfileramine	*Zanthoxylum nitidum, Z. integrifoliolum*
Aliphatics	*Peristrophe japonica, P. roxburghiana*
Alizarin	*Morinda umbellata, M. citrifolia, Rubia akane, R. lanceolata, R. linii*
Alizarin-1-methyl ether	*Morinda umbellata*
Alkaloids	*Aristolochia cucurbitifolia, A. manshuriensis. A. heterophylla, A. kaempferi, A. kankanensis, A. shimadai, Atalantia buxifolia, Caesalpinia pulcherrima, Capsella bursa-pastoris, Cephalotaxus wilsonianer, Corydalis pallida, Cryptocarya chinensis, Dodonaea viscosa, Eclipta alba, E. prostrate, Emilia sonchifolia, E. sonchifolia* var. *javanica, Liparis cordifolia, L. loeselii, Lycopodium cunninghamioides, Machilus kusanoi, M. zuihoensis, Magnolia liliflora, Neolitsea acuminatissima, Paracyclea ochiaiana, Passiflora suberosa, Solanum incanum, S. indicum, Tabernaemontana amygdalifolia, T. pandacaqui, Turpinia formosana, Veratrum formosanum, Vernonia cinerea, Zephyranthes carinata*

Component	Source
Alkaloid glycosides	*Solanum biflorum*
Alkylated benzoquinones	*Maesa tenera, M. lanceolata, M. laxiflora*
Allantoin	*Dioscorea opposita, Symphytum officinale*
Allelopathic compounds	*Bryophyllum pinnatum*
Allelopathic polyacetylene	*Solidago altissima*
Allicin	*Allium sativum, A. thunbergii, A. tuberosum*
Allistatin	*Allium sativum, A. thunbergii, A. tuberosum*
Alloaromadendrene	*Pogostemon amboinicus*
Allomatatabiol	*Actinidia callosa var. formosana, A. chinensis*
Allosecurinine	*Securinega suffruticosa*
Allylpyrocatchin	*Chamaecyparis formosensis, C. obtusa var. filicoides, C. obtusa var. formosana*
Allylpyrocatechol	*Piper betle*
Alocasin	*Alocasia macrorrhiza*
Aloe-emodin	*Cassia mimosoides*
Amaranthin	*Gomphrena globosa*
Ambroide	*Chenopodium ambrosioides*
Amellin	*Scoparia dulcis*
Amentoflavone	*Cycas revoluta, Rhus succedanea*
Amino acids	*Abutilon indicum, A. taiwanensis, Albizzia lebbeck, Aleurites fordii, A. moluccana, A. montana, Ampelopsis brevuoedybcykata, A. cantoniensis, Arachis hypogea, A. agallocha, Caryopteris incana, Chenopodium album, Clausena lansium, Cleistocalyx operculatus, Cordyline fruticosa, Cratoxylon ligustrinum, Croton crassifolius, Cucurbita moschata, Dicliptera chinensis, D. riparia, Elaeagnus morrisonensis, E. angustifolia, Eupatorium clematideum, Ficus microcarpa, Flemingia macrophylla, F. prostrata, Glochidion puberum, Ilex rotunda, Ixora chinensis, Ludwigia octovalvis, Lycium chinense, Mallotus paniculatus, M. japonicus, Mirabilis jalapa, Melastoma dodecandrum, M. septemnervium, Nymphaca tetragona, Pinellia ternata, Rhodomyrtus tomentose, Rotala rotundifolia, Saururus chinensis, Scutellaria indica, S. formosana, Smilax china, Talinum triangulare, Teucrium viscidum, Urena procumbens, Zornia diphylla*
Amorphigenin	*Derris elliptica*

Amphicoside	*Veronicastrum simadai*
Amritoside	*Psidium guajave*
Amygdalin	*Eriobotrya japonica*
Amylase	*Brousonetia papyrifera*
Amylodextrins	*Myristica cagayanensis, M. fragrans*
Amyrenol	*Sedum formosanum*
Amyrin	*Ficus carica, F. benjamina*
Anadoline	*Symphytum officinale*
Analgesic sesquiterpene dilactone	*Mikania cordata*
Ananasic acid	*Ananas comosus*
Andrographolide	*Andrographis paniculata*
Andromedotoxin	*Rhododendron simsii*
Androsin	*Pyrola morrisomensis, P. japonica*
Anemonin	*Clematis chinensis, C. florida, C. grata, Imperata cylindrica var. major, Ranunculus japonicus, R. sceleratus*
Anethole	*Agastache rugosa, Ocimum basilicum*
Anethole-D-fenchone	*Foeniculum vulgare*
Angelic acid	*Angelica keiskei*
Angelicin	*Angelica keiskei*
Anhydroderrid	*Millettia nitida, M. taiwaniana*
Anisaldehyde	*Agastache rugosa, Foeniculum vulgare*
Anneparine	*Nelumbo nucifera*
Annocherine A–B	*Annona muricata, A. cherimola, A. reticulata*
Annonaceous acetogenins	*Annona muricata, A. cherimola, A. reticulata*
Anomalin	*Peucedanum formosanum*
Anonaine	*Nelumbo nucifera*
Antictabubs	*Dianella ensifolia, D. chinensis, D. longifolia*
Anthelmintic	*Chenopodium ambrosioides*
Anthocyanidines	*Achillea millefolium, Dianella ensifolia, D. chinensis, D. longifolia, Myrica adenophora, Taraxacum officinale, T. mongolicum*

Component	Source
Anthocyanins	*Citrus maxima, C. sinensis* var. *sekken, Glehnia littoralis, Graptopetalum paraguayense, Impatiens balsamina, Ipomoea batata, I. obscura, I. stans, Passiflora suberosa, Perilla frutescens, P. frutescens* var. *crispa, P. ocymoides, Ribes formosanum, R. nigrum, Torenia concolor* var. *formosana, Tricytis formosana, Vaccinium japonium, V. myrtillus*
Anthraglycoside A	*Polygonum cuspidatum*
Anthraglycoside B	*Polygonum cuspidatum*
Anthranil acid	*Jasminum sambac*
Anthraquinon	*Hedyotis diffusa*
Anthraquinone	*Dianella ensifolia, D. chinensis, D. longifolia*
Anthraquinone glycoside	*Rhamnus formosana*
Anthraquinones	*Cassia occidentalis, C. tora, C. torosa, Rhamnus formosana, Taraxacum officinale, Ventilago leiocarpa*
Anthrone	*Polygonum multiflorum* var. *hypoleucum*
Anti-HIV protein MAP 3	*Momordica charantia*
Antineoplastic agents	*Typhonium divaricatum*
Antofine	*Ficus septica, F. superba* var. *japonic*
Apigenin	*Cassia occidentalis, C. torosa, Clinopodium umbrosum, Juncus effusus* var. *decipiens, Leucas aspera, L. chinensis, Perilla frutescens, P. frutescens* var. *crispa, P. ocymoides*
Apigenin glycosides	*Bellis perennis*
Apigenin-7-β-glucoside	*Agrimonia pilosa*
Apigenin-7-D-glucuronide	*Ruellia tuberosa*
Apigenin-7-diglucuronide	*Clerodendrum trichotomum, C. trichotomum* var. *fargesii*
Apodonta	*Lonicera kawakamii, L. confusa*
Araban	*Abelmoschus moschatus, Cayratia japonica*
Arabinan polymer	*Bupleurum chinensis, B. falcatum*
Arabinogalactan	*Opuntia dillenii*
Arabinopyranosyl	*Deutzia cordatula, D. taiwanensis, D. corymbosa, D. gracilis*
Arabinose	*Bombax malabarica, Camellia japonica* var. *hozanensis, Gnaphalium affine, G. luteoalbum* ssp. *affine, Taraxacum officinale*
Arachidic acid	*Pueraria lobata, P. montana, Viscus multinerve*

Arachidonic acid	*Euchresta formosana*
Araliosides	*A. taiwaniana, A. chinensis*
Arborinine	*Ruta graveolens*
Arbutin	*Saxifraga stolonifera*
Archangelicin	*Angelica keiskei*
Arctigenin	*Arctium lappa*
Arctiin	*Arctium lappa*
Ardisic acid	*Ardisia crenata, A. squamulosa*
Areca red	*Areca catechu*
Arecaidine	*Areca catechu*
Arecaine	*Areca catechu*
Arecolidine	*Areca catechu*
Arecoline	*Areca catechu*
Arginine	*Dioscorea opposita, Drynaria cordata*
Arginine glucoside	*Solanum lyratum*
Aricine	*Rauvolfia verticillata*
Aristofolin E	*Aristolochia cucurbitifolia, A. manshuriensis, A. heterophylla, A. kaempferi, A. kankanensis, A. shimadai*
Aristoliukine C	*Aristolochia cucurbitifolia, A. manshuriensis, A. heterophylla, A. kaempferi, A. kankanensis, A. shimadai*
Aristolochic acid	*Akebia longeracemosa, A. quinata, Aristolochia cucurbitifolia, A. manshuriensis, A. heterophylla, A. kaempferi, A. kankanensis, A. shimadai*
Aristolochic acid-Ia methyl ester	*Aristolochia cucurbitifolia, A. manshuriensis, A. heterophylla, A. kaempferi, A. kankanensis, A. shimadai*
Arjunolic acid	*Elaeagnus oldhamii, E. thunbergii, E. wilsonii, Mussaenda parviflora, Psidium guajave*
Aromadendrene	*Cinnamomum camphora*
Aromadendrene II	*Callicarpa formosana, C. japonica*
Aromatic compounds	*Mahonia japonica, M. oiwakensis*
Arsenic	*Juncus effusus* var. *decipiens*
Artabonatine B	*Annona muricata, A. cherimola, A. reticulata*
Artabonatine C–F	*Artabotrys uncinatus*
Artemisia alcohol	*Artemisia indica, A. japonica*

Component	Source
Articulain	*Equisetum ramosissimum*
Arundoin	*Lophatherum gracile*
Asarensinotannol	*Ferula assa-foetida*
Asarone	*Asarum hypogynum, A. macranthum, A. hongkongense, A. longerhizomatosum*
Ascaridol	*Chenopodium ambrosioides*
Ascorbic acid	*Achyranthes aspera var. indica, A. aspera var. rubro-fusca, Basella alba, Boehmeria densiflora, Petasites japonicus*
Aseculin	*Alyxia insularis, A. sinensis*
Ash	*Maranta arundinacea*
Asiaticoside	*Centella asiatica*
Asparagine	*Asparagus cochinchinensis, Celosia cristata, Hemerocallis fulva, Humulus scandens, Taraxacum mongolicum*
Asparagine-linked glycon	*Celosia cristata*
Asperuloside	*Paederia cavaleriei*
Aspidistrin	*Aspidistra elatior*
Astragalin	*Adiantum capillus-veneris, A. flabellulatum, Diospyros eriantha, Dumasia pleiantha, Dysosma pleiantha, Equisetum ramosissimum, Ricinus communis*
Atherospermidine	*Artabotrys uncinatus*
Aucubin	*Eucommia ulmoides, Plantago asiatica, P. major, Vitex negundo, V. rotundifolia*
Aurantiamide	*Polygonum chinense*
Aurantio-obtusin	*Cassia tora*
Auriculasin	*Millettia taiwaniana*
Auroxanthin	*Viola inconspicua ssp. nagasakiensis, V. mandshurica*
Austroinulin	*Stevia rebaudiana*
Avicine	*Zanthoxylum avicennae*
Avicularin	*Psidium guajave, Saururus chinensis*
Awobanin	*Commelina benghalensis, C. communis*
Azadarachtin	*Melia azedarach*
Azaleatin 3-rhamnosyl glucoside	*Rhododendron simsii*
Azulenen	*Cinnamomum camphora*

Bacopaside III	*Bacopa monniera*
Bacopasides A–C	*Bacopa monniera*
Bacopasoponin G	*Bacopa monniera*
Baicalein	*Scutellaria rivularis*
Baicalin	*Scutellaria rivularis*
Balsam	*Liquidambar formosana*
Banana lectin	*Musa insularimontana, M. paradisiaca*
Barium	*Juncus effusus* var. *decipiens*
Bauenyl acetate	*Ixeris tamagawaensis*
Baureny acetate	*Alyxia insularis, A. sinensis*
Befotenine	*Phyllodium pulchellum*
Behenic acid	*Angelica keiskei, Arisaema consanguineum, A. erubescens, A. vulgaris, Ipomoea pes-caprae* ssp. *brasiliensis*
Bellidifolin	*Gentiana atkinsonii, G. campestris, G. flavo-maculata*
Bellidin	*Gentiana atkinsonii, G. campestris, G. flavo-maculata*
Benihinal	*Chamaecyparis formosensis, C. obtusa* var. *filicoides, C. obtusa* var. *formosana*
Benihinol	*Chamaecyparis formosensis, C. obtusa* var. *filicoides, C. obtusa* var. *formosana*
Benihiol	*Chamaecyparis formosensis, C. obtusa* var. *filicoides, C. obtusa* var. *formosana*
Benzaldehyde	*Mahonia japonica, M. oiwakensis, Pogostemon amboinicus*
Benzene	*Zanthoxylum ailanthoides*
Benzo[a]pyrene	*Paris polyphylla*
Benzoate	*Euphorbia thymifolia*
Benzofurans	*Piper kadsura, P. kawakamii*
Benzoic acid	*Gynura japonica* var. *flava, Jasminum sambac*
Benzolacetic ester	*Daemonorops margaritae*
Benzophenones	*Cudrania cochinchinensis*
Benzopyran derivatives	*Mallotus apelta*
Benzoquinone	*Gynura japonica* var. *flava*
Benzoxanzinoid glucosides	*Acanthus ilicifolius*
Benzoxaxinones	*Coix lacryma-jobi*

Component	Source
Benzyl 2-β*O*-ᴅ-glucopyranosyl-3,6-dimethoxybenzoate	*Cassia fistula*
Benzyl 2-hydroxy-3,6-dimethoxybenzoate	*Cassia fistula*
Benzyl alcohol	*Mahonia japonica, M. oiwakensis*
Benzyl mustard oil	*Tropaeolum majus*
Benzyl isothiocyanate	*Tropaeolum majus*
Benzylacetone	*Aquilaria sinensis, A. sibebsus*
Benzylalcohol	*Jasminum sambac*
Benzylbenzoate	*Piper kadsura, P. kawakamii*
Berbamine	*Stephania cephalantha, Thalictrum fauriei*
Berberine	*Coptis chinensis, Nandina domestica, Phellodendron wilsonii, P. amurense, P. chinensis, Stephania cephalantha, Thalictrum fauriei, Toddalia asiatica*
Berberine hydrochloride	*Coptis chinensis*
Bergapten	*Ruta graveolens*
Bergaptene	*Angelica acutiloba, A. citriodor, A. hirsutiflora*
Bergaptin	*Ficus carica, F. benjamina*
Bergaten	*Angelica keiskei, A. sieboldii*
Bergenin	*Ardisia crenata, A. squamulosa, Astilbe longicarpa, Mallotus repandus, M. tiliaefolius, M. paniculatus, M. japonicus*
Besperidin	*Clinopodium laxiflorum*
Betacyanin	*Portulaca grandiflora, P. pilosa*
Betaine	*Chenopodium album, Lycium chinense*
Betanidin	*Portulaca grandiflora, P. pilosa*
Betanin	*Portulaca grandiflora, P. pilosa*
Betaxanthins	*Mirabilis jalapa*
Betelphenoal	*Piper betle*
Betonicine	*Achillea millefolium*
Betula acid	*Clinopodium laxiflorum*
Betulalbuside A	*Breynia officinalis*
Betulic acid	*Viscus multinerve*

Betulin	*Aspidixia articulata, A. liquidambaricala, Euphorbia thymifolia, E. lathyris, E. lectin, E. milli, E. neriifolia, Hypoestes purpurea, Platycodon grandiflorum*
Betulinic acid	*Adina pilulifera, A. racemose, Diospyros angustifolia, D. kaki, Scoparia dulcis*
Biflavone	*Cephalotaxus wilsonianer*
Biflavonoids	*Rhus succedanea, Selaginella delicatula*
Biotin	*Arachis hypogea, A. agallocha*
Bis (2-ethyl butyl) phthalate	*Oenanthe javanica*
Bis-bithiophene	*Tridax procumbens*
Bis-p-hydroxybenzyl-2-isobutylmalate	*Habenaira dentata*
Bisabolene	*Cinnamomum camphora, Murraya paniculata, Ocimum gratissimum*
Bisabolol constituents	*Heterotropa hayatanum, H. macrantha*
Bisbenzylisoquinoline alkaloid	*Stephania tetrandra, S. moore*
Bishordeninyl alkaloid	*Zanthoxylum nitidum, Z. integrifoliolum*
Bisphenanthrene	*Bletilla striata*
Blespirol	*Bletilla striata*
Blestrianol	*Bletilla striata*
Borneol	*Artemisia indica, A. japonica, Blumea balsamifera var. microcephala, Chrysanthemum morifolium, C. segetum, Curcuma longa, Cymbopogon nardus, Dendranthema indicum, Dendropanax pellucidopunctata, Hedyotis corymbosa, Kaempferia galanga*
Bornyl acetate	*Chryanthemum segetum, Hedyotis corymbosa, Piper arboricola*
Bornylautate	*Lindera glauca*
Bourbonene	*Luffa cylindrica*
Brachystamide B	*Piper kadsura, P. kawakamii*
Brassicasterol	*Eryngium foetidum*
Bretylium compounds	*Rhoeo spathacea*
Breyniaionosides A-D	*Breynia officinalis*
Breyniosides A-B	*Breynia officinalis*
Bromelin	*Ananas comosus*
Bronchodilator flavonoid	*Clerodendrum petasites*
Broussonetines	*Broussonetia kazinoki*

Component	Source
Bruceines	Brucea javanica
Bruceolide	Brucea javanica
Brucine	Strychnos angustiflora
Brusatol	Brucea javanica
Bryophyline	Bryophyllum pinnatum, Kalanchoe spathulata, K. pinnata, K. gracillis, K. crenata, K. tubiflora
Buadleglucoside	Dendropanax pellucidopunctata
Buddleoglucoside	Buddleja asiatica, B. formosana, Dendranthema indicum
Bufadienolides	Kalanchoe spathulata, K. pinnata, K. gracillis, K. crenata, K. tubiflora
Bufotenine	Desmodium pulchellum
Bupleuran	Bupleurum chinensis, B. falcatum
Bupleurumol	Bupleurum kaoi, B. chinense
Bursic acid	Capsella bursa-pastoris
Busamine E	Buxus microphylla
Butanolides	Litsae acutivena
Butanone	Pinus massoniana
Butein	Gnaphalium hypoleucum, G. adnatum
Butulinic acid	Hyptis rhomboides, H. martiusii, H. suaveolens
Bututic acid	Pericampylus trinervatus, P. glaucus, P. formosanus, Melia azedarach
Buxpiine	Buxus microphylla
Buxtauine	Buxus microphylla
Byakangelicin	Angelica hirsutiflora, Ruta graveolens
Byakangelicol	Angelica hirsutiflora
c-3Epi-wilsonine	Cephalotaxus wilsonianer
c-Glycoside flavonoid	Drynaria diandra
Cadinane-type sesquiterpenes	Chamaecyparis formosensis, C. obtusa var. filicoides, C. obtusa var. formosana
Cadinene	Piper betle, Podocarpus macrophyllus, Zanthoxylum ailanthoides
Cadinone	Cinnamomum camphora, Piper betle, Podocarpus macrophyllus var. nakaii, Zanthoxylum ailanthoides
Cadmium	Talinum triangulare

Caffeic acid	*Allium cepa, Bryophyllum pinnatum, Cirsium albescens, Clinopodium laxiflorum, Dichrocephala bicolor, Kalanchoe spathulata, K. pinnata, K. gracillis, K. crenata, K. tubiflora, Prunella vulgaris, Pyrrosia polydactylis, Saxifraga stolonifera, Solidago virgo-aurea, Xanthium sibiricum, X. strumarium*
Caffeic acid derivatives	*Wedelia biflora, W. chinensis*
Caffeine	*Camellia oleifera, C. sinensis, Ilex asprella, Ilex pubescens*
Caffeoylquinic acid	*Siphonostegia chinensis*
Calamene	*Acorus calamus, A. gramineus, Agastache rugosa*
Calamenol	*Acorus calamus, A. gramineus*
Calamenone	*Acorus calamus, A. gramineus*
Calcium	*Duchesnea indica, Oxalis corymbosa*
Calcium oxalate	*Achyranthes japonica, Curculigo capitulata, Sesamum indicum*
Calotropin	*Asclepias curassavica*
Calumbianadin	*Angelica keiskei*
Calycine	*Daphniphyllum calycinum*
Camelliagenins	*Camellia japonica* var. *hozanensis*
Camellin	*Camellia japonica* var. *hozanensis*
Campesterol	*Dioscorea bulbifera, Eryngium foetidum, Hibiscus tillaceus. H. esculentus, Jatropha curcas, Viscus multinerve*
Campesteryl	*Viscus multinerve*
Camphene	*Chamaecyparis formosensis, C. obtusa* var. *filicoides, C. obtusa* var. *formosana, Cinnamomum camphora, Curcuma domestica, Cymbopogon nardus, Kaempferia galanga, Lindera glauca. Liquidambar formosana, Mentha canadensis, Myristica cagayanensis, M. fragrans, Podocarpus macrophyllus* var. *nakaii, Vitex negundo, Ruta graveolens, V. rotundifolia, Zingiber officinale*
Camphor	*Artemisia indica, A. japonica, Blumea balsamifera* var. *microcephala, B. laciniata, Chryanthemum segetum, C. morifolium, C. indicum, Curcuma domestica, Dendranthema indicum, Dendropanax pellucidopunctata*
Camphorene	*Ruta graveolens*
Camptothecin	*Nothapodytes foetida, N. nimmoniana*
Camptothecine	*Camptotheca acuminata*
Canaline	*Medicago polymorpha*
Canavalia	*Canavalia ensiformis*
Canavalia gibberelin I-II	*Canavalia ensiformis*

Component	Source
Canavaline	*Astragalus sinicus, Canavalia ensiformis, Desmodium capitatum, Phellodendron wilsonii, P. amurense, P. chinensis*
Cannabiscitrin	*Myrica adenophora*
Caoutchoue	*Artocarpus heterophyllus*
Capillanol	*Artemisia capillaris*
Capillarin	*Artemisia capillaris*
Capillene	*Artemisia capillaris*
Capillin	*Artemisia capillaris*
Capillon	*Artemisia capillaris*
Caprylic	*Cymbopogon citratus*
Capsaicin	*Capsicum frutescens*
Capscin	*Capsicum frutescens*
Capsularin	*Corchorus capsularis, C. olitorius*
Carbazoles	*Clausena excavata*
Carbohydrates	*Arachis hypogea, A. agallocha, Ludwigia octovalvis, Phoenix dactylifera, Pyracantha fortuneana, Solanum indicum*
Carbon tetrachloride	*Ixeris laevigata var. oldhamii*
Carbonhydre compounds	*Mesona procumbens*
Carbonyl compounds	*Hydrangea chinensis*
Carboxylic acid	*Berchemia formosana, B. lineata, Hypericum chinense, H. patulum, Sanguisorba formosana, S. officinalis, S. minor, Viola verecunda, V. hondoensis, V. philippica*
Cardamunin	*Gnaphalium hypoleucum, G. adnatum*
Carene	*Murraya paniculata*
Carene-3	*Murraya paniculata*
Carmichaemine	*Aconitum bartletii, A. fukutomei, A. formosanum, A. kojimae, A. kojimae var. lassiocarpium, A. kojimae var. ramosum, A. yamamotoanum*
Carnaubic acid	*Chenopodium album*
Carosine	*Catharanthus rosens*
Carotenes	*Allium cepa, Blumea laciniata, Eriobotrya japonica, Gardenia jasminoides, Gnaphalium affine, G. luteoalbum ssp. affine, Taraxacum officinale, Toona sinensis*

Component	Sources
Carotenoids	*Cucurbita moschata, Diospyros eriantha, Eichhornia crassipes, Morus australis, Pseudosasa usawai, P. owatrii, Scirpus ternatanus, S. maritimus*
Carrageenan	*Crassocephalum crepidioides*
Cartharmin	*Carthamus tinctorius*
Carvacrol	*Cinnamomum camphora, Mosla punctulata, Piper betle*
Carvone	*Chrysanthemum indicum, Dendranthema indicum, Dendropanax pellucidopunctata, Luffa cylindrica*
Caryophyliene	*Agastache rugosa, Lindera glauca, Luffa cylindrica, Piper betle, P. nigrum*
Caryophyllane oxide	*Gynura japonica var. flava, Piper arboricola, Vitex cannabifolia*
Caryophyllin	*Akebia longeracemosa, A. quinata*
Caryoptyllen, α-guriunene	*Pogostemon amboinicus*
Cassiollin	*Cassia occidentalis, C. torosa*
Castalagin	*Melastoma candidum*
Castcin	*Vitex rotundifolia, V. negundo*
Catalpol	*Veronicastrum simadai*
Catalposide	*Veronicastrum simadai*
Catechin	*Areca catechu*
Catechutanic acid	*Acacia confusa, A. farnesiana, Bombax malabarica*
Catecol	*Portulaca oleracea, Toona sinensis*
Cedrol	*Jasminum hemsleyi*
Celastrol	*Celastrus kusanoi, C. hypoleucus*
Cellulose	*Bixa orellana*
Celosiaol	*Celosia argentea*
Cephalotaxine	*Cephalotaxus wilsonianer*
Cephalotaxinone	*Cephalotaxus wilsonianer*
Cepharadione B	*Piper sarmentosum, P. sanctum*
Cepharamine	*Stephania cephalantha*
Cepharanoline	*Stephania cephalantha*
Cepharanone B	*Piper sarmentosum, P. sanctum*
Cepharanthine	*Stephania cephalantha*

Component	Source
Ceramide	*Alocasia macrorrhiza, Premna obtusifolia, P. crassa, P. serratifolia, P. microphylla*
Cerotic acid	*Akebia longeracemosa, A. quinata, Artocarpus heterophyllus, Brousonetia papyrifera, Plumeria rubra* cv. *acutifolia,*
	Viscus multinerve
Cerotinic acid	*Ficus carica, F. benjamina, Plumeria rubra* cv. *acutifolia*
Cerylic alcohol	*Lactuca indica, Taraxacum officinale*
Chalcone	*Angelica keiskei*
Chalcone glucose	*Glycyrrhiza uralensis*
Chalepensin	*Ruta graveolens*
Chamaecin	*Jasminum hemsleyi*
Chamazulene	*Heterotropa hayatanum, H. macrantha*
Charantin	*Momordica charantia*
Chavibetol	*Piper betle*
Chavicine	*Piper arboricola, P. nigrum*
Chavicol	*Cymbopogon nardus, Illicium arborescens, Piper betle*
Chebulagic acid	*Phyllanthus multiflorus, P. emblica*
Cherianoine	*Annona muricata, A. cherimola, A. reticulata*
Chimaphilin	*Pyrola morrisonensis, P. japonica*
Chinamic aldehyde	*Pogostemon amboinicus*
Chingchengenamide	*Piper kadsura, P. kawakamii*
Chisulactone	*Polygala glomerata*
Chlorine	*Capsella bursa-pastoris*
Chlorgenin	*Gardenia angusta* var. *kosyunensis, G. oblongifolia*
Chlorogenic acid	*Boehmeria nivea* var. *tenacissima, Blechnum orientale, B. pyramidatum, B. amabile, Cirsium albescens, Lonicera*
	japonica, L. japonica var. *sempervillosa, Lonicera kawakamii, L. confusa, Prunus persica, Pyrrosia adnascens,*
	P. petiolosa, Senecio scandens, Sesamum indicum, Solidago virgo-aurea
Chlorophenolic acid	*Euphorbia hirta*
Cholestanes	*Blechum pyramidatum, B. orientale, B. amabile*
Cholesterol	*Hibiscus tillaceus, H. esculentus*

Choline	*Abrus cantoniensis, Aralia taiwaniana, A. chinensis, Chrysanthemum morifolium, Chryanthemum segetum, Diospyros khaki, D. opposita, Humulus scandens, Morus australis, Pinellia ternata, Solanum lyratum, Sesamum indicum, Taraxacum mongolicum, T. officinale, Trichosanthes cucumeroides*
Chromane	*Gynura formosana, G. elliptica*
Chrysanthemaxanthin	*Chrysanthemum indicum, Dendranthema indicum, Dendropanax pellucidopunctata, Senecio scandens*
Chrysanthemin	*Chrysanthemum morifolium, C. segetum, Dendranthema indicum, Dendropanax pellucidopunctata*
Chrysanthenol	*Chrysanthemum segetum*
Chrysanthenone	*Chrysanthemum segetum*
Chrysanthinin	*Chrysanthemum indicum*
Chryso-obtusin	*Cassia tora*
Chrysophanein	*Cassia fistula, Rumex acetosa*
Chrysophanic acid	*Dianella ensifolia, D. chinensis, D. longifolia, Duchesnea indica, Polygonum cuspidatum, P. multiflorum var. hypoleucum, Rumex crispus*
Chrysophanol	*Cassia fistula, C. tora, Hemerocallis fulva, Polygonum cuspidatum*
Chrysuphanic acid	*Rumex japonicus*
Chymase	*Brousonetia papyrifera*
Cis-linalool	*Osmanthus fragrans*
Cincholic acid	*Adina pilulifera, A. racemose*
Cineal	*Dendropanax pellucidopunctata*
Cineol	*Chrysanthemum indicum, Dendranthema indicum, Eucalyptus robusta, Vitex negundo*
Cineole	*Blumea balsamifera var. microcephala, B. laciniata, Cinnamomum camphora, Curcuma longa, Kaempferia galanga, Laungusa galanga, Lindera glauca, Luffa cylindrica, Piper betle, Ruta graveolens*
Cinnamyl acetate	*Cinnamomum cassia*
Cinnamic acid	*Bryophyllum pinnatum, Cinnamomum cassia, Liquidambar formosana, Lycium chinense*
Cinnamic alcohol	*Liquidambar formosana*
Cinnamic aldehyde	*Cinnamomum cassia*
Cis-dihydrodehydro-diconiferyl-9-0-β-D-glucoside	*Pteris vittata*
Citerpene ferruginol	*Taiwania cryptomeriodes*

Component	Source
Citral	*Blumea laciniata, Citrus tangerina, Cymbopogon citratus, C. nardus, Litsea cubeba, Pandanus odoratissimus var. sinensis, Zingiber officinale*
Citric acid	*Bryophyllum pinnatum, Capsella bursa-pastoris, Chaenomeles japonica, Eriobotrya japonica, Lactuca indica, Kalanchoe spathulata, K. pinnata, K. gracillis, K. crenata, K. tubiflora, Oxalis corymbosa*
Citrogellol	*Cymbopogon citratus*
Citronella	*Toddalia asiatica*
Citronellal	*Cymbopogon citratus, C. nardus*
Citronellal laurotetannine	*Litsea cubeba*
Citronellic	*Cymbopogon citratus*
Citronellol	*Cymbopogon nardus, Murraya paniculata. Plumeria rubra cv. acutifolia, Vitex negundo*
Citronenal	*Ocimum gratissimum*
Citropten	*Citrus medica var. gaoganensis*
Citrulline	*Medicago polymorpha*
Ckenaogebik A	*Clematis chinensis, C. florida*
Claulactones A-J	*Clausena excavata*
Clauseactones A-D	*Clausena excavata*
Clauszoline M	*Clausena excavata*
Clelerythrine	*Toddalia asiatica*
Clemontanos-C	*Clematis montana*
Clenodane diterpenoids	*Casearia membranacea*
Cleomin	*Cleome gynandra*
Clerodane glycosides	*Dicranopteris dichotoma*
Clerodendron	*Clerodendrum japonicum, C. kaempferi, C. trichotomum, C. trichotomum var. fargesii*
Clerodolone	*Clerodendrum trichotomum, C. trichotomum var. fargesii*
Clerosterol	*Clerodendrum japonicum, C. kaempferi*
Clovene	*Zanthoxylum ailanthoides*
Cocaine	*Diospyros eriantha*

Codeine	*Stephania cephalantha*
Codeine phosphate	*Drynaria cordata*
Coixenolide	*Coix lacryma-jobi*
Coixol	*Coix lacryma-jobi*
Colchicine	*Hemerocallis fulva*
Columbamine	*Coptis chinensis*
Commelinin flavocommelitin	*Commelina benghalensis, C. communis*
Commelnin	*Commelina benghalensis, C. communis*
Complanatine	*Lycopodium salvinioides*
Condurangin	*Humulus scandens*
Convallarin	*Smilacina formosana*
Conyzasaponins I-Q	*Conyza sumatrensis, C. blinii*
Copaene	*Artemisia lactiflora, A. princeps*
Copper	*Talinum triangulare*
Coptisine	*Coptis chinensis, Thalictrum fauriei*
Corchorin	*Corchorus capsularis, C. olitorius*
Corchoritin	*Corchorus capsularis, C. olitorius*
Corchoroside	*Corchorus capsularis, C. olitorius*
Corchotoxin	*Corchorus capsularis, C. olitorius*
Coriandrol	*Coriandrum sativum*
Corilagin	*Phyllanthus multiflorus*
Coroloside	*Corchorus capsularis, C. olitorius*
Coronaridin	*Tabernaemontana divaricata*
Corticosteroids	*Costus speciosus, Paederia foetida*
Corynantheine	*Uncaria hirsuta, U. rhynchophylla, U. kawakamii*
Corynoxeine	*Uncaria hirsuta, U. rhynchophylla, U. kawakamii*
Cosmosiin	*Agrimonia pilosa, Chryanthemum segetum, C. morifolium*
Costunolide	*Eupatorium formosanum*
Coumaric acid	*Allium cepa*

Component	Source
Coumarin	*Ageratum conyzoides, A. houstonianum, Achillea millefolium, Alternanthera nodiflora, A. sessillis, A. philoxeroides, A. sessilis, Artemisia lactiflora, A. princeps, Cinnamomum cassia, Clausena excavata, Coriandrum sativum, Eupatorium tashiroi, Flemingia macrophylla, F. prostrata, Peucedanum formosanum, Taraxacum officinale, Zanthoxylum pistaciflorum, Z. piperitum, Z. dimorphophylla*
Coumarin glycoside	*Cissus repens, C. sicyoides*
Coumaronochromones	*Euchresta formosana*
Coumaroyl triterpene lactone	*Diospyros angustifolia*
Coumesterol	*Medicago polymorpha*
Coumurrolin	*Murraya paniculata*
Crataegolic acid	*Psidium guajave*
Crigeron	*Erigeron canadensis*
Cristatin A	*Lepidagathis formosensis, L. hyalina, L. cristata*
Croalbidine	*Crotalaria albida*
Crocin	*Gardenia jasminoides*
Crotastriatine	*Crotalaria pallida*
Crotin	*Croton tiglium*
Croton resin	*Croton tiglium*
Croton oil	*Croton tiglium*
Crotonic acid	*Croton tiglium*
Crotonoside	*Croton tiglium*
Crude fiber	*Maranta arundinacea*
Crude protein	*Maranta arundinacea*
Cryptolepinone	*Sida acuta*
Cryptone	*Piper arboricola, P. nigrum*
Cryptopine	*Thalictrum fauriei*
Cryptotaenen	*Cryptotaenia canadensis, C. japonica*
Cryptoxanthin	*Eriobotrya japonica, Taraxacum mongolicum*
Cryptozaponol	*Jasminum hemsleyi*
Crystalline components	*Trachelospermum jasminoides*

Crytomeridiol	*Magnolia liliflora*
Cucurbitacin E	*Cucumis melo ssp. melo*
Cucurbitacin B	*Cucumis melo ssp. melo*
Cucurbitine	*Cucurbita moschata*
Cucurbitine sterol	*Cucurbita moschata*
Cudraxanthone B	*Cudrania cochinchinensis*
Cudraxanthone S	*Cudrania cochinchinensis*
Cumaldehyde	*Cinnamomum camphora*
Cumic alcohol	*Zanthoxylum ailanthoides*
Cumulene	*Conyza sumatrensis, C. blinii*
Curassavicin	*Asclepias curassavica*
Curcolone	*Curcuma domestica, C. zedoaria*
Curcumenal	*Curcuma zedoaria*
Curcumin	*Curcuma domestica, C. longa, C. zedoaria*
Curcumol	*Curcuma domestica, C. longa, C. zedoaria*
Curdione	*Curcuma domestica, C. zedoaria*
Curmarin	*Curcuma domestica*
Curzenene	*Curcuma domestica, C. zedoaria*
Curzerenone	*Curcuma zedoaria*
Curzemone	*Curcuma domestica*
Cyandidin-3-sophoroside	*Hibiscus rosa-sinensis*
Cyanic acid	*Nandina domestica*
Cyanidin	*Impatiens balsamina, Parthenocissus tricuspidata, Prunella vulgaris, Quisqualis indica*
Cyanidin glucoside	*Hibiscus rosa-sinensis*
Cyanidin 3-glucoside	*Rhododendron simsii, Solidago virgo-aurea*
Cyanidin 3,5-diglucoside	*Rhododendron simsii*
Cyanidin-3-gentiobioside	*Solidago virgo-aurea*
Cyanogenic glucoside	*Ageratum conyzoides, A. houstonianum*
Cyasterone ecdysones	*Akebia longeracemosa, A. quinata*
Cyaterone	*Cyathula prostrata*

Component	Source
Cycasin	*Cycas revoluta*
Cyckivuvibuxine C	*Buxus microphylla*
Cyckivuvibuxine D	*Buxus microphylla*
Cyclanoline	*Stephania hispidula, S. japonica*
Cycleanine	*Stephania cephalantha*
Cyclitol	*Trachelospermum jasminoides*
Cycloartenol	*Abrus precatorius, Lepidagathis formosensis, L. hyalina, L. cristata*
Cycloencalenol	*Melia azedarach*
Cyclomonerviol	*Nervilia taiwaniana, N. purpurea*
Cyclomulberrochromene	*Morus alba*
Cyclonerviol	*Nervilia taiwaniana, N. purpurea*
Cyclonerviol	*Nervilia taiwaniana, N. purpurea*
Cyclopropenoid	*Sida rhombifolia*
Cycloprotobuxamine A	*Buxus microphylla*
Cycloprotobuxamine C	*Buxus microphylla*
Cyclotides	*Viola diffusa, V. tricolor, V. betonicifolia*
Cyclovirobuxine D	*Buxus microphylla*
Cydoartnal	*Euphorbia atoto*
Cydohoreanine B	*Buxus microphylla*
Cylindrin	*Lophatherum gracile*
Cymbopogonol	*Cymbopogon citratus*
Cymene	*Coriandrum sativum, Agastache rugosa, Myristica cagayanensis, M. fragrans*
Cynorin	*Senecio nemorensis, S. scandens*
Cyrtophyllin	*Clerodendrum japonicum, C. kaempferi*
Cysteic acid	*Taraxacum mongolicum*
Cysteine	*Ceratopteris thalictroides, Taraxacum mongolicum*
Cystine	*Taraxacum mongolicum*

Cytisine	Sophora flavescens, S. tomentosa
Cytochrone C	Ricinus communis
D-α-Pinene	Chamaecyparis formosensis, C. obtusa var. filicoides, C. obtusa var. formosana
D-Apiose	Lemmaphyllum microphyllum
D-Borneol	Zingiber officinale
D-Camphene	Acorus calamus, A. gramineus
D-Camphor	Achillea millefolium, Chenopodium ambrosioides, Cinnamomum camphora, Prunella vulgaris
D-Candinene, comphene	Piper arboricola
D-Catechin	Acacia confusa, A. farnesiana
D-Catechol	Camellia japonica var. hozanensis
D-Fenchone	Prunella vulgaris
D-Galactose	Abelmoschus esculentus, Ocimum gratissimum
D-Galacturonic acid	Abelmoschus esculentus, Ocimum gratissimum, Plantago asiatica, P. major
D-Glucose	Bupleurum chinensis, B. falcatum, Ocimum gratissimum, Solanum incanum
D-Glucoside	Hemerocallis fulva
D-Limonene	Citrus tangerina, Schizophragma integrifolium, Tagetes erect
D-Linalool	Pandanus odoratissimus var. sinensis
D-Longifolene	Piper arboricola
D-Mannose	Ocimum gratissimum
D-Mannuronic acid	Ocimum gratissimum
D-Matrine	Sophora flavescens
D-Menthone	Schizophragma integrifolium
D-Oxymatrine	Sophora flavescens
D-Pinitol (3-O-methyl-chiroinositol)	Bougainvillea spectabilis
D-Sabinene	Litsea cubeba
D-Sophoranol	Sophora flavescens
D-Xylose	Plantago asiatica, P. major
Daechuine-S3	Paliurus ramosissimus
Daidzein	Medicago polymorpha, Pueraria lobata, P. montana
Daidzin	Pueraria lobata, P. montana

Component	Source
Dalpanol	*Derris elliptica*
Dammarane	*Pterocypsela indica, Rhus javanica*
Danmacanthal	*Morinda umbellata*
Daphnetin	*Euphorbia lathyris, E. lectin, E. milli, E. neriifolia, E. thymifolia, Hydrangea macrophylla, Wikstroemia indica*
Daphniglaucins A	*Daphniphyllum glaucescens* spp. *oldhamii*
Daphniglaucins B	*Daphniphyllum glaucescens* spp. *oldhamii*
Daphnin	*Setaria italica*
Darutin-bitter	*Siegesbeckia orientalis*
Daturodiol	*Datura metel, D. metel f.fastuosa, D. tatula*
Daturolone	*Datura metel, D. metel f.fastuosa, D. tatula*
Daucosterol	*Alyxia insularis, A. sinensis, Coleus scutellarioide var. crispipilus, C. parvifolius, Conyza sumatrensis, C. blinii, Dicliptera chinensis, D. riparia, Gossampinus malabarica, Pericampylus formosanus, P. trinervatus, P. glaucus, Salvia hayatana, S. japonica, S. roborowskii, Tetrastigma dentatum, T. formosanum, T. umbellatum, T. hemsleyanum*
Deacetylasperulosidic acid	*Galium echinocarpum*
Deacylcyanchogenin	*Cynanchum paniculatum*
Deacylmetaplexlgenin	*Cynanchum paniculatum*
Decanal	*Coriandrum sativum, Cymbopogon citratus*
Decanol	*Coriandrum sativum*
Decanoylacetaldehyde	*Houttuynia cordata*
Decylic aldehyde	*Coriandrum sativum*
Deguelin	*Derris elliptica*
Dehydroandrographolide	*Andrographis paniculata*
Dehydrodigallic acid	*Rhynchosia volubilis, R. minima*
Dehydrolindestrenolide	*Lindera aggregata, L. okoensis*
Dehydromatricaria	*Erigeron canadensis*
Dehydromatricaria ester	*Conyza sumatrensis, C. blinii*
Dehydropodophyllotoxin	*Dumasia pleiantha, Dysosma pleiantha*
Dehydrosulfurenic acid	*Cinnamomum insulari-montanum, C. kotoense, C. micranthum*
Dehydroxycubebin	*Hypoestes purpurea*

Delphin	*Commelina benghalensis, C. communis*
Delphindin	*Commelina benghalensis, C. communis, Impatiens balsamina, Medicago polymorpha, Prunella vulgaris*
Delphinidin-3-diglucoside	*Eichhhornia crassipes*
Delphinidin-3-monoglucoside	*Solanum lyratum*
Delta(5)24-stigmastadienol	*Eryngium foetidum*
Delta-3-carene	*Vitex negundo*
Delta-5-avenasterol	*Eryngium foetidum*
Demethylcarolignane E	*Hibiscus taiwanensis*
Demethylcephalotaxine	*Cephalotaxus wilsonianer*
Demethyltetrandrine	*Stephania hispidula*
Deoxyandrographolide	*Andrographis paniculata*
Deoxyelephantopin	*Elephantopus mollis, E. scaber*
Deoxymikanolide	*Mikania cordata*
Deoxypodophyllotoxin	*Dumasia pleiantha, Dysosma pleiantha*
Dephynyl methane-2-carboxylic acid	*Conyza sumatrensis, C. blinii*
Desmethoxyyangonin	*Piper sarmentosum, P. sanctum*
Desmodium alkaloids	*Desmodium triflorum*
Detetrahydroconidendrin	*Jasminum hemsleyi*
Detuydrosugiol	*Jasminum hemsleyi*
Dextrin	*Maranta arundinacea*
Digycoside	*Hedyotis diffusa*
Diacetate	*Euphorbia thymifolia*
Diacetate nicotinate	*Euphorbia thymifolia*
Diacetyltambulin	*Zanthoxylum nitidum, Z. integrifoliolum*
Diallyl sulfide	*Allium sativum, A. thunbergii, A. tuberosum*
Diandraflavone	*Drynaria diandra*
Dianthronic	*Cassia occidentalis, C. torosa*
Diarylpentanoid	*Anoectochilus formosanus*
Diazomethane	*Cycas revoluta*
Dicaffeoxylquinic acid	*Xanthium sibiricum, X. strumarium*

Component	Source
Dichrins	*Dichroa febrifuga*
Dichroidine	*Dichroa febrifuga*
Dichroines	*Dichroa febrifuga*
Dicliripariside A	*Dicliptera chinensis, D. riparia*
Dicliripariside C	*Dicliptera chinensis, D. riparia*
Dicoumarol	*Medicago polymorpha*
Dictamnine	*Zanthoxylum avicennae*
Dietary fiber	*Phoenix dactylifera*
Diethyl phthalate	*Oenanthe javanica*
Digalactosyl	*Colocasia antiquorum var. illustris, C. esculenta*
Digicirin	*Digitalis purpurea*
Digicoside	*Digitalis purpurea*
Digifolein	*Digitalis purpurea*
Digipurin	*Digitalis purpurea*
Digitonin	*Digitalis purpurea*
Digitoxigenin	*Corchorus capsularis, C. olitorius, Digitalis purpurea*
Digitoxin	*Digitalis purpurea*
Dihydro-β-ionone	*Osmanthus fragrans*
Dihydro-4-hydroxy-5-hyroxymethy-2(3H)-furanone	*Clematis chinensis, C. florida*
Dihydrocarveol	*Piper arboricola, P. nigrum*
Dihydrocherythrine	*Toddalia asiatica*
Dihydrochloride	*Ligustrum sinense*
Dihydrocyclonerviol	*Nervilia taiwaniana, N. purpurea*
Dihydroisocoumarin	*Crassocephalum crepidioides*
Dihydrokaempferol	*Morus alba*
Dihydromikanolide	*Mikania cordata*
Dihydromorin	*Morus alba*
Dihydromyricetin	*Pouteria obovata*

Dihydronepetalactol	*Actinidia callosa var. formosana, A. chinensis*
Dihydronuciferine	*Nelumbo nucifera*
Dihydrooroselol	*Angelica keiskei*
Dihydropiperlonguminine	*Piper kadsura, P. kawakamii*
Dihydrosecurinine	*Securinega suffruticosa, S. virosa*
Dihydrosterulic acid	*Euphoria longana*
Dihydroxy flavone	*Crotalaria sessiliflora*
Dihydroxy methyl anthraquinone	*Morinda citrifolia*
Dihydroxystearic acid	*Ricinus communis*
Dihydrophenanthrenes	*Spiranthes sinensis*
Dilactone	*Rhynchosia volubilis, R. minima*
Dimeric	*Rubus croceacanthus, R. lambertianus*
Dimeric guianolides	*Lactuca indica*
Dimeric acridone alkaloids	*Glycosmis citrifolia*
Dimethiodide	*Cyclea insularis, C. barbata*
Dimethoxyxanthone	*Polygala aureocauda*
Dimethy ether	*Blumea balsamifera var. microcephala, Euphorbia formosana*
Dimethyl	*Cyclea insularis, C. barbata*
Dimethylallyl ether	*Zanthoxylum nitidum, Z. integrifoliolum*
Diogenin	*Costus speciosus*
Diol	*Curcuma domestica, C. zedoaria*
Diosbulbines	*Dioscorea bulbifera*
Dioscin	*Paris polyphylla*
Dioscorecin	*Dioscorea bulbifera*
Dioscoretoxin	*Dioscorea bulbifera*
Diosgenin	*Aletris formosana, Asparagus cochinchinensis, Cissus repens, C. sicyoides, Dioscorea bulbifera, D. eriantha, D. opposita, Paris polyphylla, Solanum indicum*
Diosmetin-7-glucoside	*Chrysanthemum morifolium*
Diosmin	*Citrus medica var. gaoganensis, Toddalia asiatica, Zanthoxylum avicennae, Z. nitidum, Z. integrifoliolum*
Diospyrosonaphthoside	*Diospyros angustifolia*

Component	Source
Diospyrososide	*Diospyros angustifolia*
Dipentene	*Coriandrum sativum, Cymbopogon citratus, C. nardus, Erigeron canadensis, Liquidambar formosana, Myristica cagayanensis, M. fragrans, Pandanus odoratissimus var. sinensis, Piper arboricola, Vitex negundo*
Dipentenecitronellic acid	*Cymbopogon nardus*
Diphenhydramine	*Miscanthus sinensis var. condensatus*
Dipheny picyl	*Desmodium laxiflorum, D. gangeticum, D. multiflorum*
Diphenylamine derivative	*Pieris hieracloides, P. taiwanensis, P. formosa*
Diphenylmethane-2-carboxylic acid	*Erigeron canadensis*
Diplaziosides	*Diplazium megalophyllum, D. subsinuatum*
Diricinolein	*Ricinus communis*
Disaccharides	*Sanguisorba formosana, S. officinalis, S. minor, Trichosanthes homophylla, T. dioica*
Dispyrosooleans	*Diospyros angustifolia*
Diterpenoids	*Aralia chinensis, Cunninghamia korishii, Oreocnide pedunculata, Pieris hieracloides, P. taiwanensis, P. formosa, Podocarpus nagi, Viburnum plicatum var. formosanum, V. odoratissimum V. awabuki, V. luzonicum*
DL-(3-14C) Cysteine	*Ceratopteris thalictroides*
DL-Anabasine	*Alangium chinense*
Dodecane	*Piper sarmentosum, P. sanctum*
Dodecanol	*Angelica acutiloba, A. citriodor*
Domesticine	*Nandina domestica*
Donoxime	*Desmodium pulchellum*
Dopamine	*Musa sapientum, M. formosana, M. basjoo var. formosana, Portulaca oleracea*
Dotriacantan-1-ol	*Elephantopus mollis, E. scaber*
Dotriacontanol	*Elephantopus mollis, E. scaber*
Douminidine	*Gelsemium elegans*
Dracoalban	*Daemonorops margaritae*
Dracoresene	*Daemonorops margaritae*
Dracoresinotannol	*Daemonorops margaritae*
Drymarin A	*Drymaria diandra*

Drymarin B	*Drynaria diandra*
Drymaritin	*Drynaria diandra*
Dryocrassy formate	*Cyathea lepifera. C. podophylla*
Dubiosides D-F	*Thladiantha nudiflora*
Dulcilone	*Scoparia dulcis*
Dulciol	*Scoparia dulcis*
Dulcitol acid	*Jatropha curcas, Maytenus diversifolia*
Durantoside 1-4	*Duranta repens*
Dydimin	*Clinopodium umbrosum*
Dysoxylum	*Asplenium nidus*
E-1-(4′hydroxyphenyl)-but-1-en-3-one	*Scutellaria barbata*
Earth elements, La, Ce, Nd, Sm, Eu, Tb, Yb, Lu	*Dicranopteris dichotoma*
Ebenifoline carigorinine	*Euonymus echinatus, E. laxiflorus, E. chinensis*
Ecdysterone	*Achyranthes japonica, A. aspera var. indica, A. aspera var. rubro-fusca, A. bidentata, A. longifolia, A. ogotai, Akebia longeracemosa, A. quinata, Cyathula prostrata, Podocarpus macrophyllus var. nakaii*
Echimidine	*Symphytum officinale*
Echinopsine	*Echinops grilisii*
Ecliptine	*Eclipta alba, E. prostrate*
Edulinine	*Ruta graveolens*
Eicosenoic acid	*Cardiospermum halicacabum*
Elaterin	*Momordica charantia*
Elemene	*Agastache rugosa, Artemisia lactiflora, A. princeps*
Elemicin	*Cymbopogon citratus*
Elephantin	*Elephantopus mollis, E. scaber*
Elephantopin	*Elephantopus mollis, E. scaber*
Eleutherosides	*Acanthopanax senticosus*
Ellagic acid	*Bischofia javanica, Euphorbia formosana, Lagerstroemia subcostata, Phyllanthus multiflorus, P. emblica, Psidium guajave, Rhynchosia volubilis, R. minima, Rubus formosensis, Sanguisorba formosana, S. officinalis, S. minor*
Ellagitannins	*Rubus croceacanthus, R. lambertianus*
Elliptone	*Derris elliptica*

Component	Source
Emarginatine	*Euonymus echinatus, E. laxiflorus, E. chinensis*
Emodin	*Cassia mimosoides, C. tora, Duchesnea indica, Jasminum hemsleyi, Polygonum multiflorum var. hypoleucum, P. cuspidatum, Rhamnus formosana, Rumex crispus, R. japonicus*
Emodin monomethyl ether	*Polygonum cuspidatum*
Emodin-8-glycoside	*Muehlenbeckia platychodum, M. hastulata*
Encommiol	*Eucommia ulmoides*
Enmein	*Rabdosia lasiocarpus*
Entageric acid	*Entada phaseoloides*
Epi-narirutin	*Clinopodium laxiflorum*
Epibrassicasterol	*Nervilia taiwaniana, N. purpurea*
Epicatechin	*Acacia confusa, A. farnesiana, Ecdysanthera rosea, E. utilis, Muehlenbeckia platychodum, M. hastulata*
Epicephalotaxin	*Cephalotaxus wilsonianer*
Epifriedelanol	*Euphorbia atoto, Premna obtusifolia, P. crassa, P. serratifolia, P. microphylla*
Epifriedelanol acetate	*Bischofia javanica*
Epifriedelin	*Clerodendrum trichotomum, C. trichotomum var. fargesii*
Epifriedelinol	*Clerodendrum japonicum, C. kaempferi, Elephantopus mollis, E. scaber, Euphoria longana, Pericampylus formosanus, P. trinervatus, P. glaucus, Sapium discolor, S. sebiferum*
Epigallocatechin	*Elaeagnus morrisonensis, E. angustifolia, E. glabra, E. lanceollata*
Epiguaipyridine	*Pogostemon amboinicus*
Epihedaragenin	*Potentilla leuconta, P. multifida*
Epipriedelinol	*Vaccinium emarginatum*
Epistephanine	*Cocculus orbiculata, Stephania japonica, S. hispidula*
Epitaraxerol	*Mallotus apelta*
Epoxipiperolid	*Piper sarmentosum, P. sanctum*
Epoxyflavanone	*Atylosia scarbaeoides*
Equisetonin	*Equisetum ramosissimum*
Equisetrin	*Equisetum ramosissimum*
Ergostane derivatives	*Tubocapsicum anomalum*

Ergosterol	*Lactuca indica, Nervilia taiwaniana, N. purpurea*
Ergosterol peroxide	*Ananas comosus*
Eriodictyol-7-*O*-β-ᴅ-glucuronide	*Pyrrosia adnascens, P. petiolosa*
Erucic acid	*Tropaeolum majus*
Erycheline	*Erycibe henryi*
Erysimoside	*Corchorus capsularis, C. olitorius*
Erythrodiol	*Aspidixia articulata, A. liquidambaricala*
Esculetin	*Euphorbia lathyris, E. lectin, E. milli, E. neriifolia, E. thymifolia*
Esculin	*Alyxia insularis, A. sinensis*
Esculitin	*Saxifraga stolonifera*
Essential oil	*Asparagus cochinchinensis, Bletilla formosana, Chloranthus oldham, Cirsium albescens, Conyza sumatrensis, C. blinii, Dichrocephala integrifolia, Eriobotrya japonica, Eucalyptus robusta, Gnaphalium affine, G. luteoalbum ssp. affine, Hedyotis diffusa, Houttuynia cordata, Leonurus sibiricus, Lindera glauca, Melissa officinalis, Mosla punctulata, Plumeria rubra cv. acutifolia, Pogostemon cablin, Salvia plebeia, Sarcandra glabra, Taraxacum officinale, Tylophora ovata, Verbena officinalis, Vitex negundo, Zanthoxylum ailanthoides, Zingiber officinale*
Ether oils	*Taraxacum officinale*
Ethereal oil	*Bixa orellana*
Ethyl 4,5-dicaffeoyl quinate	*Dichrocephala bicolor*
Ethyl β-fructopyranoside	*Rosa taiwanensis, R. davurica*
Ethyl-*p*-methoxycinnamate	*Kaempferia galanga*
Ethylcinnmate	*Kaempferia galanga*
Etoposide	*Dysosma pleiantha*
Eucalyptol	*Artemisia indica, A. japonica*
Eucalyptole	*Cinnamomum camphora, Ocimum basilicum*
Euchrenone	*Euchresta formosana*
Euchretin F	*Euchresta formosana*
Euchretin M	*Euchresta formosana*
Euchretins	*Euchresta formosana*
Eugenol	*Cinnamomum camphora, C. cassia, Cymbopogon nardus, Laungusa galanga, Murraya paniculata, Myristica cagayanensis, M. fragrans, Piper betle, Pogostemon amboinicus, Psidium guajave, Vitex negundo*

Component	Source
Eugenol methyl ether	*Ocimum basilicum, Piper betle*
Eugenol toddaculine	*Toddalia asiatica*
Eugenole	*Ocimum basilicum*
Euojaponine	*Euonymus echinatus, E. laxiflorus, E. chinensis*
Eupachinilides	*Eupatorium amabile, E. lindleyanum*
Eupafolin	*Salvia plebeia*
Eupaformonin	*Eupatorium cannabinum ssp. asiaticum, E. formosanum*
Eupaformosanin	*Eupatorium formosanum, E. tashiroi*
Euparin	*Eupatorium tashiroi*
Eupatene	*Eupatorium tashiroi*
Eupatol	*Eupatorium tashiroi*
Eupatolide	*Eupatorium cannabinum ssp. asiaticum, E. formosanum, Eupatorium tashiroi*
Euphal	*Euphorbia atoto, E. lathyris, E., lectin, E. milli, E. neriifolia, E. thymifolia*
Euphorbetin	*Euphorbia lathyris, E. lectin, E. milli, E. neriifolia, E. thymifolia*
Euphorbiasteroid	*Euphorbia lathyris, E. thymifolia, E. lectin, E. milli, E. neriifolia*
Euphorbol	*Euphorbia atoto, E. lathyris, E. lectin, E. milli, E. neriifolia, E. thymifolia*
Euphorbon	*Euphorbia hirta*
Euscaphic acid	*Potentilla leuconta, P. multifida*
Evicennin	*Zanthoxylum avicennae*
Evofolin-A	*Sida acuta*
Evofolin-B	*Sida acuta*
Exoticin	*Murraya paniculata*
Extragole	*Zanthoxylum ailanthoides*
Falcalindiol	*Glehnia littoralis*
Fangchinoline	*Stephania hispidula*
Farfuomolide A	*Farfugium japonicum*
Farfuomolide B	*Farfugium japonicum*
Farnesal	*Cymbopogon citratus*
Farnesiferols	*Ferula assa-foetida*

Component	Source
Farnesol	*Cymbopogon citratus, Plumeria rubra cv. acutifolia*
Fat	*Gnaphalium affine, G. luteoalbum ssp. affine*
Fatty acids	*Alocasia cucullata, Cassia mimosoides, Clematis chinensis, C. grata, Clinopodium laxiflorum, Colocasia antiquorum var. illustris, C. esculenta, Curculigo orchioides, Diospyros khaki, Dipteracanthus repens, D. prostratus, Helianthus annuus, Lindera glauca, L. communis, Ludwigia octovalvis, Lygodium japonicum, Marsilea crenata, M. minuta, Parthenocissus tricuspidata, Phoenix dactylifera, Solanum indicum, Sida acuta, S. rhombifolia, Quisqualis indica, Ricinus communis, Sterculia nobilis, S. lychnophora, Tetrapanax papyriferus*
Febrifugin	*Hydrangea macrophylla*
Fenchone	*Blumea laciniata*
Fenicularin	*Foeniculum vulgare*
Fernene	*Adiantum capillus-veneris, A. flabellulatum*
Ferritin	*Oxalis corymbosa*
Fermadiene	*Adiantum capillus-veneris, A. flabellulatum*
Ferryl-bipyridyl	*Desmodium laxiflorum, D. gangeticum, D. multiflorum*
Ferulic acid	*Allium cepa, Bryophyllum pinnatum, Chenopodium album, Ferula assa-foetida, Kalanchoe spathulata, K. pinnata, K. gracillis, K. crenata, K. tubiflora, Sida acuta*
Feruloyltyramines	*Hibiscus syriacus*
Fetidine	*Thalictrum fauriei*
Fiber	*Pyracantha fortuneana*
Ficusin	*Ficus carica, F. benjamina*
Filcene	*Adiantum capillus-veneris, A. flabellulatum*
Filicenal	*Adiantum capillus-veneris, A. flabellulatum*
Flavan	*Daphne arisanensis, D. odora, Mariscus cyperinus*
Flavan-3-ols	*Limonium sinense*
Flavanones	*Euchresta formosana, Limonium sinense, Mariscus cyperinus*
Flavanonols	*Berchemia formosana, B. lineata*
Flavins	*Alternanthera nodiflora, A. sessillis, A. philoxeroides, A. sessilis*
Flavogallonic acid	*Rhynchosia volubilis, R. minima*

Component	Source
Flavone	*Berchemia formosana, B. lineata, Desmodium laxiflorum, D. gangeticum, D. multiflorum, Diospyros kaki, Ilex pubescens, Limonium sinense, Loropetalum chinense, Scutellaria barbata*
Flavone glucoside	*Akebia longeracemosa. A. quinata*
Flavone c-glycosides	*Viola confusa, V. yedoensis*
Flavonoid congeners	*Scutellaria barbata*
Flavonoid glycosides	*Anoectochilus formosanus, Artemisia lactiflora, A. princeps, Emilia sonchifolia, E. sonchifolia var. javanica, Kyllinga brevifolia, Ludwigia octovalvis, Pteris ensiformis, P. multifida, Rhodomyrtus tomentose, Stephania hispidula, Urena procumbens, Vernonia gratiosa, Wedelia biflora, W. chinensis, Zornia diphylla*
Flavonoid hispidulin	*Clerodendrum petasites*
Flavonoids	*Adina pilulifera, A. racemose. Abutilon indicum, A. taiwanensis, Acalypha australis, A. indica, Achillea millefolium, Ampelopsis brevuoedybcykata, A. cantoniensis, Anisomeles indica, Bauhinia championi, Blumea aromatica, B. lacera, B. lanceolaria, B. balsamifera var. microcephala, Callicarpa longissima, C. loureiri, C. nudiflora, C. pedunculata, Caryopteris incana, Chamaesyce hirta, C. thymifolia, Chloranthus oldham, Chlorophytum comosum, Chryanthemum segetum, Cichorum endivia, Citrus maxima, C. sinensis var. sekken, Clausena lansium, Clerodendrum japonicum, C. philippinum, C. kaempferi, Cleistocalyx operculatus, Conyza sumatrensis, C. blinii, Cratoxylon ligustrinum, Crotalaria sessiliflora, Curcuma longa, Dalbergia odorifeter, Desmodium caudatum, Dicranopteris dichotoma, Diospyros eriantha, Elaeagnus morrisonensis, E. angustifolia. Epimeredi indica, Eupatorium clematideum, Euphorbia atoto, Evolvulus alsinoides, Ficus microcarpa. Gnaphalium affine, G. luteoalbum ssp. affine, Gynostemma pentaphyllum, Gynura bicolor, Hedyotis corymbosa, Helicteres angustifolia. Ilex rotunda, Iris tectorum, Kaempferia galanga, Lemmaphyllum microphyllum, Lespedeza cuneata, Leucas aspera. L. chinensis, L. mollissima var. chinensis, L. lavandulaefolia, Mimosa pudica, Melastoma candidum, Miscanthus sinensis var. condensatus, Oreocnide pedunculata, Osbeckia chinensis, Phyla nodiflora, Polygonum perfoliatum, Potentilla tugitakensis, Pteris ensiformis, P. semipinnata, P. multifida, Prunus persica, Pueraria lobata, P. montana, Pyrrosia adnascens, P. petiolosa, Rhamnus formosana, Rhus chinensis, R. verniciflua, R. typhina, R. succedanea, Rosa taiwanensis, R. davurica, Rotala rotundifolia, Rubus parvifolius, Salvia plebeia. Sambucus javanica, Sapium discolor, S. sebiferum, Scutellaria formosana, Solidago virgo-aurea, Taraxacum officinale, Thalictrum fauriei, Trichosanthes homophylla, Tricytis formosana. T. dioica, Tridax procumbens, Tylophora ovata, Viscus alniformosanae, V. angulatum, V. multinerve*
Flavonoids, 5-hydroxytryptamine	*Boehmeria nivea var. tenacissima*

Flavonol glycosides	*Bellis perennis, Elaeagnus obovata, E. loureirli, E. bockii, E. macrophylla, E. glabra, E. lanceollata, Goodyera procera, G. schlechtenda, G. nankoensis, Limonium sinense*
Flavonol glycoside gallates	*Limonium sinense*
Flavonols triglycosides	*Rhamnus formosana*
Flavonols	*Limonium sinense, Scurrula loniceritolius, S. ritozonensis, S. liquidambaricolus, S. ferruginea*
Flavonone glucoside	*Glycyrrhiza uralensis*
Flavons	*Vitex rotundifolia*
Flavoxanthin	*Viola inconspicua* ssp. *nagasakiensis, V. mandshurica*
Flaxetin	*Alyxia insularis, A. sinensis*
Fluggein	*Securinega virosa*
Formic acids	*Arenga engleri, A. pinnata, A. saccharifera, Jasminum sambac*
Formonetin	*Medicago polymorpha*
Formosanatins A-D	*Euchresta formosana*
Formosanatin C	*Euchresta formosana*
Formosanin-C	*Paris arisanensis, P. formosana*
Fouquierone	*Rhus javanica*
Framine	*Phyllodium pulchellum*
Frangulin B	*Rhamnus formosana*
Fraxinella	*Melia azedarach*
Friedan-3β-ol	*Euphorbia atoto*
Friedelan-3-ol	*Glochidion lanceolarium*
Friedelan-3α-ol	*Euphorbia atoto*
Friedelan-3α-yl-acetate	*Bischofia javanica*
Friedelanol	*Mallotus apelta*
Friedelin	*Bischofia javanica, Clerodendrum trichotomum, C. trichotomum var. fargesii, C. japonicum, C. kaempferi, Diospyros angustifolia, Euphoria longana, Hemerocallis fulva, Lophatherum gracile, Mallotus apelta, Melanolepis multiglandulosa, Premna obtusifolia, P. crassa, P. serratifolia, P. microphylla, Sapium discolor, S. sebiferum, Vaccinium emarginatum*
Friedelinol	*Conyza sumatrensis, C. blinii*
Fructofuranoside	*Polygonatum falcatum, P. kingianum, P. odoratum*

Component	Source
Fructose	*Plumbago zeylanica*
Fulvoplumierin	*Plumeria rubra* cv. *acutifolia*
Fumaric acid	*Eupatorium tashiroi, Potentilla discolor, P. tugitakensis, Pyrrosia polydactylis, Sarcandra glabra*
Furanocoumarins	*Angelica hirsutiflora, Glehnia littoralis*
Furanodiene	*Curcuma domestica, C. zedoaria*
Furanodienone	*Curcuma domestica, C. zedoaria*
Furanolabdane diterpenes	*Hypoestes purpurea*
Furanosesquitepenes	*Farfugium japonicum*
Furans	*Farfugium japonicum*
Furfural, planteose	*Ocimum basilicum*
Furfuraldehyde	*Zanthoxylum avicennae*
Furfurol	*Bupleurum kaoi, B. chinense*
Furostanol saponins	*Allium sativum, A. thunbergii, A. tuberosum*
Furostanol	*Paris polyphylla*
Futoamide	*Piper kadsura, P. kawakamii*
Galactan	*Luffa cylindrica*
Galactitol	*Clerodendrum japonicum, C. kaempferi*
Galactomannan	*Cassia occidentalis, C. torosa*
Galactose	*Gnaphalium affine, G. luteoalbum* ssp. *affine, Bombax malabarica*
Galactose-specific lectin	*Pedilanthus tithymaloides, Trichosanthes homophylla, T. dioica*
Galangin	*Laungusa galanga*
Galanthamine	*Hippeastrum equestr, H. regina*
Galic acid	*Caesalpinia pulcherrima, Erigeron canadensis, Eucalyptus robusta, Euphorbia hirta, Geranium nepalense var. thunbergii, G. suzukii; Limonium sinense, Phyllanthus multiflorus, P. emblica, Potentilla discolor, P. obovata, P. tugitakensis, Psidium guajave, Rhus chinensis, R. verniciflua, R. typhina, Rhynchosia volubilis, R. minima, Ricinus communis, Sanguisorba formosana, S. officinalis, S. minor*
Galloyl flavonol glycosides	*Pemphis acidula*
Galloylglucoses	*Rhus javanica*

Galuteolin	*Equisetum ramosissimum*
Fambir-fluorescin	*Acacia confusa, A. farnesiana*
Gambirine	*Acacia confusa, A. farnesiana*
Gamma-caryophyllene	*Callicarpa formosana, C. japonica*
Gamma-linolenic acids	*Ribes formosanum, R. nigrum*
Gardenin	*Gardenia angusta var. kosyunensis, G. oblongifolia, G. jasminoides*
Gaudichaudianum	*Asplenium nidus*
Gedunin	*Melia azedarach*
Gelatin	*Bletilla striata*
Gelsemidine	*Gelsemium elegans*
Gelsemine	*Gelsemium elegans*
Gemichalcone C	*Hypericum geminiflorum*
Gemixanthone A	*Hypericum geminiflorum*
Genipin-1-β-gentiobioside	*Gardenia jasminoides*
Geniposide	*Gardenia jasminoides*
Gentianine	*Swertia randaiensis*
Gentianine	*Justicia gendarussa, J. procumbens, J. procumbens var. hayatai*
Gentianine	*Justicia gendarussa, J. procumbens, J. procumbens var. hayatai*
Gentianol	*Gentiana atkinsonii, G. campestris, G. flavo-maculata*
Gentiopicroside	*Justicia gendarussa, J. procumbens, J. procumbens var. hayatai*
Gentlanidine	*Cymbopogon citratus*
Geranic	*Chenopodium ambrosioides, Coriandrum sativum, Cymbopogon nardus, C. citratus, Murraya paniculata, Myristica*
Geraniol	*cagayanensis, M. fragrans, Ocimum basilicum, Plumeria rubra* cv. *acutifolia, Vitex negundo, Zanthoxylum*
	ailanthoides
Germacrene	*Kadsura japonica*
Germacrene B	*Callicarpa formosana, C. japonica*
Germanicyl	*Lactuca indica*
Gibberelin A	*Canavalia ensiformis*
Gibberelin A$_2$	*Canavalia ensiformis*
Gingerglycolipids A–C	*Zingiber kawagoii, Z. rhizoma*

Component	Source
Gingerol	*Zingiber officinale*
Ginsenosides Rb$_1$, Rb$_3$, Rd, Rf	*Gynostemma pentaphyllum*
Ginsenosides	*Pterocypsela indica*
Giosbulbin	*Dioscorea bulbifera*
Gitaloxigenin	*Digitalis purpurea*
Gitaloxin	*Digitalis purpurea*
Gitanin	*Digitalis purpurea*
Gitoxigenin	*Digitalis purpurea*
Gitoxin	*Digitalis purpurea*
Glaucescine	*Daphniphyllum calycinum*
Glaucine	*Thalictrum fauriei*
Globulin	*Arachis hypogea, A. agallocha*
Glochidacuminoside A-D	*Glochidion eriocarpum, G. acuminatum, G. zeylanicum*
Glochidiolide	*Glochidion eriocarpum, G. acuminatum, G. zeylanicum*
Glochidone	*Glochidion rubrum*
Glochidonol	*Glochidion lanceolarium, G. rubrum*
Glucan	*Basella rubra*
Glucobrassicin	*Clerodendrum cyrtophyllum*
Glucokinin	*Drynaria cordata*
Glucomannan	*Pleione formosana*
Glucominol	*Allium sativum, A. thunbergii, A. tuberosum*
Glucononitol	*Vitex negundo*
Glucose	*Gnaphalium affine, G. luteoalbum* ssp. *affine, Lycopus lucidus* var. *formosana, Plumbago zeylanica*
Glucosides	*Acanthopanax senticosus, Ampeloptis brevuoedybcykata, A. cantoniensis, Arachis hypogea, A. agallocha, Centella asiatica, Davallia mariesii, Desmodium laxiflorum, D. gangeticum, D. multiflorum, Dodonaea viscosa, Morinda citrifolia, Polygonatum falcatum, P. kingianum, P. odoratum, Sarcandra glabra, Serissa foetida*
Glucosides A, B, and C	*Bryophyllum pinnatum*
Glucotropaolin	*Tropaeolum majus*
Glutamine	*Dioscorea opposita, Ficus carica, F. benjamina*

Glutelin	*Alocasia cucullata*
Glyceroglycolipid	*Premna obtusifolia, P. crassa, P. serratifolia, P. microphylla*
Glycine	*Taraxacum mongolicum*
Glycogen	*Bletilla formosana*
Glycolic acid	*Asparagus cochinchinensis*
Glycone	*Hypoxis aurea*
Glycoproteins	*Celosia cristata, Oldenlandia hedyotidea, O. diffusa*
Glycoside alkaloids	*Solanum biflorum*
Glycosides	*Clerodendrum trichotomum, C. trichotomum var. fargesii, Conyza canadensis, C. dioscoridis, Corchorus capsularis, C. olitorius, Dicliptera chinensis, D. riparia, Gardenia angusta var. Kosyunensis, G. oblongifolia, Ligustrum sinense, Pyrola morrisonensis, P. japonica, Pyrrosia adnascens, P. petiolosa*
Glycosidic consituents	*Hydrocotyle sibthorpioides*
Glycovatromonoside	*Corchorus capsularis, C. olitorius*
Glycyrrhiza	*Glycyrrhiza uralensis*
Glycyrrhizic acid	*Abrus precatorius, Glycyrrhiza uralensis*
Glycyrrhizin	*Arachis hypogea, A. agallocha*
Glypenosides	*Gynostemma pentaphyllum*
Gnaphalin	*Gnaphalium hypoleucum, G. adnatum*
Gobosterin	*Arctium lappa*
Gomphrenin	*Gomphrena globosa*
Goodyerin	*Goodyera procera, G. schlechtenda. G. nankoensis*
Goodyeroside A	*Goodyera procera, G. schlechtenda. G. nankoensis*
Gracillin	*Paris polyphylla*
Gravacridonediol	*Ruta graveolens*
Gravacridonetriol	*Ruta graveolens*
Gravacridonol chlorine	*Ruta graveolens*
Gravelliferone	*Ruta graveolens*
Graveoline	*Ruta graveolens*
Graveolinine	*Ruta graveolens*
Grossypitrin	*Equisetum ramosissimum*

Component	Source
Gryptoxanthine	*Cycas revoluta*
Guaiacol	*Ficus carica, F. benjamina*
Guaiaxulene	*Ficus carica, F. benjamina*
Guaijaverin	*Psidium guajave*
Guajavolic acid	*Psidium guajave*
Guineensine	*Piper kadsura, P. kawakamii*
Gum	*Myrica adenophora*
Guvacine	*Areca catechu*
Guvacoline	*Areca catechu*
Gynuraone	*Gynura japonica var. flava*
Gypenocide	*Gynostemma pentaphyllum*
Habenariol	*Habenaira dentata*
Haemanthidien	*Zephyranthes candida*
Hamabiwalactone A	*Litsea japonica, L. hypophaea*
Hamabiwalactone B	*Litsea japonica, L. hypophaea*
Hamaudol	*Angelica hirsutiflora*
Hamoaromaline	*Stephania cephalantha*
Hananomin	*Illicium arborescens*
Harpagide acetate	*Eucommia ulmoides*
Harpagoside	*Scrophularia yoshimurae*
Harringtonine	*Cephalotaxus wilsonianer, Stephania japonica*
Hecogenin	*Cissus repens, C. sicyoides, Polygonum chinense*
Hederagenin	*Clematis chinensis, C. grata, Ilex asprella*
Helichrysoside	*Melastoma candidum*
Helioxanthin	*Hypoestes purpurea*
Helioxanthis	*Polygala glomerata*
Helminthosporin	*Cassia occidentalis, C. torosa*
Helveticoside	*Corchorus capsularis, C. olitorius*
Hemerocallin	*Hemerocallis fulva*

Hemolytic sapogenin	*Sansevieria trifasciata*
Hennadiol	*Mallotus apelta*
Hentriacontane	*Plantago asiatica, P. major, Vaccinium emarginatum*
Hepacosane	*Dicranopteris linearis*
Heptadecatrienyl	*Rhus succedanea*
Heptadecylic acid	*Aleurites fordii, A. moluccana, A. montana*
Herbacetrin	*Equisetum ramosissimum*
Hermandezine	*Thalictrum fauriei*
Hesperidin	*Citrus medica var. guoganensis, Clinopodium umbrosum, Davallia mariesii, Gossampinus malabarica, Zanthoxylum avicennae*
Hesperindin	*Citrus tangerina*
Heteromines D	*Heterostemma brownii*
Heteromines E	*Heterostemma brownii*
Heteroside	*Cassia occidentalis, C. torosa*
Heteroxylan	*Phoenix dactylifera*
Hexacosanol	*Scoparia dulcis*
Hexadecane	*Piper sarmentosum, P. sanctum*
Hexadeceonic acid	*Myristica cagayanensis, M. fragrans, Sterculia nobilis, S. lychnophora*
Hexahydromatricaria	*Erigeron canadensis*
Hexenol	*Agastache rugosa*
Hexose	*Miscanthus floridulus, M. sinensis var. condensatus, Ocimum gratissimum*
Hibiscuside	*Hibiscus syriacus*
Hibiscuwanin A-B	*Hibiscus taiwanensis*
Higerine	*Desmodium pulchellum*
Hinoguinin	*Hypoestes purpurea*
Hinokiflavone	*Cycas revoluta, Podocarpus macrophyllus var. nakaii*
Hirsuteine	*Uncaria hirsuta, U. rhynchophylla, U. kawakamii*
Hirsutine	*Uncaria hirsuta, U. rhynchophylla, U. kawakamii*
Hispidulin	*Salvia plebeia*

Component	Source
Homoarbutin	*Pyrola morrisonensis, P. japonica*
Homoarecoline	*Areca catechu*
Homoaromoline	*Stephania cephalantha*
Homocyclotirucallane	*Spiranthes sinensis*
Homocylindrocarpidine	*Tabernaemontana amygdalifolia*
Homoeriodictyol	*Viscus multinerve*
Homoeriodictyol-7-glucoside	*Viscus multinerve*
Homoplantagin	*Plantago asiatica, P. major*
Homoplantaginin	*Salvia plebeia*
Homostephanoline	*Stephania japonica*
Homotrilobine	*Cocculus orbiculata*
Hopadiene	*Adiantum capillus-veneris, A. flabellulatum*
Hopane-triterpene lactone glycosides	*Diplazium megalophyllum, D. subsinuatum*
Hormoharringtonine	*Cephalotaxus wilsonianer*
Horneol	*Chrysanthemum indicum, Coriandrum sativum*
Houttuynium	*Houttuynia cordata*
Hoyin	*Hoya carnosa*
Humulene	*Humulus scandens, Lantana camara*
Huratoxin	*Excoecaria orientalis, E. agallocha, E. kawakamii*
Hydrangeic acid	*Hydrangea macrophylla*
Hydrangenol	*Hydrangea macrophylla*
Hydrazyl	*Desmodium laxiflorum, D. gangeticum, D. multiflorum*
Hydrocinnamic acid	*Aquilaria sinensis, A. sibebsus*
Hydrocotyin	*Centella asiatica*
Hydrocotylene	*Vitex negundo*
Hydrocotylosides I–VII	*Hydrocotyle sibthorpioides*
Hydrocyanic acid	*Manihot utilissima, Passiflora suberosa, Taraxacum officinale*
Hydroquinone	*Ilex pubescens, Rhus succedanea*
Hydroxyadianthone	*Adiantum capillus-veneris, A. flabellulatum*

Hydroxyanic acid	*Chaenomeles japonica*
Hydroxybenzaldehyde	*Nothapodytes foetida, N. nimmoniana*
Hydroxybenzoic acid	*Taraxacum officinale, Vitex negundo*
Hydroxycinnamic acid	*Achillea millefolium, Solidago virgo-aurea, Taraxacum officinale*
Hydroxydaidzein	*Hibiscus syriacus*
Hydroxyephalotaxine	*Cephalotaxus wilsonianer*
Hydroxygenkwanin	*Wikstroemia indica*
Hydroxyl group	*Zanthoxylum pistaciflorum, Z. piperitum, Z. dimorphophylla*
Hydroxyleamptothecin	*Camptotheca acuminata*
Hydroxyleucine	*Deutzia cordatula, D. taiwanensis, D. corymbosa, D. gracilis*
Hydroxylinderstrenolide	*Lindera aggregata, L. okoensis*
Hydroxymethylene	*Diplazium megaphllum, D. subsinuatum*
Hyoscine	*Datura metel, D. metel f.fastuosa, D. tatula*
Hyoscyamine	*Aconitum bartletii, A. fukutomel, A. formosanum, A. kojimae, A. kojimae var. lassiocarpium, A. kojimae var. ramosum,*
	A. yamamotoanum, Datura metel, D. metel f.fastuosa, D. tatula
Hypaphorine	*Abrus precatorius*
Hypericin	*Hypericum japonicum*
Hyperin	*Dysosma pleiantha, Eupatorium tashiroi, Hypericum japonicum, Pyrola morrisonensis, P. japonica, Saururus chinensis*
Hyperoside	*Hibiscus mutabilis, Prunella vulgaris*
Hypochlorous acid	*Desmodium laxiflorum, D. gangeticum, D. multiflorum*
Hypoepistephanine	*Stephania japonica*
Hypophyllanthin	*Phyllanthus urinaria*
Hypopurin A-D	*Hypoestes purpurea*
Hypoxoside	*Hypoxis aurea*
Hystonin	*Physalis angulata*
Iffaionic acid	*Scoparia dulcis*
Illicin	*Illicium arborescens*
Imidazole	*Murdannia keisak, M. loriformis*
Imidazolylothylamine	*Solanum lyratum*
Imperatorin	*Glehnia littoralis*

Component	Source
Indican	*Clerodendrum cyrtophyllum, Indigofera tinctoria, Polygonum perfoliatum*
Indicine	*Heliotropium indicum*
Indigoferabietone	*Indigofera zollingeriana, I. longeracemosa*
Indigotin	*Indigofera tinctoria*
Indimulin	*Indigofera tinctoria*
Indirubin	*Clerodendrum cyrtophyllum, Indigofera suffruticosa*
Indole	*Mahonia japonica, M. oiwakensis*
Indolizinone	*Polygonatum falcatum, P. kingianum, P. odoratum*
Indolopyrido quinazoline	*Zanthoxylum nitidum, Z. integrifoliolum*
Indoxyl	*Indigofera tinctoria*
Ineol	*Artemisia indica, A. japonica*
Ingigo	*Clerodendrum cyrtophyllum*
Inokosterone	*Achyranthes longifolia, A. ogotai, A. japonica, A. bidentata, Blechnum orientale*
Inositol	*Aspidixia articulata, A. liquidambaricala, Lonicera macrantha, L. shintenensis, L. japonica, L. japonica var. sempervillosa, Sonchus arvensis, S. oleraceus*
Insect molting hormones	*Achyranthes bidentata*
Insulanoline	*Paracyclea gracillima*
Insularine	*Paracyclea ochiaiana, P. gracillima, Stephania japonica, S. hispidula*
Integramine	*Zanthoxylum nitidum, Z. integrifoliolum*
Inulin	*Arctium lappa, Cirsium japonicum, C. japonicum var. australe, Taraxacum mongolicum, T. officinale*
Invertase	*Plumbago zeylanica*
Iodine	*Dioscorea bulbifera*
Ipomarone	*Ipomoea batata, I. obscura, I. stans*
Iresenin	*Clematis chinensis, C. grata*
Iridoid	*Paederia cavaleriei*
Iridoid glucosides	*Hedyotis diffusa, Hemiphragma heterophyllum var. dentatum, Paederia scandens, Viburnum plicatum var. formosanum, V. odoratissimum, V. awabuki, V. luzonicum, Scrophularia yoshimurae, Wendlandia formosana*
Iridoidglycoside–nishindaside	*Vitex negundo*
Iridomyrmecin	*Actinidia callosa var. formosana, A. chinensis*

Irinotecan	*Camptotheca acuminata*
Iron	*Talinum triangulare*
Isatan B	*Clerodendrum cyrtophyllum*
Iso-chondrodendrine	*Paracyclea gracillima*
Iso-mangiferin	*Pyrrosia polydactylis, Ruta graveolens*
Iso-prenepolymer	*Clerodendrum japonicum, C. kaempferi*
Iso-rhynchophylline	*Uncaria hirsuta, U. rhynchophylla, U. kawakamii*
Iso-trilobine	*Cocculus trilobus*
Iso-trilobine	*Cocculus sarmentosus*
Isoadiantone	*Adiantum capillus-veneris, A. flabellulatum*
Isoamaranthin	*Gomphrena globosa*
Isocedrolic acid	*Jasminum hemsleyi*
Isocorydine	*Litsea cubeba*
Isocorynoxeine	*Uncaria hirsuta, U. rhynchophylla, U. kawakamii*
Isocurcumenol	*Curcuma zedoaria*
Isodeoxyelophantopin	*Elephantopus mollis, E. scaber*
Isoeugenol	*Myristica cagayanensis, M. fragrans*
Isofernene	*Adiantum capillus-veneris, A. flabellulatum*
Isoflavanoids	*Desmodium laxiflorum, D. gangeticum, D. multiflorum, Millettia taiwaniana*
Isofouquierone	*Rhus javanica*
Isoglochidiolide	*Glochidion eriocarpum, G. acuminatum, G. zeylanicum*
Isoguvacine arecotidine	*Areca catechu*
Isoharringtonine	*Cephalotaxus wilsonianer*
Isoindigo	*Clerodendrum cyrtophyllum*
Isoliquiritigene	*Glycyrrhiza uralensis*
Isoliquiritigenin	*Polygonatum falcatum, P. kingianum, P. odoratum*
Isoliquiritin	*Glycyrrhiza uralensis*
Isolobelamine	*Lobelia chinensis*
Isomangiferin	*Anemarrhena asphodeloides*
Isomenthone	*Glechoma hederacea var. grandis, Mentha canadensis*

Component	Source
Isomesityl oxide	*Cryptotaenia canadensis, C. japonica*
Isomucronulator	*Polygonatum falcatum, P. kingianum, P. odoratum*
Isoneomatatabiol	*Actinidia callosa* var. *formosana, A. chinensis*
Isoorientin	*Vaccinium emarginatum, Vitex negundo, V. rotundifolia, Viola inconspicua* ssp. *nagasakiensis, V. mandshurica*
Isopimarane-type diterpenes	*Orthosiphon aristatus, O. stamineus*
Isopimpinellin	*Angelica hirsutiflora, Ruta graveolens, Toddalia asiatica*
Isoproterenal	*Vandellia crustacea, V. cordifolia*
Isoproterenol	*Boussingaultia gracilis* var. *pseudobaselleoides*
Isopinocamphone	*Glechoma hederacea* var. *grandis*
Isoquercitrin	*Angelica keiskei, Hibiscus mutabilis, Houttuynia cordata, Hypericum japonicum, Loropetalum chinense, Morus australis, Polygonum cuspidatum, Ricinus communis, Sapium discolor, S. sebiferum, Saururus chinensis, Taxillus matsudai, T. levinei*
Isoquercitroside	*Tropaeolum majus*
Isquereitrin	*Equisetum ramosissimum*
Isoquinoline alkaloids	*Cryptocarya chinensis*
Isorhamnetin	*Dodonaea viscosa*
Isorhamnetin-3-D-rutinoside	*Goodyera procera, G. schlechtenda, G. nankoensis*
Isoricinoleic acid	*Ricinus communis*
Isorobustaside A	*Breynia officinalis*
Isorottlerin	*Mallotus tiliaefolius*
Isosafrole	*Angelica acutiloba, A. citriodor*
Isosakurannetin	*Clinopodium laxiflorum*
Isotemonidine	*Stemona tuberosa*
Isotetrandrine	*Stephania cephalantha, Thalictrum fauriei*
Isothalidenzine	*Thalictrum fauriei*
Isotrictiniin	*Phyllanthus multiflorus, P. emblica*
Isotrilobine	*Cocculus orbiculata*
Isovaleric	*Cymbopogon citratus*

Isovitexin	*Crotalaria sessiliflora*
Isoxanthanol	*Xanthium sibiricum, X. strumarium*
Jaligonic acid	*Pieris hieracloides, P. taiwanensis, P. formosa*
Jasminoidin	*Gardenia jasminoides*
Jasmine	*Jasminum sambac*
Jatrorhizine	*Coptis chinensis, Thalictrum fauriei*
Jervine	*Hemerocallis fulva, Veratrum formosanum*
Juglandic acid	*Juncus effusus var. decipiens*
Juglone	*Juncus effusus var. decipiens*
Justicidine E	*Hypoestes purpurea*
Justicin	*Gendarussa vulgaris*
Kadsurarin A	*Kadsura japonica*
Kadsurenone	*Piper kadsura, P. kawakamii*
Kadsuric acid	*Kadsura japonica*
Kadsurin	*Kadsura japonica*
Kaempferide	*Kaempferia galanga*
Kaempferitrin	*Celastrus orbiculatus, C. punctatus, C. paniculatus, Trichosanthes cucumeroides*
Kaempferol	*Hypericum japonicum, Impatiens balsamina, Kalanchoe spathulata, K. pinnata, K. gracillis, K. crenata, K. tubiflora, Kaempferia galanga, Plumeria rubra cv. acutifolia, Potentilla discolor, Ricinus communis*
Kaempferol glucoside	*Tropaeolum majus*
Kaempferol gossypetin	*Hibiscus rosa-sinensis*
Kaempferol-3-rutinoside	*Bauhinia championi, Ricinus communis*
Kaempferol-3-D-rutinoside	*Goodyera procera, G. schlechtenda, G. nankoensis*
Kaempferol-3-galactoside	*Adiantum capillus-veneris, A. flabellulatum, Bauhinia championi*
Kaempferol-3-glucosylgalactoside	*Ophiopogon japonicus*
Kaempferol-3,7-diglucoside	*Equisetum ramosissimum*
Kaempferol-7-shamnoside	*Chenopodium ambrosioides*
Kaempferol-rhamno glucoside	*Solidago virgo-aurea*
Kaempferol-rhamnoside	*Onychium japonicum*

Component	Source
Kaempferols	*Cichorum endivia, Coriandrum sativum*
Kaepferal	*Coriandrum sativum*
Kameofero	*Lindera aggregata, L. okoensis*
Karabin	*Nerium indicum*
Kaurene	*Podocarpus macrophyllus var. nakaii*
Kaxinol B (Isoprenylated flavan)	*Broussonetia kazinoki*
Ketones	*Plumeria rubra* cv. *acutifolia*
Kiganen	*Cryptotaenia canadensis, C. japonica*
Kiganol	*Cryptotaenia canadensis, C. japonica*
Kinganone	*Polygonatum falcatum, P. kingianum, P. odoratum*
Kinsenone	*Anoectochilus formosanus*
Kinsenoside	*Goodyera procera, G. schlechtenda, G. nankoensis*
Kiransin	*Akebia longeracemosa, A. quinata*
Kokusaginine	*Ruta graveolens*
Konokiol	*Magnolia liliflora*
Koumine	*Gelsemium elegans*
Kouminicine	*Gelsemium elegans*
Kouminine	*Gelsemium elegans*
Kukoamine	*Lycium chinense*
Kulinone	*Melia azedarach*
Kuraridin	*Sophora flavescens*
L-1-Leucine	*Chenopodium album*
L-Acetyl-4-isopropylidenecyclopentene	*Cinnamomum camphora*
L-Anagyrine	*Sophora flavescens*
L-Arabinose	*Bupleurum chinensis, B. falcatum, Ocimum gratissimum*
L-Baptifoline	*Sophora flavescens*
L-Cadinene	*Murraya paniculata*
L-Citrulline	*Diospyros eriantha*

L-Curcamene	*Curcuma domestica*
L-Epicatechol	*Camellia japonica* var. *hozanensis*
L-Limonene	*Chenopodium ambrosioides*
L-Linalool	*Tagetes erecta*
L-Methylcytisine	*Sophora flavescens*
L-Pimara-8,15-dien-19-oic acid	*Aralia chinensis*
L-Rhamnose	*Abelmoschus esculentus*
L-Sesamen	*Acanthopanax senticosus*
Labenzyme	*Cirsium japonicum, C. japonicum* var. *australe*
Lacerol	*Clerodendrum cyrtophyllum*
Lacnophylium	*Erigeron canadensis*
Lactic acid	*Bryophyllum pinnatum*
Lactiflorenol	*Artemisia lactiflora, A. princeps*
Lactone	*Bombax malabarica, Cleome gynandra, Litsea japonica, L. hypophaea*
Lactose	*Abelmoschus moschatus*
Lactucerol	*Sonchus arvensis, S. oleraceus*
Lactucin	*Lactuca indica*
Lagenaria D	*Lagenaria siceraria* var. *microcarpa*
Lambertianins A-D	*Rubus croceacanthus, R. lambertianus*
Lanostanoids	*Amentotaxus formosana*
Lantadene A	*Lantana camara*
Lantadene B	*Lantana camara*
Lantanotic acid	*Lantana camara*
Lantic acid	*Lantana camara*
Lappine	*Arctium lappa*
Lariciresinol-9-0-β-D-glucoside	*Pteris vittata*
Latex	*Euphorbia tirucalli*
Lathyrol	*Euphorbia thymifolia*
Lathyrol diacetate benzoate	*Euphorbia lathyris, E. lectin, E. milli, E. neriifolia*
Lathyrol diacetate nicotinate	*Euphorbia lathyris, E. lectin, E. milli, E. neriifolia*

Component	Source
Laurate	*Psidium guajave*
Lauric acid	*Melia azedarach, Myristica cagayanensis, M. fragrans, Taraxacum mongolicum*
Laurolitsine	*Cinnamomum camphora*
Lavoxanthin	*Senecio scandens*
Laxifolone A	*Euonymus echinatus, E. laxiflorus, E. chinensis*
Lead	*Talinum triangulare*
Lecithin	*Sesamum indicum*
Lectin	*Euphorbia tirucalli, Psophocarpus tetragonolobus, Urtica thunbergiana, U. dioica*
Lenrosine	*Catharanthus rosens*
Lenrosivine	*Catharanthus rosens*
Leonrine	*Leonurus artemisia*
Leonurine	*Leonurus sibiricus*
Leucine	*Dioscorea opposita, Medicago polymorpha*
Leucoanthocyanin	*Camellia japonica var. hozanensis*
Leucocyanidin	*Areca catechu, Psidium guajave*
Leucocyanidol	*Euphorbia hirta*
Levidulinase	*Amorphophallus konjac, A. rivieri*
Leviduline	*Amorphophallus konjac, A. rivieri*
Levulose	*Eriobotrya japonica*
Lignan glucosides	*Acanthus ilicifolius, Lactuca indica*
Lignan helioxanthin	*Taiwania cryptomeriodes*
Lignanoids	*Siphonostegia chinensis*
Lignans	*Bupleurum kaoi, B. chinense, Hibiscus syriacus, Hypoestes purpurea, Justicia gendarussa, J. procumbens, J. procumbens var. hayatai, Leucas aspera, L. chinensis, L. mollissima var. chinensis, L. lavandulaefolia, Piper kadsura, P. kawakamii, Pluchea indica, Rhinacanthus nasutus, Trachelospermum jasminoides*
Lignoceric acid	*Viscus multinerve*
Limettin	*Citrus medica var. gaoganensis*

Limonene	*Artemisia lactiflora, A. princeps, Atalantia buxifolia, Blumea balsamifera var. microcephala, Chrysanthemum indicum, Cinnamomum camphora, Clausena excavata, Coriandrum sativum, Conyza sumatrensis, C. blinii, Cymbopogon nardus, Dendranthema indicum, D. pellucidopunctata, Erigeron canadensis, Glechoma hederacea var. grandis, Lindera glauca, Luffa cylindrica, Medicago polymorpha, Mentha canadensis, Ocimum basilicum, Perilla frutescens, P. frutescens var. crispa, P. ocymoides, Ruta graveolens, Zanthoxylum ailanthoides, Z. avicennae*
Linalool	*Agastache rugosa, Artemisia indica, A. japonica, Atalantia buxifolia, Conyza sumatrensis, C. blinii, Coriandrum sativum, Cymbopogon citratus, C. nardus, Erigeron canadensis, Eupatorium tashiroi, Glechoma hederacea var. grandis, Hedyotis corymbosa, H. diffusa, Jasminum sambac, Litsea cubeba, Luffa cylindrica, Mahonia japonica, M. oiwakensis, Medicago polymorpha, Michelia alba, Myristica cagayanensis, M. fragrans, Ocimum basilicum, Osmanthus fragrans, Perilla frutescens, P. frutescens var. crispa, P. ocymoides, Plumeria rubra cv. acutifolia, Ruta graveolens, Vitex cannabifolia, Zanthoxylum ailanthoides, Zingiber officinale*
Linalyl beozeate	*Jasminum sambac*
Linderalactone	*Lindera aggregata. L. okoensis*
Linderane	*Lindera strychifolialinderane, L. aggregata, L. okoensis*
Linoleic acid	*Bupleurum kaoi, B. chinense, Chenopodium album, Cibotium barometz, Coix lacryma-jobi, Helianthus annuus, Jatropha curcas, Myristica cagayanensis, M. fragrans, Nothapodytes foetida, N. nimmoniana, Ricinus communis, Rubus formosensis, Tabernaemontana pandacaqui*
Linolein	*Sesamum indicum*
Linolenic acid	*Jatropha curcas, Ludwigia octovalvis, Ricinus communis, Rubus formosensis*
Linolic acid	*Angelica acutiloba, A. citriodor, Corchorus capsularis, C. olitorius*
Lipase	*Brousonetia papyrifera*
Liquiritigenin	*Glycyrrhiza uralensis, Polygonatum falcatum, P. kingianum, P. odoratum*
Liquirtin	*Glycyrrhiza uralensis*
Liriodendrin	*Alyxia insularis, A. sinensis*
Liriodenine	*Magnolia liliflora, Nelumbo nucifera*
Lithium	*Adiantum capillus-veneris, A. flabellulatum*
Litsealacton A–B	*Litsea japonica. L. hypophaea*
Lobelanidine	*Lobelia chinensis*

Component	Source
Lobelanine	*Lobelia chinensis*
Lobeline	*Lobelia chinensis*
Lobetyol	*Pratia nummularia*
Lobetyolin	*Pratia nummularia*
Lobetyolinin	*Pratia nummularia*
Loganin	*Lonicera japonica, L. japonica* var. *sempervillosa*
Loliolide	*Maytenus diversifolia, Sida acuta*
Long-chain carboxylic acid	*Corydalis pallida*
Lonicerin	*Lonicera macrantha, L. japonica, L. japonica* var. *sempervillosa, L. shintenensis, Stephania tetrandra, S. moore*
Lonicern	*Lonicera kawakamii, L. confusa*
Lonone-related compounds	*Oreocnide pedunculata*
Lucernol	*Medicago polymorpha*
Lucidin	*Morinda umbellata*
Lugrandoside	*Dicliptera chinensis, D. riparia*
Lumicaerulic acid	*Coptis chinensis*
Lupane-type triterpenoids	*Helicteres angustifolia*
Lupane triterpenes	*Viburnum plicatum* var. *formosanum, V. odoratissimum, V. awabuki, V. luzonicum*
Lupenl acetate	*Ixeris tamagawaensis*
Lupensterol	*Ricinus communis*
Lupeol	*Bombax malabarica, Diospyros angustifolia, Elephantopus mollis, E. scaber, Ficus carica, F. benjamina, Hypoestes purpurea, Plumeria rubra* cv. *acutifolia, Myrica adenophora, Salvia hayatana, S. japonica, S. roborowskii, Viscus alniformosanae, V. angulatum, V. multinerve*
Lupeol acetate	*Artocarpus altilis, Elephantopus mollis, E. scaber*
Lupulone	*Humulus scandens*
Lutecolin	*Akebia longeracemosa, A. quinata, Begonia fenicis, B. laciniata, B. malabarica*
Luteic acid	*Psidium guajave*
Lutein	*Taraxacum mongolicum*

Luteolin	*Coleus scutellarioide var. crispipilus, C. parvifolius, Crotalaria pallida, Humulus scandens, Ixeris tamagawaensis, Lonicera macrantha, L. japonica, L. japonica var. sempervillosa, L. shintenensis, Perilla frutescens var. crispa, P. ocymoides*
Luteolin 5-0-β-D-glucopyranoside	*Coleus scutellarioide var. crispipilus, C. parvifolius*
Luteolin 5-0-β-D-glucuronide	*Coleus scutellarioide var. crispipilus, C. parvifolius*
Luteolin 4′-β-D-glucoside	*Gnaphalium hypoleucum, G. adnatum*
Luteolin-7-glucoside	*Chryanthemum segetum, Glossogyne tenuifolia, Ixeris chinensis, Juncus effusus var. decipiens, Vitex rotundifolia*
Luteolin 7-methyl ether	*Coleus scutellarioide var. crispipilus, C. parvifolius*
Luteolin-7-rhamnoglucoside	*Lonicera japonica, L. japonica var. sempervillosa, L. shintenensis, L. macrantha*
Luteolin-7-β-D-glucopyranoside	*Lemmaphyllum microphyllum*
Luteolin-7-β-D-glucoside	*Agrimonia pilosa*
Luteolin-7-0-glucoside	*Ixeris tamagawaensis*
Luteolin-7-0-β-D-galactoside	*Lonicera kawakamii, L. confusa*
Luteolinidin	*Juncus effusus var. decipiens*
Luteorin	*Angelica keiskei*
Luzonoid A-G	*Viburnum plicatum var. formosanum, V. odoratissimum, V. awabuki, V. luzonicum*
Luzonoside A-D	*Viburnum plicatum var. formosanum, V. odoratissimum, V. awabuki, V. luzonicum*
Lycopene	*Stephania cephalantha, Viola inconspicua ssp. nagasakiensis, V. mandshurica*
Lycopodine	*Lycopodium salvinioides*
Lycopose	*Lycopus lucidus var. formosana*
Lycoramine	*Hippeastrum equestre, H. regina*
Lycorine	*Zephyranthes carinata, Z. candida*
Lycororine	*Hippeastrum equestre, H. regina*
Lysine	*Drynaria cordata, Hemerocallis longituba, Taraxacum mongolicum*
Lysopine	*Parthenocissus tricuspidata*
m-Phthalic acid	*Potentilla discolor*
Macelignan	*Leucas aspera, L. chinensis*
Maclurin	*Morus alba*
Macranthoin	*Siphonostegia chinensis*
Macrephyllic acid	*Podocarpus macrophyllus var. nakaii*
Macrophylline	*Senecio scandens, S. nemorensis*

Component	Source
Madolin-p	*Aristolochia cucurbitifolia, A. manshuriensis, A. heterophylla, A. kaempferi, A. kankanensis, A. shimadai*
Maesaquinone	*Maesa tenera, M. lanceolata, M. laxiflora*
Magnocurarine	*Litsea cubeba, Magnolia litiflora*
Magnoflorine	*Magnolia litiflora*
Magnolol	*Magnolia litiflora*
Makisterones	*Podocarpus macrophyllus var. nakaii*
Malic acid	*Bryophyllum pinnatum, Coriandrum sativum, Chaenomeles japonica, Eriobotrya japonica, Lactuca indica, Oxalis corymbosa*
Malloprenol	*Mallotus paniculatus, M. japonicus*
Mallorepine	*Mallotus repandus*
Mallotinin	*Mallotus repandus*
Malospicine	*Indigofera suffruticosa*
Maltase	*Solanum indicum*
Malvidin	*Impatiens balsamina, Medicago polymorpha*
Mangiferin	*Anemarrhena asphodeloides, Gentiana arisanensis*
Mannit	*Gardenia angusta var. kosyunensis, G. oblongifolia*
Mannitol	*Scoparia dulcis, Sonchus arvensis, S. oleraceus, Vernonia gratiosa*
Mannose	*Amorphophallus konjac, A. rivieri*
Markogenin	*Anemarrhena asphodeloides*
Marmesin	*Angelica hirsutiflora, Ruta graveolens*
Marmesinin	*Ruta graveolens*
Maslinic acid	*Psidium guajave*
Masperuloside	*Morinda citrifolia*
Massonianoside E	*Pinus massoniana*
Metatable acid	*Actinidia callosa var. formosana, A. chinensis*
Matricaria	*Erigeron canadensis*
Matricaria ester	*Conyza sumatrensis, C. blinii, Erigeron canadensis*
Matteucinin	*Rhododendron simsii*

Matteucinol	*Rhododendron simsii*
Maytanacine	*Maytenus emarginata, M. serrata*
Maytanbutine	*Maytenus emarginata, M. serrata*
Maytanprine	*Maytenus emarginata, M. serrata*
Maytansine	*Maytenus emarginata, M. serrata*
Maytansinol	*Maytenus diversifolia, Maytenus emarginata, M. serrata*
Maytanvaline	*Maytenus emarginata, M. serrata*
Mearnsitrin-3-*O*-α-L-rhamnoside	*Berchemia formosana, B. lineata*
Medicagenic acid	*Medicago polymorpha*
Megastigmane glucosides	*Breynia officinalis, Glochidion eriocarpum, G. acuminatum, G. zeylanicum*
Melessic acid	*Pericampylus formosanus*
Melialactone	*Melia azedarach*
Melianodiol	*Melia azedarach*
Melianol	*Melia azedarach*
Melianotriol	*Melia azedarach*
Melibiase	*Solanum indicum*
Melissic acid	*Ipomoea pes-caprae* ssp. *brasiliensis, Pericampylus trinervatus, P. glaucus*
Melissyl alcohol	*Clerodendrum japonicum, C. kaempferi*
Melodinus	*Melodinus angustifolius*
Melotoxin	*Cucumis melo* ssp. *melo*
Menisarine	*Cocculus sarmentosus*
Menisidine	*Stephania hispidula*
Menisin	*Stephania hispidula*
Menthenone	*Mentha canadensis*
Menthol	*Glechoma hederacea* var. *grandis, Luffa cylindrica, Mentha canadensis*
Menthone	*Luffa cylindrica, Mentha haplocalyx*
Menthyl acetate	*Mentha canadensis, M. haplocalyx*
Mesaconitine	*Aconitum bartletii, A. fukutomel, A. formosanum, A. kojimae, A. kojimae* var. *lassiocarpium, A. kojimae* var. *ramosum, A. yamamotoanum*
Mesityl oxide	*Cryptotaenia canadensis, C. japonica*

Component	Source
Meso-dihydro-guaiaretic acid	*Leucas aspera, L. chinensis*
Mesoinositol	*Clerodendrum trichotomum, C. trichotomum* var. *fargesii, Ficus pumila* var. *awkeotsang, Viscus alniformosanae, V. angulatum*
Metaphanine	*Stephania japonica, S. hispidula*
Methanethiol	*Asparagus cochinchinensis*
Methoxyl-camptothecin	*Camptotheca acuminata*
Methyl 3-*O*-β-glucopyranosyl-gallate	*Rosa taiwanensis, R. davurica*
Methyl 3,5-dicaffeoyl quinate	*Dichrocephala bicolor*
Methyl anthranilate	*Murraya paniculata*
Methyl benzoate	*Psidium guajave*
Methyl betulin	*Bischofia javanica*
Methyl chavicol	*Agastache rugosa, Foeniculum vulgare, Ocimum basilicum*
Methyl cinnamate	*Ocimum basilicum*
Methyl ester	*Clerodendrum calamitosum, C. cyrtophyllum, Ehretia acuminata, E. dicksonii, E. resinosa, Mussaenda parviflora, Rhynchosia volubilis, R. minima*
Methyl eugenol	*Michelia alba, Ocimum basilicum, Piper sarmentosum, P. sanctum*
Methyl heptenone	*Litsea cubeba*
Methyl isobutyl ketone	*Cryptotaenia canadensis, C. japonica*
Methyl isochondodendrine	*Stephania cephalantha*
Methyl laurate	*Osmanthus fragrans*
Methyl *p*-methoxycinnamate	*Duranta repens*
Methyl phenylethyl ether	*Pandanus odoratissimus* var. *sinensis*
Methyl salicylate	*Murraya paniculata*
Methyl-*trans*-2-decene-4,6,8-triynoate	*Ricinus communis*
Methylacetic acid	*Erigeron canadensis*
Methylendioxy-xanthone	*Polygala aureocauda*
Methylene-bishydroxy-coumarin	*Medicago polymorpha*

Methylethylacetic ester	*Michelia alba*
Methylheptenol	*Cymbopogon citratus, C. nardus, Zingiber officinale*
Methylkulonate	*Melia azedarach*
Methylmyristate	*Osmanthus fragrans*
Methylpalmintate	*Osmanthus fragrans*
Michelabine	*Michelia alba*
Michelenolide	*Eupatorium formosanum*
Mikanolide	*Mikania cordata*
Millewanin A-E	*Millettia taiwaniana*
Minerals	*Artocarpus heterophyllus, Eichhornia crassipes*
Minosine	*Mimosa pudica*
Miscathoside	*Miscanthus floridulus*
Misrathoside	*Miscanthus sinensis var. condensatus*
Molephantin	*Elephantopus mollis, E. scaber*
Molluscacides	*Canna flaccida, C. indica*
Momordicine	*Momordica charantia*
Momordin lib	*Galium echinocarpum*
Monoacetyl derivatives	*Diplazium megaphllum, D. subsinuatum*
Monoacyldigalactosyl glycerols	*Zingiber kawagoii, Z. rhizoma*
Monobornylphthalate	*Chryanthemum segetum*
Monocrotalines	*Crotalaria albida,C. sessiliflora*
Monoepoxylignan	*Pinus massoniana*
Monogalactocyl diacylglycerols	*Colocasia antiquorum var. illustris, C. esculenta*
Monoglycoside	*Impatiens balsamina, Quisqualis indica*
Monoterpenes	*Pluchea indica, Taraxacum officinale*
Monotropein	*Galium echinocarpum, Pyrola morrisonensis, P. japonica*
Moridon	*Morinda citrifolia*
Morin	*Morus alba*
Morindadiol	*Morinda citrifolia*

Component	Source
Morindanigrin	*Morinda citrifolia*
Morindin	*Morinda citrifolia, Morinda umbellata*
Morolic acid	*Adina pilulifera, A. racemose*
Morphine	*Desmodium laxiflorum, D. gangeticum, D. multiflorum, Stephania cephalantha*
Mucilage	*Aralia taiwaniana, A. chinensis, Bletilla formosana, Bombax malabarica, Liriope spicata, Pericampylus formosanus*
Mukorosside	*Sapindus mukorossi*
Mulberrin	*Morus alba*
Mulberrochromene	*Morus alba*
Munduserone	*Derris elliptica*
Munjistin	*Morinda umbellata, Rubia akane, R. lanceolata, R. linii*
Musarin	*Musa sapientum, M. formosana, M. basjoo var. formosana*
Muslinic acid	*Elaeagnus oldhamii, E. thunbergii, E. wilsonii*
Mussaenoside	*Mussaenda parviflora*
Myoinositol	*Myrica adenophora, Orthosiphon aristatus, O. stamineus, Vaccinium emarginatum*
Myrcene	*Artemisia lactiflora, A. princeps, Cymbopogon nardus, Oenanthe javanica, Zanthoxylum ailanthoides*
Myrceus	*Cymbopogon citratus*
Myrecene	*Medicago polymorpha*
Myricetin	*Myrica adenophora*
Myricetin 5-methyl ether	*Rhododendron simsii*
Myricetin	*Berchemia formosana, B. lineata, Diospyros eriantha, Myrica adenophora, M. rubra, Viola inconspicua* ssp. *nagasakiensis, V. mandshurica*
Myricyl	*Spilanthes acmella, S. acmella var. oleracea*
Myrieitrin	*Pouteria obovata*
Myristic acid	*Blumea balsamifera var. microcephala, Coix lacryma-jobi, Ipomoea pes-caprae ssp. brasiliensis, Jatropha curcas, Myristica cagayanensis, M. fragrans, Sesamum indicum, Taraxacum mongolicum*
Myristicin	*Myristica cagayanensis, M. fragrans*

Myrtillin	*Viola inconspicua* ssp. *nagasakiensis, V. mandshurica*
N,N-Dimethyltryptamine	*Phyllodium pulchellum*
N,N-Dimethyltryptamine oxide	*Phyllodium pulchellum*
n-1-Triacontanol	*Jatropha curcas*
n-Acetylgalactosamine-specific lectin	*Euphorbia heterophyllacamphol*
N-Acetylglucosamine	*Urtica thunbergiana, U. dioica*
n-Acridone	*Atalantia buxifolia*
n-Butyl-2-ethyl-butylphthalate	*Oenanthe javanica*
n-Butylidenephthalide	*Angelica acutiloba, A. citriodor*
N-Desmethylchelerythrine	*Zanthoxylum nitidum, Z. integrifoliolum*
n-Heptacosane	*Viscus multinerve*
N-Methyl platydesmin	*Ruta graveolens*
N-Methylcoclaurine	*Nelumbo nucifera*
N-Methylisococlaurine	*Nelumbo nucifera*
n-Methylmorpholine	*Cassia occidentalis, C. torosa*
n-Nonacosane	*Viscus multinerve*
n-Nonyl aldehyde	*Tagetes erecta*
n-Octacosane	*Viscus multinerve*
n-Octacosanoic acid	*Viscus multinerve*
n-Octacotanol	*Viscus multinerve*
n-Pentacosane	*Clerodendrum japonicum, C. kaempferi*
n-Pentacosanoic acid	*Viscus multinerve*
n-Tehacosanol	*Viscus multinerve*
N-*trans*-Feruloyltyramine	*Sida acuta*
n-Triacontanol	*Conyza sumatrensis, C. blinii*
n-Tricosanoic acid	*Viscus multinerve*
n-Valero-phenones-carboxylic acid	*Angelica acutiloba, A. citriodor*
Nagilactosides C-E	*Podocarpus nagi*
Nandazurine	*Nandina domestica*
Nandinine	*Nandina domestica*

Component	Source
Naphthalene glycoside	*Diospyros angustifolia*
Naphthalenes	*Farfugium japonicum, Piper arboricola*
Naphthaquinone	*Bombax malabarica, Plumbago zeylanica, Ventilago leiocarpa*
Naphthopyrones	*Cassia tora*
Naphthoquinone esters	*Rhinacanthus nasutus*
Naphthoquinones	*Ventilago leiocarpa*
Narcissin	*Berchemia formosana, B. lineata*
Narcotic compounds	*Passiflora suberosa*
Narcotic alkaloid	*Pericampylus formosanus*
Naringenin	*Potentilla discolor, Viscus multinerve*
Narirutin	*Clinopodium laxiflorum*
Nasunin	*Solanum lyratum*
Naucleoside	*Adina pilulifera, A. racemose*
Nectandrin B	*Leucas aspera, L. chinensis*
Negundoside	*Vitex negundo*
Neo-allicin	*Allium sativum, A. thunbergii, A. tuberosum*
Neo-lignans	*Magnolia liliflora*
Neoandrographolide	*Andrographis paniculata*
Neocarthamin	*Carthamus tinctorius*
Neocryuptomerin	*Podocarpus macrophyllus var. nakaii*
Neocycasin A-G	*Cycas revoluta*
Neogitogenin	*Anemarrhena asphodeloides*
Neoglucobrassicin	*Clerodendrum cyrtophyllum*
Neomatabiol	*Actinidia callosa var. formosana, A. chinensis*
Neoxanthin	*Taraxacum mongolicum*
Nepodin	*Rumex japonicus*
Nerinine	*Zephyranthes candida*
Nerioderin	*Nerium indicum*

Neriodin	*Nerium indicum*
Neriodorin	*Nerium indicum*
Nerol	*Cymbopogon citratus, C. nardus, Osmanthus fragrans*
Nerolidol	*Hedyotis corymbosa*
Nervisterol	*Nervilia taiwaniana, N. purpurea*
Nesperidin	*Clinopodium laxiflorum*
Niacin	*Achyranthes aspera var. indica, A. aspera var. rubro-fusca, Arachis hypogea, A. agallocha, Basella alba, Boehmeria densiflora, Hibiscus rosa-sinensis, Oxalis corniculata, Petasites japonicus*
Nicoteine	*Nicotiana tabacum*
Nicotelline	*Nicotiana tabacum*
Nicotiflorin	*Ricinus communis*
Nicotimine	*Nicotiana tabacum*
Nicotine	*Eclipta alba, E. prostrate, Lycopodium salvinioides, Nicotiana tabacum*
Nicotinic acid	*Celosia argentea, Lycopersicon esculentum, Solanum nigraum, S. undatum*
Nilgirine	*Crotalaria pallida*
Nimbin	*Melia azedarach*
Nimbolins	*Melia azedarach*
Niranthin	*Phyllanthus urinaria*
Nirtetralin	*Phyllanthus urinaria*
Nishindine	*Vitex negundo*
Nitidine	*Zanthoxylum avicennae, Z. nitidum, Z. integrifoliolum*
Nitrates	*Drynaria cordata*
Nitre	*Cayratia japonica, Mollugo pentaphylla*
Nitric acid	*Cayratia japonica*
Nitric oxide	*Desmodium laxiflorum, D. gangeticum, D. multiflorum*
Nitrile	*Tropaeolum majus*
Nobiletin	*Citrus maxima, C. sinensis var. sekken, C. tangerina*
Nodifloretin	*Phyla nodiflora*
Nodifloridin A	*Phyla nodiflora*
Nodifloridin B	*Phyla nodiflora*

Component	Source
Nonacosan-10-ol	*Dicranopteris linearis*
Nonacosan-10-one	*Dicranopteris linearis*
Nonacosane	*Chenopodium album, Dicranopteris linearis*
Nonan-2--ol	*Ruta graveolens*
Nonan-2-one	*Ruta graveolens*
Nonanal	*Coriandrum sativum*
Nonricinolein	*Ricinus communis*
Nonylaldehyde	*Zingiber officinale*
Nootkatin	*Jasminum hemsleyi*
Nordamnacanthal	*Morinda citrifolia*
Nordrenaline	*Portulaca oleracea*
Norepinephrine	*Musa sapientum, M. formosana, M. basjoo var. formosana*
Noreugenin	*Adina pilulifera, A. racemose*
Norkurarinone	*Sophora flavescens*
Normenisarine	*Cocculus orbiculata*
Norsecurinine	*Securinega virosa*
Norswertianolin	*Gentiana atkinsonii, G. campestris, G. flavo-maculata*
Nuciferine	*Nelumbo nucifera*
Nummularine H	*Paliurus ramosissimus*
Nuzhenide	*Ligustrum lucidum, L. pricei*
o-Benzylbezoic acid	*Conyza sumatrensis, C. blinii*
o-Nornuciferine	*Nelumbo nucifera*
Obacunone	*Phellodendron wilsonii, P. amurense, P. chinensis*
Obtusifolin	*Cassia tora, Hemerocallis fulva*
Obtusin	*Cassia tora*
Ocimene	*Ocimum basilicum, O. gratissimum*
Octacosane	*Ficus carica, F. benjamina*
Octadecanoic acid	*Sterculia nobilis, S. lychnophora*

Octadecatetraenoic acid	*Stellaria media*
Octanol	*Agastache rugosa*
Octopinic acid	*Parthenocissus tricuspidata*
Odoratin	*Eupatorium tashiroi*
Oleanalic acid	*Aspidixia articulata, A. liquidambaricala*
Oleanane flavonoids	*Adiantum capillus-veneris, A. flabellulatum*
Oleanane saponins	*Sanicula petagniodes, S. elata*
Oleanane-type triterpenoid saponins	*Hydrocotyle sibthorpioides*
Oleanane-type triterpenes	*Ludwigia octovalvis*
Oleandrin	*Nerium indicum*
Oleandrose	*Nerium indicum*
Oleanen dervatives	*Asparagus cochinchinensis*
Oleanolic acid	*Achyranthes aspera var. indica, A. aspera var. rubro-fusca, A. japonica, Aralia chinensis, Aspidixia articulata, A. liquidambaricala, Clinopodium laxiflorum, Diospyros khaki, Duranta repens, Gentiana arisanensis, Gossampinus malabarica, Glossogyne tenuifolia, Ilex asprella, Kalimeris indica, Ligustrum lucidum, L. pricei, Lysimachia ardisloides, L. capillipes, L. davurica, Prunella vulgaris, Pterocypsela indica, Randia spinoa, Viscus multinerve, V. alniformosanae, V. angulatum*
Oleic acid	*Angelica acutiloba, A. citriodor, Bixa orellana, Brucea javanica, Bupleurum kaoi, B. chinense, Chenopodium album, Coix lacryma-jobi, Corchorus capsularis, C. olitorius, Jatropha curcas, Myristica cagayanensis, M. fragrans, Ricinus communis*
Olein	*Ricinus communis, Sesamum indicum*
Oleyl alcohol	*Chenopodium album*
Oligoglycosidic compounds	*Kyllinga brevifolia*
Olitoriside	*Corchorus capsularis, C. olitorius*
Ononitol	*Medicago polymorpha*
Ophiopogenins	*Ophiopogon japonicus*
Opifriedelanol	*Mallotus apelta*
Organic acid	*Sansevieria trifasciata*
Organoids	*Ceratopteris thalictroides, Oxalis corymbosa*
Orientin	*Viola inconspicua ssp. nagasakiensis, V. mandshurica, Vitex negundo, V. rotundifolia*

Component	Source
Orienting 2″-0-xyloside	*Setaria palmifolia, S. viridis*
Orienting-7-0-glucoside	*Uraria crinita, U. lagopodioides*
Orlean	*Bixa orellana*
Orthomethylcoumaric aldehyde	*Cinnamomum cassia*
Orthosiphoni	*Orthosiphon aristatus, O. stamineus*
Osmane	*Osmanthus fragrans*
Osthenol-7-*o*-β-gentiobioside	*Glehnia littoralis*
Osthol	*Angelica hirsutiflora*
Osthole	*Murraya paniculata*
Oxalate	*Oxalis corniculata, O. corymbosa*
Oxalic acid	*Coriandrum sativum, Juncus effusus var. decipiens, Lactuca indica, Oxalis corniculata, Taraxacum officinale*
Oxoushinsunine	*Michelia alba*
Oxycanthine	*Thalictrum fauriei*
Oxychelerythrin	*Zanthoxylum nitidum, Z. integrifoliolum*
Oxycoccicyanin	*Viola inconspicua ssp. nagasakiensis, V. mandshurica*
Oxynitidine	*Zanthoxylum avicennae, Z. nitidum, Z. integrifoliolum*
Oxypurpureine	*Thalictrum fauriei*
Oxyristic acid	*Pieris hieracloides, P. taiwanensis, P. formosa*
p-Coumaric acid	*Allium cepa, Bryophyllum pinnatum*
p-Coumaroyl iridoids	*Viburnum plicatum var. formosanum, V. odoratissimum, V. awabuki, V. luzonicum*
p-Cymene	*Angelica acutiloba, A. citriodor, Chenopodium ambrosioides, Glechoma hederacea* var. *grandis, Lantana camara, Mosla punctulata, Piper betle, Piper sarmentosum, P. sanctum, Ruta graveolens*
p-Hydroxy-cinmamic acid	*Setaria palmifolia, S. viridis, Vaccinium emarginatum*
p-Hydroxyacetophenone	*Gynura formosana, G. elliptica*
p-Methoxybenzylacetone	*Aquilaria sinensis, A. sibebsus*
p-Vinylguaiacol	*Hedyotis diffusa*
p-Vinylphenol	*Hedyotis diffusa*

Paederoside	*Paederia cavaleriei, P. scandens*
Paeonin	*Cynanchum paniculatum*
Paeonol	*Cynanchum paniculatum*
Paliurines A-C	*Paliurus ramosissimus*
Paliurines F	*Paliurus ramosissimus*
Paliurines G-I	*Paliurus ramosissimus*
Palmatic acid	*Pericampylus formosanus, P. trinervatus, P. glaucus*
Palmatine	*Coptis chinensis, Phellodendron wilsonii, P. amurense, P. chinensis, Thalictrum fauriei*
Palmatate	*Sambucus chinensis, S. formosana*
Palmitic acid	*Angelica acutiloba, A. citriodor, Ajuga bracteosa, A. decumbens, A. gray, A. pygmaea, Balanophora spicata, Blumea balsamifera var. microcephala, Bupleurum kaoi, B. chinense, Chenopodium album, Cibotium barometz, Coix lacryma-jobi, Corchorus capsularis, C. olitorius, Helianthus annuus, Jatropha curcas, Lysimachia ardisloides, L. capillipes, L. davurica, Melia azedarach, Ricinus communis, Sonchus arvensis, S. oleraceus, Taraxacum mongolicum, Viscus multinerve*
Palmitine	*Bixa orellana, Sesamum indicum*
Palmitoleic	*Jatropha curcas*
Palustrine	*Equisetum ramosissimum*
Panasterone A	*Blechnum orientale*
Panaxadiol	*Gynostemma pentaphyllum*
Panaxatriol	*Gynostemma pentaphyllum*
Pandanamine	*Pandanus amaryllijolius, P. pygmaeus*
Pandanin	*Pandanus amaryllijolius, P. pygmaeus*
Pangelin	*Ruta graveolens*
Paniculatin	*Murraya paniculata*
Paniculatincoumurrayin	*Murraya paniculata*
Papain	*Ficus carica, F. benjamina*
Papaverine	*Stephania cephalantha*
Papyriflavonol A	*Brousonetia papyrifera*
Parietin	*Polygonum multiflorum var. hypoleucum*
Pariphyllin	*Paris polyphylla*

Component	Source
Parthenolide	*Eupatorium formosanum*
Passifloricins	*Passiflora foetida* var. *hispida*
Patchouli acid	*Pogostemon amboinicus*
Patchouli alcohol	*Pogostemon cablin*
Patchouli oil	*Pogostemon amboinicus*
Patchoulipyridine	*Pogostemon amboinicus*
Pectic polysaccharide	*Diospyros kaki*
Pectic acid	*Centella asiatica*
Pectic compound	*Lactuca indica*
Pectic polysaccharide	*Silene morii, S. vulgaris*
Pectins	*Musa insularimontana, M. paradisiaca, Myristica cagayanensis, M. fragrans, Plumeria rubra cv. acutifolia, Taraxacum formosanum*
Pectolinarigenin	*Clerodendrum inerme, Duranta repens*
Pectolinarin	*Cirsium japonicum, C. japonicum* var. *australe*
Peinene	*Myristica cagayanensis, M. fragrans*
Pelargonaldehyde	*Osmanthus fragrans*
Pelargonidin	*Impatiens balsamina*
Pelargonidin-3-rhamnosylglucoside	*Chloranthus oldham, Chlorophytum comosum*
Pellitorine	*Piper kadsura, P. kawakamii*
Peltutin	*Dysosma pleiantha*
Penformosin	*Peucedanum formosanum*
Pentacosyl	*Ixeris tamagawaensis*
Pentacyclic triterpenoids	*Mallotus apelta*
Pentahydroxy flavone	*Hyphea kaoi*
Pentosan	*Sesamum indicum*
Pentose	*Miscanthus sinensis* var. *condensatus, Ocimum gratissimum*
Pentoson	*Aleurites fordii, A. moluccana, A. montana*
Pentse	*Miscanthus floridulus*

Pepsin	*Ceratopteris thalictroides*
Pepsin A	*Ficus carica, F. benjamina*
Peptides	*Lycium chinense*
Peraksine	*Rauvolfia verticillata*
Pericalline	*Catharanthus rosens*
Perilladehyde	*Perilla frutescens, P. frutescens var. crispa, P. ocymoides*
Periodic acid	*Arenga engleri, A. pinnata, A. saccharifera*
Perividine	*Catharanthus rosens*
Peroxide	*Rhus javanica*
Peruvosides	*Thevetia peruviana*
Pervine	*Catharanthus rosens*
Petasiformin A	*Petasites formosanus*
Petasiphyll A	*Petasites formosanus*
Petroselenic acid	*Glehnia littoralis*
Petroselic acid	*Cryptotaenia canadensis, C. japonica*
Petroselidinic acid	*Glehnia littoralis*
Petunidin	*Medicago polymorpha*
Phantomolin	*Elephantopus mollis, E. scaber*
Phellandrene	*Coriandrum sativum, Cinnamomum cassia, Curcuma longa, Mosla punctulata, Zingiber officinale*
Phellodendrine	*Phellodendron wilsonii, P. amurense, P. chinensis*
Phellopterin	*Angelica hirsutiflora*
Phenals	*Stephania hispidula*
Phenanthraquinones	*Dendrobium moniliforme*
Phenanthrene	*Bletilla striata*
Phenanthrene derivatives	*Aristolochia cucurbitifolia, A. manshuriensis, A. heterophylla, A. kaempferi, A. kankanensis, A. shimadai*
Phenanthruindolizidine alkaloids	*Tylophora lanyuensis, T. atrofolliculta*
Phennolics	*Teucrium viscidum*

Component	Source
Phenolic	*Berchemia formosana, B. lineata, Clerodendrum philippinum, Cratoxylon ligustrinum, Desmodium capitatum, Diospyros eriantha, Elaeagnus morrisonensis, E. angustifolia, Eupatorium clematideum, Flemingia prostrata, Ilex rotunda, Ixora chinensis, Lophatherum gracile, Melastoma dodecandrum, M. septemnervium, Millettia nitida, Mimosa pudica, Mussaenda parviflora, Pandanus odoratissimus var. sinensis, Phyllanthus urinaria, Polygonum perfoliatum, Prunus persica, Psychotria rubra, Pteris ensiformis, P. multifida, Rotala rotundifolia, Rhodomyrtus tomentose, Ribes formosanum, R. nigrum, Rubus formosensis, Sarcandra glabra, Scutellaria indica, Stachytarpheta jamaicensis, Vitex negundo*
Phenolic compounds	*Breynia accrescens, B. fruticosa, Caryopteris incana, Cibotium cumingii, Cleistocalyx operculatus, Cordyline fruticosa, Croton lachnocarpus, Desmodium caudatum, Echinochloa colonum, Ficus microcarpa, Glochidion puberum, Hedyotis corymbosa, Helicteres angustifolia, Emilia sonchifolia, E. sonchifolia var. javanica, Epimeredi indica, Eucalyptus robusta, Evolvulus alsinoides, Mesona chinensis, Smilax bracteata, S. china, Urtica thunbergiana, U. dioica, Youngia japonica*
Phenolic glycosides	*Breynia officinalis, Diospyros angustifolia, Lilium formosanum, L. speciosum, Pyrola morrisonensis, P. japonica, Viscus alniformosanae, V. angulatum*
Phenols	*Graptopetalum paraguayense, Flemingia macrophylla, Ludwigia octovalvis, Urena lobata, U. procumbens, Zornia diphylla*
Phenoxy benzamine	*Vandellia crustacea, V. cordifolia*
Phenyl ethyl acetate	*Psidium guajave*
Phenyl ethyl alcohol pentosans	*Eriobotrya japonica*
Phenylalanine	*Medicago polymorpha*
Phenylethy acetate	*Pandanus odoratissimus var. sinensis*
Phenylethyl alcohol	*Plumeria rubra cv. acutifolia*
Phenylpropanoid	*Hemiphragma heterophyllum var. dentatum*
Phenylpropanoid esters	*Hibiscus taiwanensis*
Phenylpropenolyl sulfonic acid	*Petasites formosanus*
Phenylpropyl alcohol	*Cinnamomum cassia*
Phenyheptatriyne	*Bidens pilosa var. minor, B. racemosa*
Pheophorbide related compounds	*Clerodendrum calamitosum, C. cyrtophyllum*
Phetidine	*Thalictrum fauriei*

Phloroglucinal	*Mallotus tiliaefolius*
Phorbol	*Croton tiglium*
Phorbol diester	*Croton tiglium*
Phosphatidic acid	*Vigna radiata, V. angularis, V. umbellata*
Phosphatidylcholine	*Vigna radiata, V. angularis, V. umbellata*
Phosphatidylethanolamine	*Vigna radiata, V. angularis, V. umbellata*
Phospholipids	*Coix lacryma-jobi, Vigna radiata, V. angularis, V. umbellata*
Phthalic acid ester	*Pandanus odoratissimus var. sinensis*
Phyllanthine	*Phyllanthus urinaria, Securinega suffruticosa*
Phylteralin	*Phyllanthus urinaria*
Physalin	*Physalis angulata*
Physcie	*Damnacanthus indicus*
Physcim-1-gluco-rhamnodies	*Phyllodium pulchellum*
Physcion	*Cassia tora*
Phytic acid	*Vigna radiata, V. angularis, V. umbellata*
Phytin	*Sesamum indicum*
Phytoecdysteroids	*Sida rhombifolia*
Phytofluene	*Viola inconspicua ssp. nagasakiensis, V. mandshurica*
Phytogenic	*Typhonium divaricatum*
Phytolaccatoxin	*Pieris hieracloides, P. taiwanensis, P. formosa*
Phytolacine	*Pieris hieracloides, P. taiwanensis, P. formosa*
Phytosterindigitonid	*Hoya carnosa*
Phytosterines	*Taraxacum officinale*
Phytosteroids	*Drynaria cordata*
Phytosterol	*Aleurites fordii, A. moluccana, A. montana, Bixa orellana, Duchesnea indica, Gnaphalium affine, G. luteoalbum ssp. affine, Hypoxis aurea*
Picein	*Clerodendrum japonicum, C. kaempferi*
Picrolonic acid	*Paris polyphylla*
Picryldydrazyl	*Hypericum chinense, H. patulum, Polygonum plebeium, P. paleaceum*
Pierisformosides G-I	*Pieris hieracloides, P. taiwanensis, P. formosa*

Component	Source
Pimpinellin	*Toddalia asiatica*
Pinene	*Chamaecyparis formosensis, C. obtusa var. filicoides, C. obtusa var. formosana*
Pinene acid	*Cinnamomum camphora, Citrus medica var. gaoganensis, Laungusa galanga, Luffa cylindrica, Mentha canadensis, Podocarpus macrophyllus var. nakaii, Vitex negundo, V. rotundifolia*
Pinipicrin	*Biota orientalis*
Pinitol	*Bougainvillea spectabilis, Lespedeza cuneata*
Pinocarveol	*Cinnamomum camphora*
Pinoresinol-di-β-D-glucoside	*Eucommia ulmoides*
Pinoresinol-di-0-β-D-glucopyranside	*Alyxia insularis, A. sinensis*
Pipataline	*Piper kadsura, P. kawakamii*
Piperaidine	*Piper kadsura, P. kawakamii*
Piperamine	*Piper nigrum*
Piperanine	*Piper arboricola, P. kadsura, P. kawakamii*
Piperic acid	*Piper arboricola*
Piperidine alkaloids	*Lobelia nummularia, L. laxiflora, Microcos paniculata*
Piperine	*Piper arboricola, P. kadsura, P. kawakamii, P. nigrum*
Piperitone	*Cymbopogon nardus, Piper arboricola, Zanthoxylum ailanthoides*
Piperlonguminine	*Piper kadsura, P. kawakamii*
Piperolactam A	*Piper sarmentosum, P. sanctum*
Piperonal	*Piper arboricola, P. nigrum*
Pirolatin	*Pyrola morrisonensis, P. japonica*
Plant insulin	*Drynaria cordata*
Plantagin	*Plantago asiatica, P. major*
Plantasan	*Plantago asiatica, P. major*
Plantenolic acid	*Plantago asiatica, P. major*
Plastoguinoae	*Polygonum cuspidatum*
Placodosides	*Platycodon grandiflorum*
Platycodigenic acid	*Platycodon grandiflorum*
Platycodigenin	*Platycodon grandiflorum*

Platycodonin	*Platycodon grandiflorum*
Platyconin	*Platycodon grandiflorum*
Platynecic acid	*Crotalaria sessiliflora*
Plectranthin	*Plectranthus amboinicus, Rabdosia lasiocarpus*
Plumbagin	*Plumbago zeylanica*
Plumieric acid	*Plumeria rubra* cv. *acutifolia*
Plumieride	*Plumeria rubra* cv. *acutifolia*
Podocarpane-type trinorditerpenes	*Taiwania cryptomerioides*
Podocarpene	*Podocarpus macrophyllus* var. *nakaii*
Podophyllotoxin	*Dysosma pleiantha*
Podototarin	*Podocarpus macrophyllus* var. *nakaii*
Pogostol	*Pogostemon amboinicus*
Pogostone	*Pogostemon amboinicus, P. cablin*
Poliumonside	*Dicliptera chinensis, D. riparia*
Polyacetylenes	*Bidens pilosa* var. *minor, B. racemosa*
Polyacetylene glycosides	*Pratia nummularia*
Polycyclic quaternary alkaloids	*Daphniphyllum glaucescens* spp. *oldhamii*
Polycyclic compounds	*Pometia pinnata*
Polydain	*Polygonum cuspidatum*
Polygalacic acid	*Platycodon grandiflorum, Solidago virgo-aurea*
Polyine	*Glehnia littoralis*
Polyketides	*Passiflora foetida* var. *hispida*
Polypeptides	*Spinacia oleracea*
Polyphenolic compounds	*Mussaenda pubescens, Pouteria obovata*
Polyphenols	*Melissa officinalis, Pouzolzia elegans, P. pentandra, P. zeylanica, Camellia oleifera, C. sinensis*
Polyphyllin D	*Paris polyphylla*
Polysaccharides	*Achyranthes bidentata, Allium cepa, A. sativum, A. thunbergii, A. tuberosum, Asparagus cochinchinensis, Arenga engleri, A. pinnata, A. saccharifera, Blumea riparia* var. *megacephala, Cinnamomum insulari-montanum, C. kotoense, C. micranthum, Coix lacryma-jobi, Dicranopteris dichotoma, Glehnia littoralis, Gnaphalium affine, G. luteoalbum* ssp. *affine, Lycium chinense, Ophiopogon japonicus, Polygala aureocauda*

Component	Source
Pomolic acid	*Debregeasia edulis, D. salicifolia*
Pomolic acid methyl ester	*Debregeasia edulis, D. salicifolia*
Ponasterone	*Podocarpus macrophyllus var. nakaii*
Populnin	*Equisetum ramosissimum*
Portulal	*Portulaca grandiflora, P. pilosa*
Potassium	*Achyranthes longifolia, A. ogotai, Cayratia japonica*
Potassium compounds	*Drynaria cordata*
Potassium nitrate	*Drynaria cordata, Gossampinus malabarica*
Potassium oxide	*Desmodium triuetrum*
Potassium salt	*Portulaca oleracea*
Potassium quisqualate	*Quisqualis indica*
Precatorine	*Abrus precatorius*
Prenyl chalcone	*Hypericum geminiflorum*
Prenyl flavanones	*Macaranga tanarius*
Prinsepiol	*Prinsepia scandens*
Pristimesin	*Celastrus kusanoi, C. hypoleucus*
Proanthocyanidin A_1	*Ecdysanthera rosea, E. utilis*
Proanthocyanidin A_2	*Ecdysanthera rosea, E. utilis*
Proanthocyanidins	*Ecdysanthera rosea, E. utilis*
Procumbenetin	*Tridax procumbens*
Procurcumenol	*Curcuma domestica, C. zedoaria*
Procyanidin B_2	*Ecdysanthera rosea, E. utilis, Melastoma candidum*
Prodelphinidin B-2	*Myrica rubra*
Prodocarpus flavones	*Podocarpus macrophyllus var. nakaii*
Prometaphanine	*Stephania hispidula, S. japonica*
Pronase	*Ceratopteris thalictroides*
Pronuciferine	*Nelumbo nucifera*
Prosapogenin	*Platycodon grandiflorum*

Protacatechuic acid	*Allium cepa*
Protease	*Brousonetia papyrifera, Plumbago zeylanica*
Protein	*Achyranthes aspera var. indica, A. aspera var. rubro-fusca, Artocarpus heterophyllus, Blumea riparia var. megacephala, Bombax malabarica, Cassia mimosoides, Celosia cristata, Coix lacryma-jobi, Colocasia antiquorum var. illustris, C. esculenta, Euryale ferox, E. chinese, Lemmaphyllum microphyllum, Lycopersicon esculentum, Pyracantha fortuneana, Rubus formosensis*
Protein arachine	*Arachis hypogea, A. agallocha*
Protoisorubosides	*Allium sativum, A. thunbergii, A. tuberosum*
Protoanemomin	*Clematis chinensis, C. grata, Clematis gouriana ssp. lishanensis, Ranunculus japonicus, R. sceleratus*
Protocatechuic acid	*Aralia taiwaniana, A. chinensis, Cirsium albescens, Potentilla discolor, P. tugitakensis, Taxillus matsudai, T. levinei*
Protohypericin	*Hypericum japonicum*
Protopine	*Thalictrum fauriei*
Protostephanine	*Stephania hispidula, S. japonica, Stemona tuberosa*
Protoveratrine	*Hemerocallis fulva, Veratrum formosanum*
Prunellin	*Prunella vulgaris*
Prunia	*Miscanthus sinensis var. condensatus*
Prunioside A	*Spiraea prunifolia var. pseudoprunifolia*
Pseudoaconitine	*Aconitum bartletii, A. fukutomel, A. formosanum, A. kojimae, A. kojimae var. lassiocarpium, A. kojimae var. ramosum, A. yamamotoanum*
Pseudojervine	*Hemerocallis fulva*
Pseudoprotopine	*Zanthoxylum nitidum, Z. integrifoliolum*
Pseudopurpurin	*Rubia akane, R. lanceolata, R. linii*
Psidiolic acid	*Psidium guajave*
Psoralen	*Ficus carica, F. benjamina, Glehnia littoralis, Ruta graveolens*
Psoralene	*Angelica hirsutiflora, A. keiskei*
Psychopharmacological properties	*Leucas mollissima var. chinensis, L. lavandulaefolia*
Ptaquiloside	*Hypolepis tenuifolia*
Pterocarpanoid	*Crotalaria pallida*
Puerarin	*Pueraria lobata, P. montana*
Puerarin-xyloside	*Pueraria lobata, P. montana*

Component	Source
Punicalagin	*Terminalia catappa*
Punicalin	*Terminalia catappa*
Punicic acid	*Diplocyclos palmatus*
Puriniums	*Heterostemma brownii*
Purpureal glycosides	*Digitalis purpurea*
Purpurin	*Rubia akane, R. lanceolata, R. linii*
Purpuroxanthin	*Morinda umbellata*
Putranjivadione	*Euonymus echinatus, E. laxiflorus, E. chinensis*
Putranjivain A	*Euphorbia jolkini*
Pyranoanthocyanins	*Citrus maxima, C. sinensis var. sekken*
Pyridine	*Murdannia keisak, M. loriformis*
Pyrimidines	*Heterostemma brownii*
Pyrocatechic tannin	*Blumea balsamifera var. microcephala*
Pyrocatechol	*Portulaca oleracea*
Pyrocatechol equivalent	*Urtica thunbergiana, U. dioica*
Pyrogallol	*Ranunculus japonicus, R. sceleratus*
Pyrrolidine	*Pandanus amaryllifolius, P. pygmaeus*
Pyrrolidine alkaloids	*Broussonetia kazinoki, Lobelia chinensis*
Pyrrolizidine alkaloids	*Liparis keitaoensis, Symphytum officinale*
Pyrropetioside	*Pyrrosia adnascens, P. petiolosa*
Quaternary alkaloids	*Phellodendron wilsonii, P. amurense, P. chinensis*
Quercet	*Potentilla discolor*
Quercetin	*Allium cepa, Astilbe longicarpa, Begonica fenicis, B. laciniata, B. malabarica, Biota orientalis, Corchorus aestuand, Coriandrum sativum, Euchresta formosana, Euphoria longana, Geranium nepalense var. thunbergii, G. suzukii, Hibiscus mutabilis, H. rosa-sinensis, Hypericum japonicum, Impatiens balsamina, Kyllinga brevifolia, Lonicera kawakamii, L. confusa, Pemphis acidula, Phyllanthus multiflorus, P. emblica, Plumeria rubra cv. acutifolia, Potentilla discolor, Psidium guajave, Ricinus communis, Scurrula loniceritolius, S. ritozonensis, S. liquidambaricolus, S. ferruginea, Trichosanthes homophylla, T. dioica*

Quercetin-3-galactoside	*Rumex acetosa*
Quercetin-3-*O*-(6″-galloyl)-β-D-glucoside	*Taxillus matsudai, T. levinei*
Quercetin-3-*O*-β-D-glucuronide	*Taxillus matsudai, T. levinei*
Quercetin-3-*O*-rutinoside	*Dichrocephala bicolor*
Quercetin-4-glucoside	*Hibiscus mutabilis*
Quercetin-monomethylether	*Tamarix juniperina, T. chinensis*
Quercetol	*Vaccinium emarginatum*
Quercetrin	*Euphoria longana*
Quercimeritrin	*Hibiscus mutabilis*
Quercitol	*Euphorbia hirta*
Quercitrin	*Dicranopteris linearis, Euphorbia hirta, Houttuynia cordata, Hypericum japonicum, Loropetalum chinense, Saururus chinensis, Scurrula loniceritolus, S. ritozonensis, S. liquidambaricolus, S. ferruginea*
Queritin	*Ficus carica, F. benjamina*
Quillaic acid	*Thladiantha nudiflora*
Quindolinone	*Sida acuta*
Quine	*Stephania cephalantha*
Quinic acid	*Solidago virgo-aurea*
Quinochalone	*Carthamus tinctorius*
Quinoline	*Murdannia keisak, M. loriformis*
Quinoline alkaloids	*Melicope semecarpifolia*
Quinonoid terpenoid	*Gynura japonica var. flava*
Quinoric acid	*Adina pilulifera, A. racemose*
r-(+)-Deoxytylophorinidine	*Tylophora lanyuensis, T. atrofolliculta*
r-Decanolactone	*Osmanthus fragrans*
r-Glutamyl-valyl-glutamic acid	*Juncus effusus var. decipiens*
r-Guanidinobutyric acid	*Trichosanthes cucumeroides*
r-Hydroxyglutamic acid	*Hemerocallis longituba*
r-Linolenic acid	*Stellaria media*
Radicamines A–B	*Lobelia chinensis*
Raffinose	*Lycopus lucidus var. formosana*

Component	Source
Ranunculin	*Ranunculus japonicus, R. sceleratus*
Rapanone	*Ardisia sieboldii*
Raunescine	*Rauvolfia verticillata*
Rauwelline	*Rauvolfia verticillata*
Rauwolfia A	*Rauvolfia verticillata*
Rebaudiosides	*Stevia rebaudiana*
Rellosimine	*Rauvolfia verticillata*
Relroncine	*Crotalaria sessiliflora*
Repandusinic acids	*Mallotus repandus*
Repandusinin	*Mallotus repandus*
Reptoside	*Eucommia ulmoides*
Rerpinenol	*Zanthoxylum ailanthoides*
Reserpine	*Rauvolfia verticillata*
Resins	*Artocarpus heterophyllus, Bixa orellana, Caesalpinia pulcherrima, Curculigo capitulata, C. orchioides, Dodonaea viscosa, Eucommia ulmoides, Excoecaria orientalis, E. agallocha, E. kawakamii, Ficus carica, F. benjamina, Gnaphalium affine, G. luteoalbum ssp. affine, Lemmaphyllum microphyllum, Lycopus lucidus var. formosana, Myristica cagayanensis, M. fragrans*
Retrofractamides A	*Piper kadsura, P. kawakamii*
Retrofractamides B	*Piper kadsura, P. kawakamii*
Retrofractamides D	*Piper kadsura, P. kawakamii*
Reynoutrin	*Ricinus communis*
Rhalidasine	*Thalictrum fauriei*
Rhamnopyranosy	*Deutzia cordatula, D. taiwanensis, D. corymbosa, D. gracilis*
Rhannazin	*Viscus multinerve*
Rhannazin-3-glucoside	*Viscus multinerve*
Rhamnosan	*Abelmoschus moschatus*
Rhamnose	*Camellia japonica var. hozanensis, Euphorbia hirta, Ficus carica, F. benjamina, Ilex asprella*
Rhein	*Cassia tora, Hemerocallis fulva, Ruta graveolens*

Rhinacanthin-M, N, Q	Rhinacanthus nasutus
Rhodexin A	Rhodea japonica
Rhynchophylline	Uncaria hirsuta, U. rhynchophylla, U. kawakamii
Ribalinidin	Ruta graveolens
Riboflavin	Achyranthes aspera var. indica, A. japonica, A. aspera var. rubro-fusca, Arachis hypogea, A. agallocha, Basella alba, Boehmeria densiflora, Hibiscus rosa-sinensis, Ipomoea batata, I. obscura, I. stans, Lycopersicon esculentum, Petasites japonicus, Solanum nigraum, S. undatum
Riccionidin A	Rhus javanica
Ricinine	Ricinus communis
Ricinoleic acid	Ricinus communis
Ricinolein	Ricinus communis
Rivalosides A-B	Galium echinocarpum
Rivalosides C-E	Galium echinocarpum
Robinin	Rauvolfia verticillata
Robustol	Grevillea robusia
Roemerine	Nelumbo nucifera
Rollicosin	Rollinia mucosa
Romucosine H	Annona muricata, A. cherimola, A. reticulata
Rosmarinic acid	Clinopodium laxiflorum, Coleus scutellarioide var. crispipilus, C. parvifolius, Mentha canadensis
Rotenoids	Derris elliptica, Millettia nitida, M. pachycarpa, M. taiwaniana
Rotenone	Derris elliptica, Millettia nitida, M. pachycarpa, M. taiwaniana
Rotenone acid	Derris elliptica
Rottlerin	Mallotus tiliaefolius
Rovidine	Catharanthus rosens
Rubiadin	Damnacanthus indicus
Rubiadin	Morinda umbellata
Rubiadin-1-methyl ether	Damnacanthus indicus, Morinda citrifolia
Rubiadin-2-methyl ether	Morinda umbellata
Rubichloric acid	Morinda citrifolia
Rubrofusarin nor-rubrofusarin	Cassia tora

Component	Source
Ruoperol	*Hypoxis aurea*
Rutacridone	*Ruta graveolens*
Rutacultin	*Ruta graveolens*
Rutalinidin	*Ruta graveolens*
Rutamarin	*Ruta graveolens*
Rutarin	*Ruta graveolens*
Rutin	*Angelica keiskei, Cirsium albescens, Coriandrum sativum, Goodyera procera, G. schlechtenda, G. nankoensis, Hibiscus mutabilis, Hydrangea macrophylla, Lonicera kawakamii, L. confusa, Mallotus paniculatus, M. japonicus, Muehlenbeckia platychodum, M. hastulata, Prunella vulgaris, Ricinus communis, Viola inconspicua ssp. nagasakiensis, V. mandshurica*
Rutinoside	*Viola inconspicua ssp. nagasakiensis, V. mandshurica*
s-Guaiazulene	*Artemisia lactiflora, A. princeps*
s-Isopetasin	*Petasites formosanus*
s-Petasin	*Petasites formosanus*
s-Quaiazulene	*Murraya paniculata*
Sabinene	*Curcuma longa, Zanthoxylum ailanthoides*
Saccharase	*Solanum indicum*
Safflomin	*Carthamus tinctorius*
Safflower yellow	*Carthamus tinctorius*
Safro eugenol	*Illicium arborescens*
Safrol	*Angelica acutiloba, A. citriodor*
Safrole	*Cinnamomum camphora, Myristica cagayanensis, M. fragrans*
Saikogenin	*Bupleurum kaoi, B. chinense*
Saikosaponins	*Bupleurum chinensis, B. falcatum*
Sacifoline	*Michelia alba*
Salicyclic acid	*Scoparia dulcis*
Salicylates	*Salix warburgii*
Salicylic acid	*Polygonatum falcatum, P. kingianum, P. odoratum*

Saltpeter	*Mollugo pentaphylla*
Saluianin	*Salvia coccinea*
Samarangenin B	*Limonium sinense*
Samatine	*Rauvolfia verticillata*
Sambubiosides	*Vaccinium japonium, V. myrtillus*
Santamarine	*Eupatorium formosanum*
Santhine	*Camellia oleifera, C. sinensis*
Santhotoxin	*Angelica keiskei*
Sapogenins	*Bupleurum chinensis, B. falcatum*
Saponaretin	*Hibiscus sabdariffa, Viola inconspicua ssp. nagasakiensis, V. mandshurica*
Saponartin-4-0-glucoside	*Uraria crinita, U. lagopodioides*
Saponins	*Achyranthes aspera var. indica, A. aspera var. rubro-fusca, Adina pilulifera, A. racemose, Akebia longeracemosa, A. quinata, Albizzia lebbeck, Alternanthera nodiflora, A. sessillis, A. philoxeroides, Aralia taiwaniana, A. chinensis, Aristolochia cucurbitifolia, A. manshuriensis, A. heterophylla, A. kaempferi, A. kankanensis, A. shimadai, Bellis perennis, Bupleurum kaoi, B. chinense, Chenopodium ambrosioides, Clematis montana, C. chinensis, C. florida, Codonopsis kawakami, Deutzia cordatula, D. taiwanensis, D. corymbosa, D. gracilis, Dioscorea bulbifera, Fatsia polycarpa, F. japonica, Gomphrena globosa, Gynura japonica var. flava, Gynostemma pentaphyllum, Hibiscus sabdariffa, Lonicera japonica, L. japonica var. sempervillosa, Luffa cylindrica, Mollugo pentaphylla, Paris polyphylla, Phytolacca acinosa, P. americana, P. japonica, Pterocypsela indica, Randia spinoa, Sapindus mukorossi, Solanum nigraum, S. undatum, Thladiantha nudiflora, Vernonia cinerea, Viola inconspicua ssp. nagasakiensis, V. mandshurica*
Saptoxin	*Alocasia macrorrhiza*
Sarcostin	*Cynanchum paniculatum*
Sarmentine	*Piper kadsura, P. kawakamii*
Sarolactone	*Hypericum japonicum*
Sarracine	*Senecio nemorensis, S. scandens*
Sarsasapogenin	*Anemarrhena asphodeloides, Asparagus cochinchinensis*
Sativol	*Medicago polymorpha*
Saturated fatty acid	*Rubus formosensis*
Savinin	*Ruta graveolens*
Scandenolide	*Mikania cordata*
Scandoside	*Galium echinocarpum, Paederia cavaleriei*

Component	Source
Schisantherin A-E	*Schisandra arisanensis*
Sciadopitysin	*Podocarpus macrophyllus var. nakaii*
Scopanol	*Scoparia dulcis*
Scoparon	*Artemisia capillaris*
Scopletin	*Alyxia insularis, A. sinensis*
Scopolamine	*Datura metel, D. metel f. fastuosa, D. tatula*
Scopoletin	*Coriandrum sativum, Erycibe henryi, Ilex pubescens, Murraya paniculata, Nerium indicum, Nothapodytes foetida, N. ninmoniana, Ruta graveolens, Sida acuta*
Scorodose	*Allium bakeri, A. scorodoprasum*
Scuevolin	*Scaevola sericea*
Scutellarein	*Duranta repens, Scutellaria formosana, S. indica*
Scutellarein-7-0-glucuronides, 4'-methyl scutellarein	*Clerodendrum inerme*
Scutellarin	*Clerodendrum inerme, Scutellaria barbata*
Secoiridoid glucosides	*Adina pilulifera, A. racemose*
Secoisolaricinresinol	*Hibiscus taiwanensis*
Securinine	*Securinega suffruticosa*
Securinol	*Securinega suffruticosa*
Securitinine	*Securinega suffruticosa*
Sedanonic acid	*Angelica acutiloba, A. citriodor*
Sedoheptose	*Sedum lineare, S. sempervivoides, S. morrisonense*
Semi-α-carotenoids	*Murraya paniculata*
Semialactone	*Rhus javanica*
Sempervirine	*Gelsemium elegans*
Septicine	*Ficus septica, F. superba var. japonic*
Sequiterpene	*Curcuma domestica*
Sequoyitol	*Nephrolepis auriculata*
Serecionine	*Emilia sonchifolia, E. sonchifolia var. javanica*

Serine	*Medicago polymorpha, Taraxacum mongolicum*
Serotonin	*Musa sapientum, M. formosana, M. basjoo var. formosana*
Serpentinine	*Rauvolfia verticillata*
Serratenediol	*Lycopodium salvinioides*
Sesamin	*Sesamum indicum*
Sesamol	*Sesamum indicum*
Sesquijasmine	*Jasminum sambac*
Sesquiterpenes	*Euonymus echinatus, E. laxiflorus, E. chinensis, Farfugium japonicum, Gossampinus malabarica, Jasminum sambac,*
	Petasites formosanus, Pluchea indica, Taiwania cryptomeriodes
Sesquiterpene alcohols	*Blumea balsamifera var. microcephala, Curcuma zedoaria*
Sesquiterpene esters	*Celastrus kusanoi, C. hypoleucus*
Sesquiterpene glucosides	*Taraxacum officinale*
Sesquiterpene hydroperoxides	*Pogostemon cablin*
Sesquiterpene lactine	*Carpesium divaricatum*
Sesquiterpene lacton	*Tithonia diversifolia*
Sesquiterpene lactones	*Cichorum endivia, Eupatorium formosanum*
Sesquiterpenoids	*Eupatorium amabile, E. lindleyanum*
Shanzhiside	*Mussaenda parviflora*
Shibuol	*Diospyros kaki*
Shikimetin	*Illicium arborescens*
Shikimic acid	*Ricinus communis*
Shikimintoxin	*Illicium arborescens*
Shimadoside A	*Kalimeris indica*
Shisonin	*Solanum lyratum*
Shonzhiside	*Gardenia jasminoides*
Siaresinolic acid	*Randia spinoa*
Sigmateryl-D-glucoside	*Euphoria longana*
Sikimin	*Illicium arborescens*
Silenan	*Silene morii, S. vulgaris*
Silicic acid	*Desmodium triquetrum*

Component	Source
Silicon dioxide	*Eichhornia crassipes*
Siliptinone	*Plumbago zeylanica*
Sinapic acid	*Allium cepa, Sida acuta*
Sinensols G-H	*Spiranthes sinensis*
Sinetirucallol	*Spiranthes sinensis*
Sinodiosgenin	*Dioscorea opposita*
Siosakuranetin	*Clinopodium umbrosum*
Sitostanyl formate	*Cyathea lepifera, C. podophylla*
Sitosterols	*Centella asiatica, Chenopodium album, Clematis chinensis, C. grata, Deutzia cordatula, D. taiwanensis, D. corymbosa, D. gracilis, Elaeagnus oldhamii, E. thunbergii, E. wilsonii; Eucalyptus robusta, Ficus carica, F. benjamina, Glycine javanica, Gnaphalium affine, G. luteoalbum ssp. affine, Hedyotis corymbosa, Hibiscus tillaceus, H. esculentus, Ilex rotunda, Ipomoea pes-caprae ssp. brasiliensis, Marsilea crenata, M. minuta, Nothapodytes foetida, N. nimmoniana, Psychotria rubra, Talinum paniculatum, T. patens*
Sitosteryl glucopyranosid	*Elaeagnus oldhamii; E. thunbergii, E. wilsonii*
Sitosteryl-β-D-glucoside	*Nothapodytes foetida, N. nimmoniana*
Sitosteryl-o-β-D-glucoside	*Spilanthes acmella, S. acmella var. oleracea*
Skimmianine	*Illicium arborescens, Ruta graveolens, Toddalia asiatica, Zanthoxylum avicennae, Z. integrifoliolum, Z. nitidum*
Skullcap flavones	*Scutellaria rivularis*
Solamargine	*Solanum incanum*
Solanidine	*Solanum indicum*
Solanigrines	*Solanum nigraum, S. undatum*
Solanine	*Capsicum frutescens, Solanum aculeatissimum, S. indicum, S. lyratum*
Solanocapsine	*Solanum incanum*
Solasodine	*Solanum aculeatissimum, S. indicum, S. lyratum, S. verbascifolium*
Soranjudiol	*Morinda citrifolia*
Sotelsulflavone	*Cycas revoluta*

Soyacerebroside — *Lysimachia ardisloides, L. capillipes, L. davurica*

Spathulenol — *Artemisia lactiflora, A. princeps*

Sphingolipid — *Conyza sumatrensis, C. blinii*

Spilanthol — *Spilanthes acmella, S. acmella* var. *oleracea*

Spinasterols — *Platycodon grandiflorum*

Spinosic acid A — *Randia spinoa*

Spinosic acid B — *Randia spinoa*

Spiradin A-D, F-G — *Spiraea formosana*

Spiraine — *Spiraea formosana*

Spirostanol — *Paris polyphylla*

Sponbaneous — *Symphytum officinale*

Squalene — *Abrus precatorius, Taraxacum officinale*

Squamolone — *Artabotrys uncinatus*

Stachydrine — *Chrysanthemum morifolium, C. segetum, Solanum lyratum*

Stachyose — *Lycopus lucidus* var. *formosana*

Staminane type diterpenes — *Orthosiphon aristatus, O. stamineus*

Stansins 1-5 — *Ipomoea batata, I. obscura, I. stans*

Starch — *Cibotium cumingii, Colocasia antiquorum* var. *illustris, C. esculenta, Davallia mariesii, Euryale ferox, E. chinensis*

Stearic acid — *Bupleurum kaoi, B. chinense, Coix lacryma-jobi, Jatropha curcas, Melia azedarach, Myristic cagayanensis, M. fragrans, Pericampylus formosanus, P. formosanus, P. glaucus, P. trinervatus, Premna obtusifolia, P. crassa, P. serratifolia, P. microphylla, Ricinus communis, Sonchus arvensis, S. oleraceus, Taraxacum mongolicum, Viscus multinerve*

Stearidonic — *Ribes formosanum, R. nigrum*

Stearin — *Bombax malabarica, Ricinus communis, Sesamum indicum*

Stearodiricinolein — *Ricinus communis*

Stearoptene — *Pandanus odoratissimus* var. *sinensis*

Stemhanoline — *Stephania hispidula*

Stemholine — *Stephania hispidula*

Stemondidine — *Stemona tuberosa*

Stemonine — *Stemona tuberosa*

Component	Source
Stephanine	*Stephania hispidula, S. japonica*
Stephanoline	*Stephania japonica, S. tetrandrae, S. moore*
Stepinonine	*Stephania japonica*
Steponine	*Stephania japonica, S. hispidula*
Steroids	*Gynura japonica var. flava, Pluchea indica, Solanum biflorum*
Steroidal	*Asparagus cochinchinensis, Paris polyphylla, Saurauia oldhamii, S. tristyla var. oldhamii*
Steroidal sapogenins	*Cissus repens, C. sicyoides*
Steroidal saponin	*Polygonatum falcatum, P. kingianum, P. odoratum*
Steroidal saponins	*Allium sativum. A. thunbergii. A. tuberosum, Anemarrhena asphodeloides, Lilium formosanum, L. speciosum, Paris lancifolia*
Sterols	*Chenopodium album, Cucumis melo ssp. melo, Gynostemma pentaphyllum, Hibiscus tillaceus. H. esculentus, Luffa cylindrica, Momordica charantia, Scirpus ternatanus, S. maritimus, Toona sinensis, Urena lobata*
Sterols linoleyl acetate	*Conyza sumatrensis, C. blinii*
Sterric acid	*Corchorus capsularis, C. olitorius*
Steviolbioside	*Stevia rebaudiana*
Stevioside	*Stevia rebaudiana*
Stigmas-4-ene-3,6-dione	*Polygonum chinense*
Stigmast-5-ene-3β	*Jatropha curcas*
Stigmast-7-en-3β-ol	*Clinopodium laxiflorum*
Stigmasta-5,11(12)-diene-3β-01	*Lepidagathis formosensis, L. hyalina, L. cristata*
Stigmastane-3,6-dione	*Polygonum chinense*
Stigmasterol	*Adina pilulifera, A. racemose, Aletris formosana, Angelica hirsutiflora, Anisomeles indica, Bauhinia championi, Cirsium japonicum, C. japonicum var. australe, Clerodendrum japonicum, C. kaempferi, Curcuma zedoaria, Dioscorea bulbifera, Dodonaea viscosa, Drynaria cordata, Elephantopus mollis, E. scaber, Eryngium foetidum, Glehnia littoralis, Hibiscus tillaceus. H. esculentus, Lysimachia ardisloides, L. capillipes, L. davurica, Morinda umbellata, Nervilia taiwaniana, N. purpurea, Ophiopogon japonicus, Randia spinoa, Spilanthes acmella, S. acmella var. oleracea, Viscus multinerve*
Stilbenes	*Scirpus ternatanus, S. maritimus*
Stilbenoids	*Bletilla striata*

Strearyl palmitate	*Ixeris tamagawaensis*
Strearyl strearate	*Ixeris tamagawaensis*
Strophanthidin	*Corchorus capsularis, C. olitorius*
Strospeside	*Digitalis purpurea*
Strumaroside	*Xanthium sibiricum, X. strumarium*
Strychnine	*Strychnos angustiflora*
Styrene	*Piper arboricola*
Styrylpyrones	*Goniothalamus amuyon*
Suberenon	*Ruta graveolens*
Succinates	*Drynaria cordata*
Succine acid	*Eriobotrya japonica*
Succinic acid	*Bryophyllum pinnatum, Drynaria cordata, Eupatorium tashiroi, Hemerocallis longituba, Geranium nepalense var. thunbergii, G. suzukii, Maytenus diversifolia, Sarcandra glabra*
Suchilactone	*Polygala glomerata*
Sucrose	*Eriobotrya japonica*
Sugars	*Maranta arundinacea*
Sugar alcohols	*Talinum paniculatum, T. patens*
Sugiol	*Jasminum hemsleyi, Podocarpus nagi*
Superoxide dismutase	*Rubus hirsutus*
Stupinin	*Tournefortia sarmentosa*
Swertianmarin	*Gentiana atkinsonii, G. campestris, G. flavo-maculata, Swertia randaiensis*
Swertianolin	*Gentiana atkinsonii, G. campestris, G. flavo-maculata*
Swertism	*Desmodium capitatum*
Sylvine	*Piper arboricola*
Symphytine	*Symphytum officinale*
Syringareinol	*Acanthopanax senticosus*
Syringaresinol	*Hibiscus syriacus, Siphonostegia chinensis*
Syringic acid	*Maytenus diversifolia, Sida acuta*
Syringin	*Lonicera japonica, L. japonica var. sempervillosa*
Tabernaemontanin	*Tagetes erecta*

Component	Source
Tagetone	*Tagetes erecta*
Tagitinin C	*Tithonia diversifolia*
Talatisamine	*Aconitum bartletii, A. fukutomei, A. formosanum, A. kojimae, A. kojimae var. lassiocarpium, A. kojimae var. ramosum, A. yamamotoanum*
Tannic acid	*Camellia oleifera, C. sinensis, Coriandrum sativum, Erigeron canadensis, Rhus chinensis, R. verniciflua, R. typhina, Vigna radiata, V. angularis, V. umbellata, Vitex negundo*
Tannins	*Albizzia lebbeck, Alternanthera nodiflora, Alternanthera philoxeroides, A. sessilis, Aralia taiwaniana, A. chinensis, Bixa orellana, B. orellana Caesalpinia pulcherrima, Callicarpa longissima, C. loureiri, C. nudiflora, C. pedunculata, Cassia mimosoides, Centella asiatica, Cibotium cumingii, Cleome gynandra, Clerodendrum philippinum, Conyza sumatrensis, C. blinii, Curculigo capitulata, C. orchioides, Desmodium triquetrum, Dioscorea bulbifera, Diospyros eriantha, Dodonaea viscosa, Epimeredi indica, Ficus microcarpa. F. hispida, Geranium nepalense var. thunbergii, G. suzukii, Helicteres angustifolia, Ilex cornuta, Lagerstroemia subcostata, Lantana camara, Laungusa galanga, Lonicera japonica, L. japonica var. sempervillosa, L. shintenensis, Macaranga tanarius, Mallotus paniculatus, M. japonicus, Melastoma septemnervium, M. dodecandrum, Mesona chinensis, Morinda citrifolia, Osbeckia chinensis, Potentilla tugitakensis, Punica granatum, Ranunculus japonicus, Rhodomyrtus tomentose, Ronunculus sceleratus, Rubus croceacanthus, R. lambertianus, R. parvifolius, Rumex acetosa, R. crispus, Sarcandra glabra, Scoparia dulcis, Scutellaria formosana, Serissa foetida, Ternstroemia gymnanthera, Verbena officinalis, Vitis thunbergii, Youngia japonica*
Taraligenin	*Aralia taiwaniana, A. chinensis*
Taraverone	*Crossostephium chinense*
Taraxacerin	*Taraxacum mongolicum*
Taraxacin	*Taraxacum mongolicum*
Taraxasterol	*Balanophora spicata, Eupatorium tashiroi, Sonchus arvensis, S. oleraceus, Taraxacum mongolicum*
Taraxerol	*Acanthopanax trifoliatus, Crossostephium chinense, Euphorbia atoto, E. hirta, Jatropha curcas, Myrica adenophora, Taraxacum mongolicum*
Taraxerol-3-β-O-tridecyl-ether	*Derris trifoliata*
Taraxerone	*Euphorbia atoto, E. hirta, Mallotus apelta*
Taraxeryl acetate	*Crossostephium chinense, Ficus pumila var. awkeotsang*

Taraxevol	*Lophatherum gracile*
Taraxol	*Taraxacum mongolicum*
Taraxsteryl acetate	*Cirsium japonicum, C. japonicum var. australe*
Tartaric acid	*Chaenomeles japonica, Eriobotrya japonica, Oxalis corymbosa, Sonchus arvensis, S. oleraceus*
Taxane diterpenoids	
Taxoids	*Taxus mairei*
Taxumairols X-Z	*Taxus mairei*
	Taxus mairei
Tazettin	*Zephyranthes candida*
Tazettine	*Hippeastrum equestr, H. regina*
Tectoquinone	*Morinda umbellata*
Tectoridin	*Belamcanda chinensis*
Tectorigenin	*Belamcanda chinensis, Euchresta formosana*
Tergallic acid	*Rhynchosia volubilis, R. minima*
Terpene	*Cymbopogon nardus, Liquidambar formosana*
Terpeneol	*Erigeron canadensis*
Terpenes	*Farfugium japonicum*
Terpenic glucosides	*Breynia officinalis*
Terpenoids	*Evodia meliaefolia*
Terpenylacetate	*Vitex rotundifolia*
Terpinene	*Coriandrum sativum, Mosla punctulata*
Terpinenol -4β-caryophyllene	*Artemisia indica, A. japonica*
Terpineol	*Cinnamomum camphora, Myristica cagayanensis, M. fragrans*
Terpinolen	*Coriandrum sativum*
Terpinolene	*Cryptotaenia canadensis, C. japonica, Oenanthe javanica*
Terpinyl acetate	*Vitex cannabifolia*
Teserpine	*Rauvolfia verticillata*
Tetracyclic triterpene acids	*Rosa taiwanensis, R. davurica*
Tetradecane	*Piper sarmentosum, P. sanctum*
Tetradecanol	*Angelica acutiloba, A. citriodor*

Component	Source
Tetraflavonoid	*Cephalotaxus wilsonianer*
Tetrahydroflavone	*Xanthium sibiricum, X. strumarium*
Tetrahydroxyflavone	*Crotalaria sessiliflora*
Tetrahydroisoquinoline alkaloids	*Mucuna macrocarpa, M. nigricans, M. puriens*
Tetrahydropyran	*Piper sarmentosum, P. sanctum*
Tetrameric	*Rubus croceacanthus, R. lambertianus*
Tetrandrine	*Stephania cephalantha, S. hispidula, S. tetrandrae, S. moore*
Tetranortriterpenoids	*Severinia buxifolia*
Tetrasaccharide glycosides	*Ipomoea batata, I. obscura, I. stans*
Tetratriacontane	*Lonicera kawakamii, L. confusa*
Thalfoetidine	*Thalictrum fauriei*
Thalidezine	*Thalictrum fauriei*
Thalpine	*Thalictrum fauriei*
Theasaponin	*Camellia japonica var. hozanensis*
Theobromine	*Camellia oleifera, C. sinensis*
Theophylline	*Camellia oleifera, C. sinensis, Ilex asprella*
Theveside	*Thevetia peruviana*
Thevetin A	*Thevetia peruviana*
Thevetin B	*Thevetia peruviana*
Theviridoside	*Thevetia peruviana*
Thiamin	*Achyranthes aspera var. indica, A. aspera var. rubro-fusca, Arachis hypogea, A. agallocha, Basella alba, Boehmeria densiflora, Hibiscus rosa-sinensis, Ipomoea batata, I. obscura, I. stans, Lycopersicon esculentum, Petasites japonicus*
Thiitic simidine	*Thalictrum fauriei*
Thujone	*Biota orientalis*
Thymine	*Allium cepa, Nothapodytes foetida, N. nimmoniana*
Thymohydroquinone	*Mosla punctulata*
Thymol	*Eucalyptus robusta, Ocimum gratissimum*

Thymol derivatives	*Carpesium divaricatum*
Tiglic acid	*Croton tiglium*
Tigogenin	*Costus speciosus*
Tinosporin	*Tinospora tuberculata*
Tocopherols	*Rubus formosensis*
Toddafolactone	*Toddalia asiatica*
Toddalinine	*Toddalia asiatica*
Tohogenol	*Lycopodium salvinioides*
Toluene	*Piper arboricola*
Tomentogenin	*Cynanchum paniculatum*
Toodalia	*Toddalia asiatica*
Toosendanin	*Melia azedarach, Toona sinensis*
Toralacton	*Cassia tora*
Tormentic acid	*Debregeasia edulis, D. salicifolia, Potentilla leuconta, P. multifida*
Torosachrysone	*Cassia occidentalis, C. torosa*
Totaradiol	*Podocarpus nagi*
Totaral	*Podocarpus nagi*
Totarol	*Podocarpus macrophyllus var. nakaii, P. nagi*
Toxalbumin	*Aleurites fordii, A. moluccana, A. montana*
Toxopherol	*Viola inconspicua ssp. nagasakiensis, V. mandshurica*
Toxyloxanthone C	*Cudrania cochinchinensis*
trans-β-Farnesene	*Artemisia lactiflora, A. princeps*
trans-β-Ocimene	*Cryptotaenia canadensis, C. japonica*
trans-Caryophyllene	*Artemisia lactiflora, A. princeps*
trans-Linalool oxide	*Osmanthus fragrans*
Trehalase	*Hemerocallis fulva*
Trehalose	*Selaginella uncinata*
Triacontanic acid	*Lysimachia ardisloides, L. capillipes, L. davurica*
Triacontanol	*Elephantopus mollis, E. scaber*
Trichosanic acid	*Trichosanthes cucumeroides*

Component	Source
Tricin	*Medicago polymorpha, Miscanthus floridulus, M. sinensis var. condensatus, Setaria palmifolia, S. viridis*
Tricin-7-0-β-D-glucoside	*Setaria palmifolia, S. viridis*
Tricosyl	*Ixeris tamagawaensis*
Trifolirhizin	*Sophora flavescens*
Trigalloylgallic acid	*Rhynchosia volubilis, R. minima*
Triglycerides	*Coix lacryma-jobi*
Triglycoside	*Kyllinga brevifolia*
Trigonelline	*Abrus precatorius, Artocarpus altilis, Astragalus sinicus, Monochoria vaginalis, Nothapodytes foetida, N. nimmoniana, Quisqualis indica, Solanum lyratum*
Trihydroxyanthraquinone monomethyl ether	*Crotalaria sessiliflora*
Trihydroxyisoflavone	*Crotalaria sessiliflora*
Trilobamine	*Cocculus orbiculata*
Trilobine	*Cocculus orbiculata, C. sarmentosus, C. trilobus, Stephania cephalantha*
Trimeric	*Rubus croceacanthus, R. lambertianus*
Tripeptide	*Juncus effusus var. decipiens*
Triptolide	*Tripterygium wilfordii*
Triricinolein	*Ricinus communis*
Triterpene glycosides	*Fatsia polycarpa, F. japonica*
Triterpene saponin	*Kalimeris indica*
Triterpenes	*Conyza sumatrensis, C. blinii, Crateva adansonii ssp. formosensis, Debregeasia edulis, D. salicifolia, Euonymus echinatus, E. laxiflorus, E. chinensis, Melanolepis multiglandulosa, Mussaenda pubescens, Pluchea indica, Psidium guajave, Pterocypsela indica, Rhus javanica, Sedum lineare, S. sempervivoides, S. morrisonense*
Triterpenes amyrenone	*Sedum formosanum*
Triterpenoid	*Adiantum capillus-veneris, A. flabellulatum, Akebia longeracemosa, A. quinata, Derris trifoliata, Euonymus echinatus, E. laxiflorus, E. chinensis, Helwingia formosana, H. japonica ssp. formosana, Gentiana scabrida, G. scabrida var. horaimontana, G. lutea, Lepidagathis formosensis, L. hyalina, L. cristata, Marsdenia formosana, Mussaenda parviflora, Vernonia cinerea, Viscus multinerve*
Triterpenoid daponins	*Conyza sumatrensis, C. blinii*

Triterpenoid glycosides	*Bellis perennis*
Triterpenoid sampnins	*Achyranthes longifolia, A. ogotai, A. kojimae, A. kojimae var. lassiocarpium, A. kojimae var. ramosum, A. yamamotoanum, Adenophora stricta, A. triphylla, A. tetrophylla, Bupleurum chinensis, B. falcatum, Dianthus chinensis, Dumasia villosa, D. truncata, Glycyrrhiza uralensis, Lysimachia ardisloides, L. capillipes, L. davurica, L. mauritiana, L. davurica, L. simulans, Maesa tenera, M. lanceolata, M. laxiflora, M. perluria var. formosana, M. japonica, Mussaenda pubescens*
Tropolone	*Jasminum hemsleyi*
Trypsin	*Ceratopteris thalictroides*
Tryptanthrin	*Clerodendrum cyrtophyllum*
Tryptophane	*Allium bakeri, A. scorodoprasum, Drynaria cordata, Pratia nummularia*
Tumerone	*Curcuma longa, C. zedoaria*
Turpentine	*Chamaecyparis formosensis, C. obtusa var. filicoides, C. obtusa var. formosana*
Turpinionoside B	*Breynia officinalis*
Tylocrebrine	*Ficus septica, F. superba var. japonic*
Tylophoridicines C-F	*Tylophora lanyuensis, T. atrofolliculta*
Tylophorine	*Ficus septica, F. superba var. japonic*
Tylophorinidine	*Tylophora lanyuensis, T. atrofolliculta*
Tylophorinine	*Tylophora lanyuensis, T. atrofolliculta*
Tyrosine	*Dioscorea opposita, Dolichos lablab, Ficus carica, F. benjamina*
Udosaponin B	*Hydrocotyle sibthorpioides*
Umbellferone	*Clausena excavata, Coriandrum sativum, , Deutzia cordatula, D. taiwanensis, D. corymbosa. D. gracilis, Ruta graveolens*
Umbelliferon	*Pinus massoniana*
Uncinine	*Artabotrys uncinatus*
Undecan-2-ol	*Ruta graveolens*
Uracil	*Nothapodytes foetida, N. nimmoniana*
Urbenine	*Coptis chinensis*
Urease	*Canavalia ensiformis, Ficus carica, F. benjamina*
Uronic acid	*Bischofia javanica, Clinopodium laxiflorum, C. umbrosum, Debregeasia edulis, D. salicifolia, Diospyros khaki, Ocimum gratissimum*

Component	Source
Ursolic acid	*Duranta repens, Hedyotis pinifolia, Ilex asprella, I. pubescens, I. rotunda, Ligustrum lucidum, L. pricei, Nerium indicum, Orthosiphon aristatus, O. stamineus, Plantago asiatica, P. major; Potentilla leuconta, P. multifida, Prunella vulgaris, Rauvolfia verticillata, Rhododendron simsii, Salvia hayatana, S. japonica, S. roborowskii, Solanum incanum*
Usaramine	*Crotalaria pallida*
Ushinsunine	*Michelia alba*
Usigtoercin	*Hypericum japonicum*
Usnic acid	*Hedyotis uncinella*
Ussonic acid	*Prunella vulgaris*
Uvaol	*Debregeasia edulis, D. salicifolia, Osmanthus fragrans*
Valencene	*Pogostemon amboinicus*
Valerianic acid	*Melia azedarach*
Valeric acid	*Luffa cylindrica*
Vallarine acid	*Centella asiatica*
Vanillic acid	*Chenopodium album, Dicliptera chinensis, D. riparia, Opuntia dillenii, Sida acuta*
Vanillin	*Ferula assa-foetida, Gynura formosana, G. elliptica, G. japonica var. flava*
Venoterpine	*Camptotheca acuminata*
Ventilagolin	*Ventilago leiocarpa*
Veratramine	*Veratrum formosanum*
Verbenalin	*Verbena officinalis*
Verbenalol	*Verbena officinalis*
Veronicoside	*Veronicastrum simadai*
Verproside	*Veronicastrum simadai*
Vertiaflavone	*Thevetia peruviana*
Vibsane diterpenoids	*Viburnum plicatum var. formosanum, V. odoratissimum, V. awabuki, V. luzonicum*
Vinblastine	*Catharanthus rosens*
Vincristine	*Catharanthus rosens*
Vindolinine	*Catharanthus rosens*
Violanin	*Taraxacum mongolicum, Viola inconspicua ssp. nagasakiensis, V. mandshurica*

Violaxanthin	*Rumex acetosa*
Violutin	*Viola inconspicua* ssp. *nagasakiensis, V. mandshurica*
Violutoside	*Viola inconspicua* ssp. *nagasakiensis, V. mandshurica*
Viroallosecurinine	*Securinega virosa*
Virosecurinin	*Securinega virosa*
Virosine	*Securinega virosa*
Vitamin A	*Basella rubra, Bixa orellana, Capsella bursa-pastoris, Foeniculum vulgare, Ipomoea batata, I. obscura, I. stans, Luffa cylindrica, Lycopersicon esculentum, Medicago polymorpha, Musa sapientum, M. formosana, M. basjoo var. formosana, Rumex crispus, Sesamum indicum, Taraxacum officinale*
Vitamin B	*Basella rubra, Citrus tangerina, Luffa cylindrica, Musa sapientum, M. formosana, M. basjoo var. formosana, Sesamum indicum, Taraxacum officinale, Toona sinensis*
Vitamin B$_1$	*Colocasia antiquorum* var. *illustris, C. esculenta, Ipomoea batata, I. obscura, I. stans, Prunella vulgaris, Gnaphalium affine, G. luteoalbum* ssp. *affine*
Vitamin B$_2$	*Colocasia antiquorum* var. *illustris, C. esculenta, Ipomoea batata, I. obscura, I. stans*
Vitamin C	*Agrimonia pilosa, Basella rubra, Blumea laciniata, Chaenomeles japonica, Hemerocallis fulva, Ipomoea batata, I. obscura, I. stans, Luffa cylindrica, Musa sapientum, M. formosana, M. basjoo var. formosana, Oxalis corymbosa, O. corniculata, Pinus taiwanensis, Prunella vulgaris, Pseudosasa usawai, P. owatrii, Ricinus communis, Rubus hirsutus, Rumex acetosa, R. japonicus, Taraxacum officinale, Toona sinensis*
Vitamin E	*Lactuca indica, Medicago polymorpha, M. polymorpha, Musa sapientum, M. formosana, M. basjoo var. formosana, Pseudosasa usawai, P. owatrii, Rubus hirsutus*
Vitamin K	*Agrimonia pilosa, Medicago polymorpha, Prunella vulgaris*
Vitamins	*Ananas comosus, A. keiskei, Cucurbita moschata, Juncus effusus* var. *decipiens, Lemmaphyllum microphyllum, Phoenix dactylifera, Pyracantha fortuneana, Solanum nigraum, S. undatum*
Vitexicarpin	*Vitex rotundifolia*
Vitexin	*Crotalaria pallida, Hibiscus sabdariffa, Rumex acetosa, Uraria crinita, U. lagopodioides, Viola inconspicua* ssp. *nagasakiensis, V. mandshurica, Zanthoxylum nitidum, Z. integrifoliolum*
Vitexin 2″-0-glucoside	*Setaria palmifolia, S. viridis*
Vitexin 2″-oxyloside	*Setaria palmifolia, S. viridis*
Vitexin-7-0-glucoside	*Uraria crinita, U. lagopodioides*

Component	Source
Vitexin-0-syloside	Crotalaria pallida
Vitricine	Vitex rotundifolia
Volatile oils	Citrus medica var. sarcodactylis, Acorus calamus, A. gramineus, Blumea aromatica, B. lacera, B. lanceolaria, Chenopodium ambrosioides, Chloranthus spicatus, Cleome gynandra, Duchesnea indica, Gardenia angusta var. kosyunensis, G. oblongifolia, Gendarussa vulgaris, Hedychium coronarium, Heterotropa taitonensis, Ilex pubescens, Ipomoea pes-caprae ssp. brasiliensis, Kyllinga brevifolia, Laungusa galanga, Pandanus odoratissimus var. sinensis, Ranunculus japonicus
Volatile compounds	Juniperus formosana, Mahonia japonica, M. oiwakensis
Volatile constituents	Hippobroma longiflora, Leea guineensis
Volatile substances	Callicarpa formosana, C. japonica
Vomifoliol	Ilex pubescens, Sida acuta
Wedelolactone	Wedelia biflora, W. chinensis
Wedelosin	Wedelia biflora, W. chinensis
Wighteone	Cudrania cochinchinensis
Wikstroemin	Wikstroemia indica
Wilsonine	Cephalotaxus wilsonianer
Wogmoside	Scutellaria rivularis
Wogonin	Scutellaria rivularis
Wood lignans	Juniperus formosana
Woodwardic acid	Blechnum orientale
Worenine	Coptis chinensis
Xanathoxylin	Sapium discolor, S. sebiferum
Xanthanol	Xanthium sibiricum, X. strumarium
Xanthinin	Xanthium sibiricum, X. strumarium
Xanthoangelol	Angelica keiskei
Xanthones	Gentiana atkinsonii, G. campestris, G. flavo-maculata, Polygala aureocauda

Xanthones-6,7-dihydroxy-1,3-dimethoxyxanthone	*Viola inconspicua* ssp. *nagasakiensis, V. mandshurica*
Xanthophyll	*Viola inconspicua* ssp. *nagasakiensis, V. mandshurica*
Xanthoria	*Cassia occidentalis, C. torosa*
Xanthotoxin	*Angelica hirsutiflora, Ruta graveolens*
Xanthotoxin A	*Ruta graveolens*
Xanthumin	*Xanthium sibiricum, X. strumarium*
Xeaxanthin	*Viola inconspicua* ssp. *nagasakiensis, V. mandshurica*
Xylene	*Piper arboricola*
Xylose	*Gnaphalium affine, G. luteoalbum* ssp. *affine*
Yatanaside	*Brucea javanica*
Yatanine	*Brucea javanica*
Yatanoside	*Brucea javanica*
Yejuhualactone	*Chrysanthemum indicum, Dendranthema indicum*
z-Piperolide	*Piper sarmentosum, P. sanctum*
Zanthones	*Hypericum geminiflorum*
Zanthonitrile	*Zanthoxylum nitidum, Z. integrifoliolum*
Zeaxanthin	*Lycium chinense, Taraxacum mongolicum*
Zeaxanthine	*Cycas revoluta*
Zederone	*Curcuma domestica, C. zedoaria*
Zerumbone	*Curcuma zedoaria*
Zinc	*Talinum triangulare*
Zingiberales	*Canna flaccida, C. indica*
Zingiberen	*Curcuma longa*
Zingiberene	*Alpinia speciosa, A. zerumbet, Curcuma zedoaria, Zingiber officinale*
Zingiberol	*Alpinia speciosa, A. zerumbet, Zingiber officinale*
Zizyphine A-type cyclopeptice alkaloids	*Paliurus ramosissimus*

Appendix 2

List of English and Scientific Names

English Name	Scientific Name
	Alyxia insularis, A. sinensis
A Li Teng	*Glochidion acuminatum, G. eriocarpum, G. zeylanicum*
Abacus plant	*Abutilon taiwanensis*
Abutilon	*Achyranthes ogotai, A. longifolia*
Achyranthes	*Aconitum kojimae, A. kojimae* var. *lassiocarpium, A. fukutomel, A. yamamotoanum*
Acomite	*Ophioglossum vulgatum*
Adder's tongue	*Adina racemose, A. pilulifera*
Adina	*Vigna angularis*
Adzuki bean	*Zanthoxylum ailanthoides*
Ailanthus prickly ash	*Bryophyllum pinnatum, Kalanchoe pinnata, K. gracillis, K. spathulata, K. crenata, K. tubiflora*
Airplant	*Melicope semecarpifolia*
Alani	*Ficus sarmentosa* var. *nipponica*
Alishan fig	*Alternanthera philoxeroides*
Alligator alternanthera	*Aquilaria sibebsus, A. agallocha*
Aloe wood	*Pilea microphylla, P. rotundinucula*
Aluminum plant	*Hippeastrum regina, H. equestre*
Amaryllis	*Amentotaxus formosana*
Amentotaxus	*Bletilla striata*
Amethyst orchid	*Bixa orellana*
Anatto tree	*Angelica hirsutiflora, A. citriodor, A. acutiloba, A. keiskei*
Angelica	*Ficus septica*
Angular fruit fig	*Jasminum sambac*
Arabian jasmine	*Fatsia polycarpa, A. chinensis*
Aralia	*Urena lobata*
Aramina	*Viola philippica*
Arrow leaf violet	*Maranta arundinacea*
Arrow root	*Artemisia indica*
Artemisia	*Biota orientalis*
Artorvitae leaves	*Arisaema consanguineum*
Arum	

Asafetida	*Ferula assa-faoetida*
Asian toddalia	*Toddalia asiatica*
Asian persimmon	*Diospyros khaki*
Asiatic butterfly bush	*Buddleja asiatica*
Asiatic wormwood	*Artemisia princeps var. orientalis*
Asparagus	*Asparagus cochinchinensis*
Asthma herb	*Euphorbia hirta*
Asthma plant	*Chamaesyce thymifolia, C. hirta*
Autumn maple tree	*Bischofia javanica*
Ba Jiao Feng Gen	*Alangium chinense*
Bachelor's buttons	*Goodyera nankoensis, G. globosa*
Bai Tong Su	*Claoxylon polot*
Ballon vine	*Cardiospermum halicacabum*
Banana	*Musa formosana, M. sapientum, M. paradisiaca*
Barbate cyclea	*Cyclea insularis*
Barbed skullcap	*Scutellaria barbata*
Barker's garlic	*Allium bakeri*
Barnyard grass	*Echinochloa colonum*
Basil	*Ocimum basilicum*
Bastard agrimony	*Ageratum houstonianum, A. conyzoides*
Bayberry	*Myrica adenophora*
Beach naupaka	*Scaevola sericea*
Bearing runners tournefortia	*Tournefortia sarmentosa*
Beautybush	*Callicarpa loureiri, C. nudiflora, C. longissima, C. pedunculata*
Beech silver-top	*Glehnia littoralis*
Begonia	*Begonia malabarica, B. laciniata*
Bei Xian	*Cyathula prostrata*
Bellflower	*Adenophora triphylla, A. stricta*
Betel nut palm	*Areca catechu*
Betel pepper	*Piper betle*

English Name	Scientific Name
Betony	*Stachys sieboldii*
Big hydrangea	*Hydrangea macrophylla*
Bilberry	*Vaccinium myrtillus*
Bird's nest fern	*Asplenium nidus*
Bird's nest orchid	*Nervilia purpurea, N. taiwaniana* Ying
Bitter melon	*Momordica charantia, Manihot utilissima* Pohl.
Bitterwort	*Gentiana lutea*
Bittersweet	*Celastrus hypoleucus, C. kusanoi*
Black currant	*Ribes nigrum*
Black musli	*Curculigo orchioides, C. capitulata*
Black pepper	*Piper nigrum*
Black fruit passion	*Passiflora suberosa*
Black nightshade	*Solanum nigraum*
Black maidenhair	*Adiantum flabellulatum*
Blackberry	*Rubus croceacanthus, R. hirsutus*
Blackberry lily	*Belamcanda chinensis*
Blue pig ear	*Vandellia crustacea, V. cordifolia*
Blueberry	*Vaccinium emarginatum*
Bluestem	*Schizophragma integrifolium*
Bluets	*Hedyotis pinifolia*
Blumea camphor	*Blumea aromatica, B. balsamifera* var. *microcephala, B. lacera, B. lanceolaria, B. riparia* var. *megacephala, B. laciniata*
Bojer's spurge	*Euphorbia milli*
Boneset	*Eupatorium cannabinum* ssp. *asiaticum, E. lindleyanum, E. tashiroi*
Borduega	*Debregeasia salicifolia, D. edulis*
Boston ivy	*Parthenocissus tricuspidata*
Bottle tree	*Sterculia lychnophora*
Bottle gourd	*Lagenaria siceraria* var. *microcarpa*
Bottlebrush orchid	*Goodyera procera*
Brake	*Pteris vittata, P. multifida, P. ensiformis*

Brazilian plume	*Justicia procumbens* var. *hayatai*
Bread fruit tree	*Artocarpus altilis*
Breynia	*Breynia fruticosa, B. accrescens*
Bridal wreath	*Spiraea prunifolia* var. *pseudoprunifolia*
Bugleweed	*Lycopus lucidus* var. *formosana, Ajuga pygmaea, A. decumbens*
Bupleurum	*Bupleurum kaoi*
Bur. Marigold	*Bidens pilosa* var. *minor*
Burdock	*Arctium lappa*
Burford's holly	*Ilex cornuta*
Butterfly weed	*Asclepias curassavica*
Calica flower	*Aristolochia elegans*
Callicarpa	*Callicarpa formosana*
Camellia	*Camellia japonica* var. *hozanensis*
Camphor tree	*Cinnamomum camphora, C. kotoense*
Campion	*Silene morii, S. vulgaris*
Canadian mint	*Mentha canadensis*
Candlenut	*Aleurites fordii*
Cantaloupe	*Cucumis melo* ssp. *melo*
Cape jasmine	*Gardenia jasminoides, G. angusta* var. *kosyunensis*
Caper spurge	*Euphorbia lathyris*
Capitate bushmint	*Hyptis rhomboides*
Carrot fern	*Onychium japonicum*
Carymobse hedyotis	*Hedyotis corymbosa*
Cassia bark tree	*Cinnamomum cassia*
Castor bean	*Ricinus communis*
Cat's claw	*Uncaria rhynchophylla, U. kawakamii*
Cat's whiskers	*Orthosiphon aristatus, O. stamineus*
Cellar fungus	*Heterotropa taitonensis, Hyphea kaoi*
Ceylon leadwort	*Plumbago zeylanica*
Ceylon spinach	*Basella alba, B. rubra*

English Name	Scientific Name
Cha Gen Zi Ma	*Oreocnide pedunculata*
Chammy hop seed bush	*Dodonaea viscosa*
Chase tree	*Vitex cannabifolia*
Chinese firethorn	*Pyracantha fortuneana*
Chi Pao	*Thladiantha nudiflora*
Chickweed	*Stellaria media, Drynaria diandra, D. fortunei*
Chin Cui Zi	*Nothapodytes nimmoniana, N. foetida*
China berry tree	*Melia azedarach*
China fir	*Cunninghamia konishii*
China root	*Smilax china*
China wood oil tree	*Aleurites moluccana, A. montana*
Chinese creeper	*Mikania cordata*
Grass jelly	*Mesona chinensis*
Chinese aloe wood	*Aquilaria sinensis*
Chinese banyan tree	*Ficus microcarpa*
Chinese box orange	*Severinia buxifolia*
Chinese camellia	*Camellia sinensis*
Chinese carnation	*Dianthus chinensis*
Chinese chive	*Allium thunbergii*
Chinese clematis	*Clematis chinensis*
Chinese climber	*Heterostemma brownii*
Chinese cork tree	*Phellodendron chinensis*
Chinese culver's root	*Veronicastrum simadai*
Chinese dianella	*Dianella chinensis*
Chinese elderberry	*Sambucus chinensis*
Chinese fevervine	*Paederia scandens*
Chinese giant hyssop	*Agastache rugosa*
Chinese hare's ear	*Bupleurum chinensis*
Chinese holly	*Ilex rotunda*

Chinese honeysuckle	*Uraria crinita*
Chinese hydrangea	*Hydrangea chinensis*
Chinese indigo	*Polygonum perfoliatum, P. paleaceum*
Chinese leucas	*Leucas mollissima var. chinensis, L. chinensis*
Chinese lobelia	*Lobelia chinensis*
Chinese mayapple	*Dysosma pleiantha*
Chinese mahogany	*Toona sinensis*
Chinese mild vetch	*Astragalus sinicus*
Chinese milkwort	*Polygala glomerata*
Chinese mosla	*Mosla punctulata*
Chinese mother wort	*Leonurus artemisia*
Chinese oak	*Cyclobalanopsis stenophylla*
Chinese orange	*Citrus sinensis var. sekken*
Chinese privet	*Ligustrum sinense*
Chinese quinine	*Dichroa febrifuga*
Chinese resurrection plant	*Selaginella uncinata, S. delicatula*
Chinese St. John's wort	*Hypericum chinense*
Chinese statice	*Limonium sinense*
Chinese stargrass	*Aletris formosana*
Chinese strawberry	*Myrica rubra*
Chinese sumac	*Rhus chinensis*
Chinese tamarisk	*Tamarix chinensis*
Chinese thimble tree	*Euonymus chinensis*
Chinese water lily	*Euryale chinese*
Chinese wedelia	*Wedelia chinensis*
Chinese yam	*Dioscorea opposita*
Chive	*Allium scorodoprasum*
Chloranthus	*Chloranthus spicatus, C. oldham*
Chou Huang Jing Zi	*Premna crassa, P. microphylla, P. obtusifolia, P. serratifolia*
Chrysanthemum	*Chrysanthemum morifolium, C. indicum, Dendranthema indicum*

English Name	Scientific Name
Chrysanthemum flower tree	*Bauhinia championi*
Chung Wei Ma Lan	*Semnostachya longespicata*
Ci Luo Shi	*Maytenus diversifolia, M. emarginata, M. serrata*
Citronella	*Cymbopogon nardus*
Cleavers	*Galium echinocarpum*
Clematis	*Clematis grata*
Clerodendrum	*Clerodendrum cyrtophyllum, C. petasites, C. philippinum*
Cleyera	*Ternstroemia gymnanthera*
Climbing fern	*Lygodium japonicum*
Club moss	*Lycopodium cunninghamioides, L. salvinioides*
Cocklebur	*Xanthium sibiricum*
Cockscomb	*Celosia cristata*
Codonopsis	*Codonopsis kawakami*
Coffea arabica	*Bredia scandens, B. rotundifolia, B. oldhamii*
Coffee senna	*Cassia occidentalis*
Coin penny wort	*Hydrocotyle formosana, H. sibthorpioides, H. nepaleniss*
Coleus	*Coleus parvifolius, C. scutellarioides var. crispipilus*
Common blue beard	*Caryopteris incana*
Common bugleweed	*Ajuga bracteosa*
Common comfrey	*Symphytum officinale*
Common fig	*Ficus carica*
Common glochidion	*Glochidion rubrum*
Common Indian mulberry	*Morinda umbellata*
Common indigo	*Indigofera tinctoria, I. suffruticosa, I. trifoliata, I. longeracemosa, I. zollingeriana*
Common lantana	*Lantana camara*
Common melastoma	*Melastoma candidum*
Common paper mulberry	*Brousonetia papyrifera*
Common rue	*Ruta graveolens*
Common rush	*Juncus effusus var. decipiens*

Common St. Paul's wort	*Siegesbeckia orientalis*
Common tree fer	*Cyathea lepifera*
Copper leaf	*Acalypha australis*
Coralberry	*Ardisia crenata*
Coriander	*Coriandrum sativum*
Cork tree	*Phellodendron amurense, P. wilsonii*
Corn chrysanthemum	*Chrysanthemum segetum*
Cotton	*Gossampinus malabarica*
Cotton rose	*Hibiscus mutabilis*
Cotton tree	*Bombax malabarica*
Couchgrass	*Cyathea podophylla*
Cramp bark	*Viburnum odoratissimum, V. plicatum var. formosanum, V. awabuki, V. luzonicum*
Crape jasmine	*Tabernaemontana divaricata, T. amygdalifolia, T. pandacaqui*
Crape myrtle	*Lagerstroemia subcostata*
Creat	*Andrographis paniculata*
Creeping fig	*Ficus pumila var. awkeotsang*
Creeping gentian	*Crawfurdia fasciculata*
Creeping orchid	*Habenaria repens*
Creeping St. John's wort	*Hypericum geminiflorum*
Crepe ginger	*Costus speciosus*
Croton	*Croton lachnocarpus, C. tiglium*
Cuban bast	*Hibiscus tillaceus*
Cubebs	*Litsea hypophaea, L. cubeba, L. acutivena*
Cudweed	*Gnaphalium adnatum, G. hypoleucum, G. affine, Salvia plebeia*
Culantro	*Eryngium foetidum*
Custard apple	*Artabotrys uncinatus*
Cypress vine	*Ipomoea quamoclit*
Da Chi	*Crisium suzukii*
Da Yi Sou Gai Ju	*Diplazium subsinuatum, D. megalophyllum*
Da Tzu Da Chi	*Crisium arisanense, C. arisanense*

English Name	Scientific Name
Da Yi Tian Gin Ba	*Elemingia macrophylla*
Da Chi	*Crisium moril*
Dandelion	*Taraxacum officinale, T. mongolicum*
Date palm	*Phoenix dactylifera*
Day flower	*Commelina benghalensis, C. communis*
Day lily	*Hemerocallis longituba, H. fulva*
Dense flowered false nettle	*Boehmeria densiflora*
Deutzia	*Deutzia corymbosa, D. cordatula, D. gracilis*
Devil pepper	*Rauvolfia verticillata*
Devil's tongue	*Amorphophallus konjac*
Di Ma Huang	*Mollugo pentaphylla*
Dianella	*Dianella longifolia, D. ensifolia*
Dichondra	*Dichondra micrantha*
Ding Gui Cao	*Zornia diphylla*
Dock	*Rumex crispus*
Dog rose	*Rosa davurica*
Dong Feng Ju Gen	*Atalantia buxifolia*
Dooryard weed	*Plantago asiatica*
Dragon root	*Arisaema vulgaris*
Du Jing Shan	*Maesa laxiflora, M. lanceolata*
Duck foot	*Urena procumbens*
Duck foot mugwort	*Artemisia lactiflora*
East Indian lotus	*Nelumbo nucifera*
East Indian walnut	*Albizzia lebbeck*
Eggplant	*Solanum lyratum*
Elderberry	*Sambucus javanica*
Elephant ear	*Alocasia cucullata*
Endive	*Cichorium endivia, C. endivia*
English daisy	*Bellis perennis*

Entada	*Entada phaseoloides*
Erh Se Yeh Shan Heh Tou	*Dumasia pleiantha, D. truncata, D. bicolor, D. villosa ssp. bicolor*
Eucommia	*Eucommia ulmoides*
Euonymus	*Euonymus echinatus*
Euphorbia	*Euphorbia atoto, E. jolkini*
Eusolex	*Taxillus levinei*
Evergreen artemisia	*Artemisia capillaris*
Everlasting	*Dipteracanthus repens, D. prostratus, Gnaphalium luteoalbum ssp. affine*
Fairy fern	*Daphniphyllum calycinum, D. glaucescens spp. oldhamii*
False Jerusalem cherry	*Solanum capsicastrum*
False mallow	*Malvastrum coromandelianum*
False staghorn fern	*Dicranopteris dichotoma, D. linearis*
Fame flower	*Talinum patens, T. triangulare, T. paniculatum*
Felt fern	*Pyrrosia petiolosa, P. adnascens, P. polydactylis*
Feng-Qi grass	*Adenostemma lavenia*
Fetterbush	*Pieris hieracloides*
Fevervine	*Paederia foetida, P. cavaleriei*
Field aster	*Kalimeris indica*
Field pansy	*Viola tricolor*
Field sow thistle	*Sonchus arvensis*
Fig	*Ficus superba var. japonica, F. virgata, F. religiesa*
Figwort	*Scrophularia yoshimurae, Hemiphragma heterophyllum var. dentatum*
Fiji longan	*Pometia pinnata*
Finger citron	*Citrus medica var. sarcodactylis*
Fireweed	*Erechtites valerianaefolia*
Fishwort	*Houttuynia cordata*
Five-leave chaste tree	*Vitex negundo*
Fleabane	*Conyza blinii, C. canadensis, C. dioscoridis, C. sumatrensis, Erigeron canadensis*
Florence fennel	*Foeniculum vulgare*
Fo-ti	*Polygonum multiflorum var. hypoleucum*

English Name	Scientific Name
Fong Feng Cao	*Anisomeles indica*
Formosa palm	*Arenga saccharifera*
Formosan actinidia	*Actinidia callosa* var. *formosana*
Formosan elderberry	*Sambucus formosana*
Formosan fig tree	*Ficus formosana*
Formosan raspberry	*Rubus formosensis*
Formosan supple jack	*Berchemia formosana*
Foxglove	*Digitalis purpurea*
Fragrant orchid	*Haraella retrocalla*
Frangipani	*Plumeria rubra* cv. *acutifolia*
Fringe orchid	*Habenaira dentata*
Frogfruit	*Phyla nodiflora*
Fruit fig tree	*Ficus wightiana*
Galanga	*Kaempferia galanga*
Gambir	*Uncaria hirsuta*
Garden balsam	*Impatiens balsamina*
Garden burnet	*Sanguisorba minor*
Garden portulaca	*Portulaca grandiflora*
Garden sorrel	*Rumex acetosa*
Gardenia	*Gardenia oblongifolia*
Garlic	*Allium tuberosum, A. sativum*
Garu	*Excoecaria agallocha, E. kawakamii, E. orientalis*
Gentian	*Gentiana scabrida, G. scabrida* var. *horaimontana, G. atkinsonii, G. arisanensis, G. flavo-maculata, G. campestris*
Geranium	*Geranium suzukii, G. suzukii* var. *hayatanum, G. nepalense* var. *thunbergii*
Germander	*Teucrium viscidum*
Gesnerias	*Hemiboea bicornuta*
Ghost plant	*Graptopetalum paraguayense*
Giant miscanthus	*Miscanthus floridulus*
Giant taro	*Alocasia macrorrhiza*

Ginger	*Zingiber officinale, Z. kawagoii, Z. rhizoma*
Ginger lily	*Alpinia oxyphylla, A. speciosa, Hedychium coronarium*
Globe thistle	*Echinops grilisii*
Glochidion	*Glochidion puberum*
Gold bean	*Psophocarpus tetragonolobus*
Gold button	*Spilanthes acmella var. oleracea, S. acmella*
Gold silver nightshade	*Solanum aculeatissimum*
Gold thread	*Coptis chinensis*
Golden canna	*Canna flaccida*
Golden dewdrops	*Duranta repens*
Golden hinoki cypress	*Chamaecyparis obtusa var. filicoides*
Golden smoke	*Corydalis pallida*
Golden St. John's wort	*Hypericum patulum*
Golden-hair grass	*Pogonatherum paniceum, P. crinitum*
Goldenrod	*Solidago altissima, S. virgo-aurea*
Gotu Kola	*Centella asiatica*
Gou Gan Cai	*Dicliptera chinensis*
Gou Gan Cai	*Dicliptera riparia* Nees
Gou Teng Diao	*Rhynchoglossum holglossum*
Gourian clematis	*Clematis gouriana* ssp. *lishanensis*
Grape ivy	*Cissus sicyoides, C. repens*
Gravel root	*Eupatorium clematideum*
Gravida	*Annona muricata, A. reticulata*
Green gentian	*Swertia randaiensis*
Green penny fern	*Lemmaphyllum microphyllum*
Ground ivy	*Glechoma hederacea var. grandis*
Ground cherry	*Physalis angulata*
Ground mulberry	*Ranunculus sceleratus*
Guacatonga	*Casearia membranacea*
Guava	*Psidium guajave*

English Name	Scientific Name
Hairy bougainvillea	*Bougainvillea spectabilis*
Hairy crabweed	*Fatoua pilosa*
Hairy clerodendrum	*Clerodendrum trichotomum, C. trichotomum var. fargesii*
Hairy elephant's foot	*Elephantopus mollis*
Hairy holly	*Ilex pubescens*
Hairy purslane	*Portulaca pilosa*
Happy tree	*Camptotheca acuminata*
Hare's ear	*Bupleurum falcatum*
Heal-all	*Prunella vulgaris*
Heart-shaped leaf liparis	*Liparis loeselii, L. cordifolia, L. keitaoensis*
Hedge euphorbia	*Euphorbia neriifolia*
Hei Jao Ku Sha Pu	*Davallia mariesii*
Helwingia	*Helwingia japonica* ssp. *formosana*
Hibiscus	*Hibiscus esculentus*
Hill buckwheat	*Polygonum chinense*
Hill gooseberry	*Rhodomyrtus tomentose*
Himalayan paris	*Paris polyphylla*
Holly	*Ilex asprella*
Holly mangrove	*Acanthus ilicifolius*
Honeysuckle	*Lonicera macrantha, L. kawakamii, L. apodonta, Uraria lagopodioides*
Hong Si Shar	*Peristrophe roxburghiana*
Hops	*Humulus scandens*
Horse field euchresta	*Euchresta formosana*
Horsetail	*Equisetum ramosissimum*
Huaang Hua Jia Zhu Tao	*Thevetia peruviana*
Huang Nhi Cha	*Cratoxylon ligustrinum*
Huo Tan Mu	*Microcos paniculata*
Hyacinth bean	*Dolichos lablab*
African potato	*Hyptis suaveolens*

India abutilon	*Abutilon indicum*
India heliotrope	*Heliotropium indicum*
Indian dammacanthus	*Damnacanthus indicus*
Indian madder	*Rubia akane, R. lanceolata, R. linii*
Indian night shade	*Solanum indicum*
Indian oleander	*Nerium indicum*
Indian privet	*Vitex rotundifolia*
Indian pulchea	*Pluchea indica*
Iris	*Iris tectorum*
Iron plant	*Aspidistra elatior*
Ironweed	*Vernonia gratiosa, V. cinerea*
Jack bean	*Canavalia ensiformis*
Jack fruit tree	*Artocarpus heterophyllus*
Jack-in-the-pulpit	*Aristolochia heterophylla*
Jamaica vervain	*Stachytarpheta jamaicensis*
Japanese anise	*Illicium arborescens*
Japanese aralia	*Fatsia japonica*
Japanese artemisa	*Artemisia japonica*
Japanese blueberry	*Vaccinium japonium*
Japanese callicarpa	*Callicarpa japonica*
Japanese chaff flower	*Achyranthes japonica*
Japanese cleyera	*Cleyera japonica*
Japanese coltsfoot	*Petasites japonicus*
Japanese curly dock	*Rumex japonicus*
Japanese Da Chi	*Crisium japonicum var. australe, C. japonicum var. takaoense*
Japanese honeysukle	*Lonicera japonica, L. japonica var. sempervillosa*
Japanese hornwort	*Cryptotaenia canadensis*
Japanese knotweed	*Polygonum cuspidatum*
Japanese lily turf	*Ophiopogon japonicus*
Japanese Liu Shan	*Cryptotaenia japonica*

English Name	Scientific Name
Japanese mahonia	*Mahonia japonica*
Japanese mallotus	*Mallotus japonicus*
Japanese Mu Fang	*Cocculus sarmentosus*
Japanese Mu Guo	*Chacnomeles japonica*
Japanese pokeberry	*Phytolacca japonica*
Japanese radish	*Ranunculus japonicus*
Japanese raspberry	*Rubus parvifolius*
Japanese sage	*Salvia japonica*
Japanese silver grass	*Miscanthus sinensis var. condensatus*
Japanese snail seed	*Cocculus trilobus*
Japanese St. John's wort	*Hypericum japonicum*
Japanese stephania	*Stephania japonica*
Japanese velvet plant	*Gynura japonica var. flava*
Japanese wintergreen	*Pyrola japonica*
Jascobinia	*Justicia gendarussa*
Jasmin orange	*Murraya paniculata*
Jessamine	*Gelsemium elegans*
Jewel orchid	*Anoectochilus formosanus*
Jian Dao Gu	*Ixeris tamagawaensis*
Jimsonweed	*Datura tatula, D. metel f. fastuosa, D. metel*
Jin Guo Lan	*Tinospora tuberculata*
Jiu Jie Cha	*Sarcandra glabra*
Job's tears	*Coix lacryma-jobi*
Joint flowered knotweed	*Polygonum plebeium*
Juniper tamarisk	*Tamarix juniperina*
Jute	*Corchorus capsularis, C. olitorius*
Kikio root	*Platycodon grandiflorum*
Kitamura	*Kitamura forma*
Koda tree	*Ehretia resinosa, E. dicksonii, E. acuminata*

Kong Xin Hua	*Maesa perlaria* var. *formosana, M. tenera*
Kosam seed	*Brucea javanica*
Kudzu vine	*Pueraria lobata*
Kudzu	*Pueraria montana*
Ladies' tresses	*Spiranthes sinensis*
Lamb of tartary	*Cibotium cumingii*
Lamb's quarter	*Chenopodium album*
Lantern seedbox	*Ludwigia octovalvis*
Lantern tridax	*Tridax procumbens*
Lanyu fig	*Ficus pedunculosa* var. *mearnsii*
Lanyu tylophora	*Tylophora lanyuensis*
Lao Theung	*Cyclea barbata*
Large-leaf abacus plant	*Glochidion lanceolarium*
Large-leaf elaeagnus	*Elaeagnus macrophylla*
Leather flower	*Clematis florida*
Lemon	*Citrus medica* var. *gaoganensis*
Lemon balm	*Melissa officinalis*
Lemon grass	*Cymbopogon citratus*
Lemonade berry	*Rhus javanica* var. *roxburghiana*
Leng Chi Cao	*Elatostema edule*
Leopard plant	*Farfugium japonicum*
Lesser melastoma	*Melastoma septemnervium, M. dodecandrum*
Lettuce	*Lactuca indica*
Li Tou Jian	*Typhonium divaricatum*
Liao Ge Wang	*Wendlandia formosana, W. indica*
Licorice	*Glycyrrhiza uralensis*
Lily of the valley	*Rhodea japonica*
Lily turf	*Liriope spicata*
Linear stonecrop	*Sedum formosanum*
Lizard's tail	*Saururus chinensis*

English Name	Scientific Name
Lo Haing Fa	*Astilbe longicarpa*
Lobelia	*Lobelia laxiflora*
Lobelia	*Lobelia nummularia*
Long Chuan Hua	*Ixora chinensis*
Long leaf chaff flower	*Achyranthes bidentata*
Long leaf wurrus	*Flemingia macrophylla*
Long-headed sedge	*Mariscus cyperinus*
Longan fruit	*Euphoria longana*
Loose-flowered euonymus	*Euonymus laxiflorus*
Loosestrife	*Lysimachia ardisioides, L. mauritiana, L. simulans*
Loquat	*Eriobotrya japonica*
Lu Huai Hua	*Rhynchosia volubilis, R. minima*
Luffa sponge	*Luffa cylindrica*
Ma Jia Zi Ye	*Paliurus ramosissimus*
Ma Sang Ye	*Coriaria japonica ssp. intermedia, C. intermedia*
Macarabga	*Macaranga tanarius*
Madagascar periwinkle	*Catharanthus rosens, C. torosa*
Maidenhair fern	*Adiantum capillus-veneris*
Malaruhat	*Cleistocalyx operculatus*
Mallotus	*Mallotus apelta*
Manila leea	*Leea guineensis*
Marble leaf	*Peristrophe japonica*
Marble vine	*Diplocyclos palmatus*
Marvel of Peru	*Mirabilis jalapa*
Matrimony vine	*Lycium chinense*
Meadow hedyotis	*Hedyotis uncinella*
Mei Leng Chow	*Canna indica*
Melia-leaf evodia	*Evodia meliaefolia*
Mesona	*Mesona procumbens*

English name	Scientific name
Mexican fire plant	*Euphorbia heterophylla*
Mexican marigold	*Tagetes erecta*
Mexican mint	*Plectranthus amboinicus*
Mexican sunflower	*Tithonia diversifolia*
Milk bush	*Euphorbia tirucalli*
Milk fig tree	*Ficus erecta* var. *beecheyana*
Milkwort	*Polygala aureocauda*
Milletia	*Millettia nitida, M. taiwaniana, M. pachycarpa*
Minimum light	*Schefflera octophylla*
Mint	*Epimeredi indica*
Mist flower	*Eupatorium amabile*
Mistletoe	*Viscus angulatum, V. multinerve, V. alniformosanae*
Mock jute	*Corchorus aestuand*
Mollotus	*Mallotus repandus, M. paniculatus, M. tiliaefolius*
Molucca mallotus	*Melanolepis multiglandulosa*
Money plant	*Epipremnum pinnatum*
Moonwort fern	*Botrychium lanuginosum, B. daucifolium*
Mountain jasmine	*Jasminum hemsleyi* Yamamoto
Mountain orange	*Citrus maxima*
Mulberry tree	*Morus alba*
Mung bean	*Vigna radiata*
Murrogen	*Cryptocarya chinensis*
Musk mallow	*Abelmoschus esculentus*
Nagi podocarp	*Podocarpus nagi*
Nan Jan	*Laungusa galanga*
Narrow leaf alternanthera	*Alternanthera nodiflora*
Narrow-leaf rattlebox	*Crotalaria similis, C. sessiliflora*
Narrow leaf sida	*Sida acuta*
Nasturtium	*Tropaeolum majus*
Native bloody leaf	*Achyranthes aspera* var. *indica*

English Name	Scientific Name
Night blooming cactus	*Epiphyllum oxypetalium, Hylocereus undatus*
Nightshade	*Solanum undatum, S. abutiloides, Tubocapsicum anomalum*
Noble bottle tree	*Sterculia nobilis*
Noni	*Morinda citrifolia*
Nutmeg	*Myristica cagayanensis, M. fragrans*
Oechids	*Hippobroma longiflora*
Officinal breynia	*Breynia officinalis*
Oil tea	*Camellia oleifera*
Oldham elaeagnus	*Elaeagnus oldhamii*
Oldham saurauia	*Saurauia oldhamii*
Oleaster	*Elaeagnus glabra, E. lanceollata, E. thunbergii, E. obovata, E. morrisonensis, E. bockii, E. loureirli, E. wilsonii*
Orange honeysuckle	*Lonicera confusa*
Orange leaf pothos	*Pothos chinensis*
Orchid tree	*Bauhinia purpurea, B. variegata*
Oregon grape	*Mahonia oiwakensis*
Oriental bittersweet	*Celastrus punctatus, C. paniculatus, C. orbiculatus*
Oriental cudrania	*Cudrania cochinchinensis*
Oriental hammock fern	*Blechum amabile, B. orientale, B. pyramidatum*
Oriental hawksbeard	*Youngia japonica*
Ornamental sweet potato vine	*Ipomoea stans*
Ovate leaf tylophora	*Tylophora ovata*
Pagoda flower	*Clerodendrum paniculatum, C. japonicum*
Pai Pan Feng Ho	*Dendropanax pellucidopunctata*
Palm grass	*Setaria palmifolia, S. italica, S. viridis*
Palm sedge	*Carex baccans*
Panpienchi	*Pteris semipinnata*
Pansy	*Viola inconspicua ssp. nagasakiensis, V. confusa, V. diffusa, V. betonicifolia, V. hondoensis*
Paris	*Paris lancifolia, P. arisanensis, P. formosana*
Pata de gallina	*Lepidagathis hyalina, L. formosensis, L. cristata*

Patchouli	*Pogostemon cablin, P. amboinicus*
Pawpaw	*Goniothalamus amuyon*
Peach	*Prunus persica*
Peanut, groundnut	*Arachis hypogea*
Peanut grass	*Atylosia scarbaeoides*
Pepper	*Piper kadsura, P. kawakamii, P. arboricola*
Pepper vine	*Ampelopsis cantoniensis*
Peppermint	*Mentha haplocalyx* Briq.
Perennial lespedeza	*Lespedeza cuneata*
Perfume herb	*Procris laevigata*
Perilla	*Perilla frutescens, P. frutescens var. crispa, P. ocymoides*
Persimmon	*Diospyros angustifolia, D. eriantha*
Petroleum plant	*Euphorbia thymifolia*
Petunia	*Ruellia tuberosa*
Pigmy water lily	*Nymphaea tetragona*
Pilose agrimony	*Agrimonia pilosa*
Pine	*Pinus massoniana*
Pineapple	*Ananas comosus*
Pinellia	*Pinellia pedatisecta*
Ping	*Marsilea crenata, M. minuta*
Pink plant	*Eclipta alba, E. prostrata*
Piper	*Piper sanctum, P. sarmentosum* Roxb.
Pittosporum	*Pittosporum pentandrum*
Plantain	*Musa basjoo var. formosana, M. insularimontana, Plantago major*
Plume thistle	*Cirsium albescens*
Poisonous wood nettle	*Laportea pterostigma*
Pokeberry	*Phytolacca acinosa, P. americana*
Polka dot plant	*Hypoestes purpurea*
Pomegranate	*Punica granatum*
Potato yam	*Dioscorea bulbifera*

English Name	Scientific Name
Potentilla	*Potentilla tugitakensis, P. leuconta*
Pouteria	*Pouteria obovata*
Pratia	*Pratia nummularia*
Prayers beads	*Abrus cantoniensis*
Prickly ash	*Zanthoxylum integrifoliolum, Z. ailanthoides, Z. avicennae, Z. dimorphophylla, Z. pistaciflorum, Z. piperitum*
Prickly chaff-flower	*Achyranthes aspera var. rubro-fusca*
Prickly-pear cactus	*Opuntia dillenii*
Privet	*Ligustrum pricei*
Purple heart	*Setcreasea purpurea*
Purple-leaf spiderwort	*Rhoeo spathacea*
Purslane	*Portulaca oleracea*
Qing Jiu Gang	*Desmodium triquetrum, D. triflorum, D. capitatum, D. pulchellum, D. multiflorum, D. laxiflorum, D. caudatum, D. sequax*
Qiu Ju Cao	*Dichrocephala bicolor*
Quail grass	*Celosia argentea*
Rabbit milkweed	*Ixeris chinensis*
Ragwort	*Senecio scandens, S. nemorensis*
Railroad vine	*Ipomoea pes-caprae ssp. brasiliensis*
Rangoon creeper	*Quisqualis indica*
Rat tail willow	*Justicia procumbens*
Red azalea	*Rhododendron simsii*
Red cluster pepper	*Capsicum frutescens*
Red Japanese hibiscus	*Abelmoschus moschatus*
Red magnolia	*Magnolia liliflora*
Red psychotria	*Psychotria rubra*
Red tassel flower	*Emilia sonchifolia, E. sonchifolia var. javanica*
Redbird cactus	*Pedilanthus tithymaloides*
Reef pemphis	*Pemphis acidula*
Rhinacanthus	*Rhinacanthus nasutus*
Rhodesian kudzu	*Glycosmis citrifolia, Glycine tabacina, G. tomentella, G. javanica*

Rice bean	*Vigna umbelbita*
Rice paper tree	*Tetrapanax papyriferus*
Rosary pea	*Abrus percatorius*
Rose of China	*Hibiscus rosa-sinensis*
Rose of Sharon	*Hibiscus syriacus*
Roselle	*Hibiscus sabdariffa*
Rough elephant's foot	*Elephantopus scaber*
Rough-leaf stem fig	*Ficus hispida*
Round brack tick clover	*Phyllodium pulchellum*
Round-leaf rotala	*Rotala rotundifolia*
Ruiren	*Prinsepia scandens*
Russian olive	*Elaeagnus angustifolia*
Rusty leaf mucuna	*Mucuna macrocarpa*
Sacred bamboo	*Nandina domestica*
Sacred garlic pear	*Crateva nurvala, C. adansonii* ssp. *formosensis*
Safflower	*Carthamus tinctorius*
Sago palm	*Cycas revoluta*
Salad burnet	*Sanguisorba officinalis*
San Leng	*Scirpus ternatanus, S. maritimus*
Sanicle	*Sanicula elata, S. petagniodes*
Sappan wood	*Caesalpinia pulcherrima*
Sasagrass	*Lophatherum gracile*
Scarlet kadsura	*Kadsura japonica*
Schisandra	*Schisandra arisanensis*
Screwpine	*Pandanus pygmaeus, P. amaryllifolius, P. odoratissimus* var. *sinensis*
Senna	*Cassia mimosoides, C. fistula*
Sensitive plant	*Mimosa pudica*
Serissa	*Serissa foetida. S. japonica*
Serrated arum	*Arisaema erubescens*
Sesame seed	*Sesamum indicum*

English Name	Scientific Name
Seythian lamb	*Cibotium barometz*
Sha Tang Mu	*Acronychia pedunculata*
Shan Ci Gu	*Pleione formosana*
Shan Zhi Ma	*Helicteres angustifolia*
Shepherd's purse	*Capsella bursa-pastoris*
Shi Gee Son	*Aspidixia articulata, A. liquidambaricala*
Shi Hu	*Dendrobium moniliforme*
Shiny bramble	*Zanthoxylum nitidum*
Short-leaf kyllinga	*Kyllinga brevifolia*
Showy milletha	*Millettia speciosa*
Shrubby false nettle	*Boehmeria nivea var. tenacissima*
Shu Don Gua	*Saurauja tristyla var. oldhamii*
Shu Qu Cao	*Glossogyne tenuifolia*
Shui Xian Cao	*Oldenlandia diffusa, O. hedyotidea*
Siberian ginseng	*Acanthopanax senticosus*
Siberian motherwort	*Leonurus sibiricus f. albiflora*
Siberian yarrow	*Achillea millefolium*
Sicklepod	*Cassia tora*
Sida	*Sida rhombifolia*
Silk oak	*Grevillea robusia*
Silver stone glorybower	*Clerodendrum calamitosum*
Silvery messerschmidia	*Messerschmidia argentea*
Siphonostegia	*Siphonostegia chinensis*
Skullcap	*Scutellaria rivularis, S. indica, S. javanica var. playfairi*
Slender pitted seed	*Bothriospermum tenellium*
Small everlasting	*Microglossa pyrifolia*
Small evolvulus	*Evolvulus alsinoides*
Small-leaf mulberry	*Morus australis*
Small paper mulberry	*Broussonetia kazinoki*

English	Scientific
Smilax	*Smilax bracteata*
Smoketree	*Turpinia formosana*
Snail seed	*Cocculus orbiculata*
Snake grape	*Ampelopsis brevuoedybcykata*
Snake plant	*Sansevieria trifasciata*
Snake root	*Aristolochia cucurbitifolia*
Snake strawberry	*Duchesnea indica*
Snake-gourd	*Trichosanthes cucumeroides*
Snaker root	*Aristolochia manshuriensis, A. shimadai, A. kaempferi, A. kankanensis*
Soap berry	*Sapindus mukorossi*
Solomon's seal	*Polygonatum falcatum, P. kingianum, P. odoratum*
Sophora	*Sophora tomentosa, S. flavescens*
Sour creeper	*Ecdysanthera rosea*
Southern yew	*Podocarpus macrophyllus var. nakaii*
Sow thistle	*Sonchus oleraceus L.*
Spiceberry	*Ardisia sieboldii, A. squamulosa*
Spicebush	*Lindera okoensis, L. strychifolia, L. communis, L. glauca*
Spider plant	*Chlorophytum comosum*
Spider wisp	*Cleome gynandra*
Spinach	*Spinacia oleracea*
Spiny randia	*Randia spinoa*
Spiny cocklebur	*Xanthium strumarium*
Splash-of-white	*Mussaenda pubescens*
Spreading hedyotis	*Hedyotis diffusa*
St. John's lily	*Crinum asiaticum*
St. Paul's wort	*Erycibe henryi*
Staghorn summac	*Rhus typhina*
Star grass	*Hypoxis aurea*
Star jasmine	*Trachelospermum jasminoides*
Star sky alternanthera	*Alternanthera sessilis*

English Name	Scientific Name
Stemona	*Stemona tuberosa*
Stephania	*Stephania cephalantha, S. hispidula, S. tetrandra*
Stevia	*Stevia rebaudiana*
Stinging nettle	*Gonostegia pentandra, G. hirta, Urtica dioica, U. thunbergiana*
Stonecrop	*Sedum lineare, S. morrisonense, S. sempervivoides*
Stout camphor tree	*Cinnamomum micranthum*
Strap flower	*Loropetalum chinense*
Strawberry geranium	*Saxifraga stolonifera*
Striped plectranthus	*Rabdosia lasiocarpus*
Striped supple jack	*Berchemia lineata*
Strychnine	*Strychnos angustiflora*
Sugar cane	*Saccharum officinarum*
Sugar palm	*Arenga engleri*
Sumac	*Rhus semialata var. roxburghiana, R. microphylla, R. succedanea, R. verniciflua*
Sunflower	*Helianthus annuus*
Sung Chi Sheng	*Taxillus matsudai*
Supplejack	*Berchema racemosa*
Swamp mahogany	*Eucalyptus robusta*
Sweet acacia	*Acacia farnesiana*
Sweet basil	*Ocimum gratissimum*
Sweet broom wort	*Scoparia dulcis*
Sweet cassava	*Jatropha curcas*
Sweet flag	*Acorus calamus, A. gramineus*
Sweet gum tree	*Liquidambar formosana* Hance
Sweet olive	*Osmanthus fragrans*
Sweet potato	*Ipomoea batatas, I. obscura*
Sweet tea vine	*Gynostemma pentaphyllum*
Sword fern	*Nephrolepis auriculata*
Sword wound weed	*Ixeris laevigata*

Ta Yeh Sang Chih Sheng	Scurrula ritozonensis, S. liquidambaricolus, S. loniceriolius, S. ferruginea
Taihoku mussaenda	Mussaenda parviflora
Taiwan acacia	Acacia confusa
Taiwan acomite	Aconitum formosanum
Taiwan amethyst orchid	Bletilla formosana
Taiwan black currant	Ribes formosanum
Taiwan buckthorn	Rhamnus formosana
Taiwan burnet	Sanguisorba formosana
Taiwan cinnamon	Cinnamomum insulari-montanum
Taiwan coltsfoot	Petasites formosanus
Taiwan cotton rose	Hibiscus taiwanensis
Taiwan Cu Fei	Cephalotaxus wilsonianer
Taiwan cypress	Chamaecyparis formosensis
Taiwan dandelion	Taraxacum formosanum
Taiwan euphorbia	Euphorbia formosana
Taiwan fir	Taiwania cryptomeriodes
Taiwan five-leaf akebia	Akebia longeracemosa
Taiwan helwingia	Helwingla formosana
Taiwan hogfennel	Peucedanum formosanum
Taiwan Juang Yang	Buxus microphylla
Taiwan juniper	Juniperus formosana
Taiwan Leng Chi Cao	Elatostema lineolatum var. majus
Taiwan lily	Lilium formosanum
Taiwan Ma Lan	Goldfussia psilostachys
Taiwan meadow	Thalictrum fauriei
Taiwan mountain onion	Veratrum formosanum
Taiwan Ma Lan	Goldfussia formosanus
Taiwan Mu	Aralia taiwaniana
Taiwan Pai Lan	Eupatorium formosanum
Taiwan pieris	Pieris taiwanensis, P. formosa

English Name	Scientific Name
Taiwan pine	*Pinus taiwanensis*
Taiwan rose	*Rosa taiwanensis*
Taiwan sage	*Salvia hayatana*
Taiwan Shan Chen	*Melodinus angustifolius* Hayata
Taiwan skullcap	*Scutellaria formosana*
Taiwan slat vine	*Pericampylus formosanus, P. trinervatus*
Taiwan Solomon's seal	*Smilacina formosana*
Taiwan Sou Su	*Deutzia taiwanensis*
Taiwan spirea	*Spiraea formosana*
Taiwan taro	*Colocasia formosana*
Taiwan toad lily	*Tricytis formosana*
Taiwan velvert plant	*Gynura formosana*
Taiwan wampee	*Clausena lansium, C. excavata*
Taiwan wild grape	*Vitis thunbergii*
Tallow tree	*Sapium sebiferum*
Tan Gen	*Dalbergia odorifera*
Tangerin orange	*Citrus tangerina*
Taro with black vein	*Colocasia antiquorum* var. *illustris*
Taro	*Colocasia esculenta*
Teng Sha Chi	*Boussingaultia gracilis* var. *pseudobaselleoides*
Ternate pinellia	*Pinellia ternata*
Texas sage	*Salvia coccinea* Juss. ex Murr.
Thatch grass	*Imperata cylindrica* var. *major*
Thick-leaf croton	*Croton crassifolius*
Thistle	*Cirsium japonicum, C. japonicum* var. *australe*
Three-leaf Acanthopanax	*Acanthopanax trifoliatus*
Thyme-leaved gratiola	*Bacopa monniera*
Ti plant	*Cordyline fruticosa*
Tian Hu Sui	*Hydrocotyle asiatica*

English Name	Scientific Name
Water murdannia	*Murdannia keisak, M. loriformis*
Wax begonia	*Begonia fenicis*
Wax plant	*Hoya carnosa*
Wax tree	*Ligustrum lucidum*
Wax weed	*Hypolepis tenuifolia*
Wedelia	*Wedelia biflora*
Wedgelet fern	*Balanophora spicata, Tetrastigma hemsleyanum*
Weed passion flower	*Passiflora foetida var. hispida*
West India chickweed	*Drynaria cordata*
Whipping fig	*Ficus benjamina*
White bolly gum	*Neolitsea acuminatissima*
White champac	*Michelia alba*
White justicia	*Gendarussa vulgaris*
White zephylily	*Zephyranthes carinata, Z. candida*
Wild basil	*Clinopodium laxiflorum, C. umbrosum*
Wild cashina	*Rollinia mucosa*
Wild copper leaf	*Acalypha indica*
Wild ginger	*Heterotropa hayatanum, H. macrantha, Asarum macranthum, A. longerhizomatosum, A. hypogynum, A. hongkongense*
Wild hops	*Flemingia prostrata*
Wild lettuce	*Pterocypsela indica*
Wild machilus	*Machilus zuihoensis, M. kusanoi*
Wild turmeric	*Curcuma zedoaria*
Willow	*Salix warburgii*
Winter crookneck squash	*Cucurbita moschata*
Wintergreen	*Pyrola morrisonensis*
Wire vine	*Muehlenbeckia hastulata, M. platychodum*
Wishbone plant	*Torenia concolor var. formosana*
Wolf tooth	*Potentilla discolor*
Wolfsbane	*Aconitum bartletii*

Wood nettle	*Laportea moroides*
Wood sorrel	*Oxalis corymbosa, O. corniculata*
Wormseed goose foot	*Chenopodium ambrosioides*
Wrinkle fruit leaf flower	*Phyllanthus multiflorus, P. urinaria, P. emblica*
Wu Shui Ge	*Pouzolzia elegans, P. pentandia, P. zeylanica*
Xiang Ju	*Crossostephium chinense*
Xiao Tzu Da Ch	*Crisium ferum, C. kawakamii*
Xu Chang Qing	*Cynanchum paniculatum*
Xue Feng Teng	*Ventilago leiocarpa*
Ya She Cao	*Monochoria vaginalis*
Yan Gan Cao	*Carpesium divaricatum*
Yellow crotalaria	*Crotalaria pallida*
Yellow rattan palm	*Daemonorops margaritae*
Yellow stem fig	*Pericampylus glaucus*
Yellow vine	*Tripterygium wilfordii*
Yew	*Taxus mairei*
Yi Ye Qiu	*Securinega virosa, S. suffruticosa*
Yuan Hua	*Daphne odora, D. arisanensis*
Zhi Mu	*Anemarrhena asphodeloides*
Zhu Ye Lian	*Pollia secundiflora*
Zou You Cao	*Tetrastigma umbellatum, T. formosanum, T. dentatum*

List of Scientific and Common Names

Scientific Name	Common Name
Abelmoschus esculentus	Musk mallow
Abelmoschus moschatus	Red Japanese hibiscus
Abrus cantoniensis	Prayer beads
Abrus precatorius	Rosary pea
Abutilon indicum	India abutilon
Abutilon taiwanensis	Abutilon
Acacia confusa	Taiwan acacia
Acacia farnesiana	Sweet acacia
Acalypha australis	Copper leaf
Acalypha indica L.	Wild copper leaf
Acanthopanax senticosus	Siberian ginseng
Acanthopanax trifoliatus	Three-leaf acanthopanax
Acanthus ilicifolius L.	Holly mangrove
Acer buerferianum Miq.	Trident maple
Achillea millefolium L.	Siberian yarrow
Achyranthes aspera L. var. *indica*	Native bloody leaf
Achyranthes aspera L. var. *rubro-fusca*	Prickly chaff-flower
Achyranthes bidentata	Long leaf chaff flower
Achyranthes japonica	Japanese chaff flower
Achyranthes longifolia	Achyranthes
Achyranthes ogotai	Achyranthes
Aconitum bartletii	Wolfsbane
Aconitum formosanum	Taiwan acomite
Aconitum fukutomel	Aconite
Aconitum kojimae	Aconite
Aconitum kojimae var. *lassiocarpium*	Aconite
Aconitum kojimae Ohwi var. *ramosum*	Aconite
Aconitum yamamotoanum Ohwi	Aconite
Acorus calamus	Sweet flag

Acorus gramineus	Sweet flag
Acronychia pedunculata	Sha Tang Mu (English name not available)
Actinidia callosa	Fomosan actinidia
Adenophora stricta	Bellflower
Adenophora triphylla	Bellflower
Adenostemma lavenia	Feng-Qi grass
Adiantum capillus-veneris	Maidenhair fern
Adiantum flabellulatum	Black maidenhair
Adina pilulifera	Adina
Adina racemose	Adina
Agastache rugosa	Chinese giant hyssop
Ageratum conyzoides	Bastard agrimony
Ageratum houstonianum	Bastard agrimony
Agrimonia pilosa	Pilose agrimony
Ajuga bracteosa	Common bugleweed
Ajuga decumbens	Bugleweed
Ajuga pygmaea	Bugleweed
Akebia longeracemosa	Taiwan five-leaf akebia
Alangium chinense	Ba Jiao Feng Gen (English name not available)
Albizzia lebbeck	East Indian walnut
Aletris formosana	Chinese stargrass
Aleurites fordii	Candlenut
Aleurites moluccana	China wood oil
Aleurites montana	China wood oil tree
Allium bakeri	Barker's garlic
Allium sativum	Garlic
Allium scorodoprasum	Chive
Allium thunbergii	Chinese chive
Allium tuberosum	Garlic
Alocasia cucullata	Elephant ear

Scientific Name	Common Name
Alocasia macrorrhiza	Giant taro
Alpinia oxyphylla	Ginger lily
Alpinia speciosa	Ginger lily
Alternanthera nodiflora	Narrow-leaf alternanthera
Alternanthera philoxeroides	Alligator alternanthera
Alternanthera sessilis	Star sky alternanthera
Alyxia insularis	A Li Teng (English name not available)
Alyxia sinensis	A Li Teng (English name not available)
Amentotaxus formosana	Amentotaxus
Amorphophallus konjac	Devil's tongue
Ampelopsis brevuoedybcykata	Snake grape
Ampelopsis cantoniensis	Pepper vine
Ananas comosus	Pineapple
Andrographis paniculata	Creat
Anemarrhena asphodeloides	Zhi Mu (English name not available)
Angelica acutiloba	Angelica
Angelica citriodor	Angelica
Angelica hirsutiflora	Angelica
Angelica keiskei	Angelica
Anisomeles indica	Fong Feng Cao (English name not available)
Annona muricata	Gravida
Annona reticulata	Gravida
Anoectochilus formosanus	Jewel orchid
Aquilaria agallocha	Aloe wood
Aquilaria sibebsus	Aloe wood
Aquilaria sinensis	Chinese Aloe wood
Arachis hypogea	Peanut, groundnut
Aralia chinensis	Aralia
Aralia taiwaniana	Taiwan Mu

Arctium lappa	Burdock
Ardisia crenata	Coralberry
Ardisia sieboldii	Spiceberry
Ardisia squamulosa	Spiceberry
Areca catechu	Betel nut palm
Arenga engleri	Sugar palm
Arenga saccharifera	Formosa palm
Arisaema consanguineum	Arum
Arisaema erubescens	Serrated arum
Arisaema vulgaris	Dragon root
Aristolochia cucurbitifolia	Snake root
Aristolochia elegans	Calica flower
Aristolochia heterophylla	Jack-in-the-pulpit
Aristolochia kaempferi	Snake root
Aristolochia kankanensis	Snake root
Aristolochia manshuriensis	Snake root
Aristolochia shimadai	Snake root
Arrabotrys uncinatus	Custard apple
Artemisia capillaris	Evergreen artemisia
Artemisia indica	Artemisia
Artemisia japonica	Japanese artemisia
Artemisia lactiflora	Duck foot mugwort
Artemisia princeps	Asiatic wormwood
Artocarpus altilis	Bread fruit tree
Artocarpus heterophyllus	Jack fruit tree
Asarum hongkongense	Wild ginger
Asarum hypogynum	Wild ginger
Asarum longerhizomatosum	Wild ginger
Asarum macranthum	Wild ginger
Asclepias curassavica	Butterfly weed

Scientific Name	Common Name
Asparagus cochinchinensis	Asparagus
Aspidistra elatior	Iron plant
Aspidixia articulata	Shi Gee Son (English name not available)
Aspidixia liquidambaricala	Shi Gee Son (English name not available)
Asplenium nidus	Bird's nest fern
Astilbe longicarpa	Lo Haing Fa (English name not available)
Astragalus sinicus	Chinese mild vetch
Atalantia buxifolia	Dong Feng Ju Gen (English name not available)
Atylosia scarbaeoides	Peanut grass
Bacopa monniera	Thyme-leaved gratiola
Balanophora spicata	Wedgelet fern
Basella alba	Ceylon spinach
Basella rubra	Ceylon spinach
Bauhinia championi	Chrysanthemum flower tree
Bauhinia purpurea	Orchid tree
Bauhinia variegata	Orchid tree
Begonia fenicis	Wax begonia
Begonia laciniata	Begonia
Begonia malabarica	Begonia
Belamcanda chinensis	Blackberry lily
Bellis perennis	English daisy
Berchemia formosana	Formosan supplejack
Berchemia lineata	Striped supplejack
Berchema racemosa	Supplejack
Bidens pilosa L. var. *minor*	Bur. marigold
Biota orientalis (L.)	Artorvitae leaves
Bischofia javanica	Autumn maple tree
Bixa orellana	Anatto tree
Blechum amabile	Oriental hammock fern

Blechum orientale	Oriental hammock fern
Blechum pyramidatum	Oriental hammock fern
Bletilla formosana	Taiwan amethyst orchid
Bletilla striata	Amethyst orchid
Blumea aromatica	Blumea camphor
Blumea balsamifera var. microcephala	Blumea camphor
Blumea lacera	Blumea camphor
Blumea laciniata	Blumea camphor
Blumea lanceolaria	Blumea camphor
Blumea riparia var. megacephala	Blumea camphor
Boehmeria densiflora	Dense flowered false nettle
Boehmeria nivea var. tenacissima	Shrubby false nettle
Bombax malabarica	Cotton tree
Bothriospermum tenellium	Slender pitted seed
Botrychium daucifolium	Moonwort fern
Botrychium lanuginosum	Moonwort fern
Bougainvillea spectabilis	Hairy bougainvillea
Boussingaultia gracilis var. pseudobaselleoides	Teng Sha Chi (English name not available)
Bredia oldhamii	Coffea arabica
Bredia rotundifolia	Coffea arabica
Bredia scandens	Coffea arabica
Breynia accrescens	Breynia
Breynia fruitcosa	Breynia
Breynia officinalis	Officinal breynia
Broussonetia kazinoki	Small paper mulberry
Broussonetia papyrifera	Common paper mulberry
Brucea javanica	Kosam seed
Bryophyllum pinnatum	Air plant
Buddleja asiatica	Asiatic butterfly bush
Bupleurum chinensis	Chinese hare's ear

Scientific Name	Common Name
Bupleurum falcatum	Hare's ear
Bupleurum kaoi	Bupleurum
Buxus microphylla	Taiwan Juang Yang
Caesalpinia pulcherrima	Sappan wood
Callicarpa formosana	Callicarpa
Callicarpa japonica	Japanese callicarpa
Callicarpa longissima	Beautybush
Callicarpa loureiri	Beautybush
Callicarpa nudiflora	Beautybush
Callicarpa pedunculata	Beautybush
Camellia japonica var. hozanensis Camellia	Frost Queen
Camellia oleifera	Oil tea
Camellia sinensis	Chinese camellia
Camptotheca acuminata	Happy tree
Canavalia ensiformis	Jack bean
Canna flaccida	Golden canna
Canna indica	Mei Leng Chow
Capsella bursa-pastoris	Shepherd's purse
Capsicum frutescens	Red cluster pepper
Cardiospermum halicacabum	Ballon vine
Carex baccans	Palm sedge
Carpesium divaricatum	Yan Gan Cao (English name not available)
Carthamus tinctorius	Safflower
Caryopteris incana	Common blue beard
Casearia membranacea	Guacatonga
Cassia fistula	Senna
Cassia mimosoides	Senna
Cassia occidentalis	Coffee senna
Cassia tora	Sicklepod

Cassia torosa	Madagascar periwinkle
Catharanthus rosens	Madagascar periwinkle
Cayratia japonica	Tree bine
Celastrus hypoleucus	Bittersweet
Celastrus kusanoi	Bittersweet
Celastrus orbiculatus	Oriental bittersweet
Celastrus paniculatus	Oriental bittersweet
Celastrus punctatus	Oriental bittersweet
Celosia argentea	Quail grass
Celosia cristata	Cockscomb
Centella asiatica	Gotu kola
Cephalotaxus wilsonianer	Taiwan Cu Fei
Ceratopteris thalictroides	Water fern
Chacnorneles japonica	Japanese Mu Guo
Chamaecyparis formosensis	Taiwan cypress
Chamaecyparis obtusa var. *filicoides*	Golden hinoki cypress
Chamaesyce hirta	Asthma plant
Chamaesyce thymifolia	Asthma plant
Chenopodium album	Lamb's quarter
Chenopodium ambrosioides	Wormseed goose foot
Cichorium endivia	Endive
Chloranthus oldham	Chloranthus
Chloranthus spicatus	Chloranthus
Chlorophytum comosum	Spider plant
Chrysanthemum indicum	Chrysanthemum
Chrysanthemum morifolium	Chrysanthemum
Chrysanthemum segetum	Corn chrysanthemum
Cibotium barometz	Scythian lamb
Cibotium cumingii	Lamb of tartary
Cichorum endivia	Endive

Scientific Name	Common Name
Cinnamomum cassia	Cassia bark tree
Cinnamomum camphora	Camphor tree
Cinnamomum insulari-montanum	Taiwan cinnamon
Cinnamomum kotoense	Camphor tree
Cinnamomum micranthum	Stout camphor tree
Cirsium albescens	Plume thistle
Cirsium japonicum	Thistle
Cirsium japonicum var. *australe*	Thistle
Cissus repens	Grape ivy
Cissus sicyoides	Grape ivy
Citrus maxima	Mountain orange
Citrus medica var. *gaoganensis*	Lemon
Citrus medica var. *sarcodactylis*	Finger citron
Citrus sinensis var. *sekken*	Chinese orange
Citrus tangerina	Tangerin orange
Claoxylon polot	Bai Tong Su (English name not available)
Clausena excavata	Taiwan wampee
Clausena lansium	Taiwan wampee
Cleistocalyx operculatus	Malaruhat
Clematis chinensis	Chinese clematis
Clematis florida	Leather flower
Clematis gouriana subsp. *lishanensis*	Gourian clematis
Clematis grata	Clematis
Clematis henryi	Virgin's bower
Clematis lasiandra	Virgin's bower
Clematis montana	Virgin's bower
Cleome gynandra	Spider wisp
Clerodendrum calamitosum	Silver stone glorybower
Clerodendrum cyrtophyllum	Clerodendrum

Clerodendrum inerme	Tube flower
Clerodendrum japonicum	Pagoda flower
Clerodendrum kaempferi	Tube flower
Clerodendrum paniculatum	Paagoda flower
Clerodendrum petasites	Clerodendrum
Clerodendrum philippinum	Clerodendrum
Clerodendrum trichotomum	Hairy clerodendrum
Clerodendrum trichotomum var. *fargesii*	Hairy clerodendrum
Cleyera japonica	Japanese cleyera
Clinopodium laxiflorum	Wild basil
Clinopodium umbrosum	Wild basil
Cocculus orbiculata	Snail seed
Cocculus sarmentosus	Japanese Mu Fang
Cocculus trilobus	Japanese snail seed
Codonopsis kawakami	Codonopsis
Coix lacryma-jobi	Job's tears
Coleus parvifolius	Coleus
Coleus scutellarioides var. *crispipilus*	Coleus
Colocasia antiquorum var. *illustris*	Taro with black vein
Colocasia esculenta	Taro
Colocasia formosana	Taiwan taro
Commelina benghalensis	Day flower
Commelina communis	Day flower
Conyza blinii	Fleabane
Conyza canadensis	Fleabane
Conyza dioscoridis	Fleabane
Conyza sumatrensis	Fleabane
Coptis chinensis	Gold thread
Corchorus aestuand	Mock jute
Corchorus capsularis	Jute

Scientific Name	Common Name
Corchorus olitorius	Jute
Cordyline fruticosa	Ti plant
Coriandrum sativum	Coriander
Coriaria intermedia	Ma Sang Ye (English name not available)
Coriaria japonica subsp. *intermedia*	Ma Sang Ye (English name not available)
Corydalis pallida	Golden smoke
Costus speciosus	Crepe ginger
Crateva adansonii subsp. *formosensis*	Sacred garlic pear
Crateva nurvala	Sacred garlic pear
Cratoxylon ligustrinum	Huang Nhi Cha (English name not available)
Crawfurdia fasciculata	Creeping gentian
Crinum asiaticum	St. John's lily
Cirsium albescens	Tzu Kai Cao (English name not available)
Cirsium arisanense	Da Tzu Da Chi (English name not available)
Cirsium ferum	Xiao Tzu Da Chi (English name not available)
Cirsium japonicum var. *australe*	Japanese Da Chi
Cirsium japonicum var. *takaoense*	Japanese Da Chi
Cirsium kawakamii	Xiao Tzu Da Chi (English name not available)
Cirsium moril	Da Chi (English name not available)
Cirsium suzukii	Da Chi (English name not available)
Crossostephium chinense	Xiang Ju (English name not available)
Crotalaria pallida	Yellow crotalaria
Crotalaria sessiliflora	Narrow-leaf rattlebox
Crotalaria similis	Narrow-leaf rattlebox
Croton crassifolius	Thick-leaf croton
Croton lachnocarpus	Croton
Croton tiglium	Croton
Cryptocarya chinensis	Murrogen

Scientific Name	Common Name
Cryptotaenia canadensis	Japanese hornwort
Cryptotaenia japonica	Japanese Liu Shan
Cucumis melo subsp. *melo*	Cantaloupe
Cucurbita moschata	Winter crookneck squash
Cudrania cochinchinensis	Oriental cudrania
Cunninghamia konishii	China fir
Curculigo capitulata	Black musli
Curculigo orchioides	Black musli
Curcuma domestica	Turmeric
Curcuma longa	Turmeric
Curcuma zedoaria	Wild turmeric
Cyathea lepifera	Common tree fern
Cyathea podophylla	Couchgrass
Cyathula prostrata	Bei Xian (English name not available)
Cycas revoluta	Sago palm
Cyclea barbata	Barbate cyclea
Cyclea insularis	Lao Theung
Cyclobalanopsis stenophylla	Chinese oak
Cymbopogon citratus	Lemon grass
Cymbopogon nardus	Citronella
Cynanchum paniculatum	Xu Chang Qing (English name not available)
Cyperus alternifolius	Umbrella plant
Daemonorops margaritae	Yellow rattan palm
Dalbergia odorifera	Tan Gen (English name not available)
Damnacanthus indicus	Indian damnacanthus
Daphne arisanensis	Yuan Hua (English name not available)
Daphne odora	Yuan Hua (English name not available)
Daphniphyllum calycinum	Fairy fern
Daphniphyllum glaucescen spp. *oldhamii*	Fairy fern
Datura metel	Jimsonweed

Scientific Name	Common Name
Datura metel f. *fastuosa*	Jimsonweed
Datura tatula	Jimsonweed
Davallia mariesii	Hei Jao Ku Sha Pu (English name not available)
Debregeasia edulis	Borduega
Debregeasia salicifolia	Borduega
Dendranthema indicum	Chrysanthemum
Dendrobium moniliforme	Shi Hu (in Chinese)
Dendropanax pellucidopunctata	Pai Pan Feng Ho (English name not available)
Derris elliptica	Trifoliate jewelvine
Derris trifoliata	Trifoliate jewelvine
Desmodium capitatum	Qing Jiu Gang (English name not available)
Desmodium caudatum	Qing Jiu Gang (English name not available)
Desmodium laxiflorum	Qing Jiu Gang (English name not available)
Desmodium multiflorum	Qing Jiu Gang (English name not available)
Desmodium pulchellum	Qing Jiu Gang (English name not available)
Desmodium sequax	Qing Jiu Gang (English name not available)
Desmodium triflorum	Qing Jiu Gang (English name not available)
Desmodium triquetrum	Qing Jiu Gang (English name not available)
Deutzia cordatula	Deutzia
Deutzia corymbosa	Deutzia
Deutzia gracilis	Deutzia
Deutzia taiwanensis	Taiwan Sou Su
Dianella chinensis	Chinese dianella
Dianella ensifolia	Dianella
Dianella longifolia	Dianella
Dianthus chinensis	Chinese carnation
Dichondra micrantha	Dichondra
Dichrocephala bicolor	Qiu Ju Cao (English name not available)
Dichroa febrifuga	Chinese quinine

Dicliptera chinensis — Gou Gan Cai (English name not available)
Dicliptera riparia — Gou Gan Cai (English name not available)
Dicranopteris dichotoma — False staghorn fern
Dicranopteris linearis — False staghorn fern
Digitalis purpurea — Foxglove
Dioscorea bulbifera — Potato yam
Dioscorea opposita — Chinese yam
Diospyros angustifolia — Persimmon
Diospyros eriantha — Persimmon
Diospyros khaki — Asian persimmon
Diplazium megaphllum — Da Yi Sou Gai Ju (English name not available)
Diplazium subsinuatum — Da Yi Sou Gai Ju (English name not available)
Diplocyclos palmatus — Marble vine
Dipteracanthus prostratus — Everlasting
Dipteracanthus repens — Everlasting
Dodonaea viscosa — Chammy hop seed bush
Dolichos lablab — Hyacinth bean
Drynaria cordata — West India chickweed
Drynaria diandra — Chickweed
Drynaria fortunei — Chickweed
Duchesnea indica — Snake strawberry
Dumasia bicolor — Shan Heh Tou (English name not available)
Dumasia pleiantha — Shan Heh Tou (English name not available)
Dumasia truncata — Shan Heh Tou (English name not available)
Dumasia villosa ssp. *bicolor* — Shan Heh Tou (English name not available)
Duranta repens — Golden dewdrops
Dysosma pleiantha — Chinese mayapple
Ecdysanthera rosea — Sour creeper
Echinochloa colonum — Barnyard grass
Echinops grilsii — Globe thistle

Scientific Name	Common Name
Eclipta alba	Pink plant
Eclipta prostrata	Pink plant
Ehretia acuminata	Koda tree
Ehretia dicksonii	Koda tree
Ehretia resinosa	Koda tree
Eichhornia crassipes	Water hyacinth
Elaeagnus angustifolia	Russian olive
Elaeagnus bockii	Oleaster
Elaeagnus glabra	Oleaster
Elaeagnus lanceollata	Oleaster
Elaeagnus loureirli	Oleaster
Elaeagnus macrophylla	Large-leaf elaeagnus
Elaeagnus morrisonensis	Oleaster
Elaeagnus obovata	Oleaster
Elaeagnus oldhamii	Oldham elaeagnus
Elaeagnus thunbergii	Oleaster
Elaeagnus wilsonii	Oleaster
Elatostema edule	Leng Chi Cao (English name not available)
Elatostema lineolatum var. majus	Taiwan Leng Chi Cao (English name not available)
Elemingia macrophylla	Da Yi Tian Gin Ba (English name not available)
Elephantopus mollis	Hairy elephant's foot
Elephantopus scaber	Rough elephant's foot
Emilia sonchifolia	Red tassel flower
Emilia sonchifolia var. javanica	Red tassel flower
Entada phaseoloides	Entada
Epimeredi indica	Mint
Epiphyllum oxypetalium	Night blooming cactus
Epipremnum pinnatum	Money plant
Equisetum ramosissimum	Horsetail

Erechtites valerianaefolia	Fireweed
Erigeron canadensis	Fleabane
Eriobotrya japonica	Loquat
Erycibe henryi	St. Paul's wort
Eryngium foetidum	Culantro
Eucalyptus robusta	Swamp mahogany
Euchresta formosana	Horse field euchresta
Eucommia ulmoides	Eucommia
Euonymus chinensis	Chinese thimble tree
Euonymus echinatus	Euonymus
Euonymus laxiflorus	Loose-flowered euonymus
Eupatorium amabile	Mist flower
Eupatorium cannabinum ssp. *asiaticum*	Boneset
Eupatorium clematideum	Gravel root
Eupatorium formosanum	Taiwan Pai Lan
Eupatorium lindleyanum	Boneset
Eupatorium tashiroi	Boneset
Euphorbia atoto	Euphorbia
Euphorbia formosana	Taiwan euphorbia
Euphorbia heterophylla	Mexican fire plant
Euphorbia hirta	Asthma herb
Euphorbia jolkini	Euphorbia
Euphorbia lathyris	Caper spurge
Euphorbia milli	Bojer's spurge
Euphorbia neriifolia	Hedge euphorbia
Euphorbia thymifolia	Petroleum plant
Euphorbia tirucalli	Milk bush
Euphoria longana	Longan fruit
Euryale chinese	Chinese water lily
Euryale ferox	Water lily

Scientific Name	Common Name
Evodia meliaefolia	Melia-leaf evodia
Evolvulus alsinoides	Small evolvulus
Excoecaria agallocha	Garu
Excoecaria kawakamii	Garu
Excoecaria orientalis	Garu
Farfugium japonicum	Leopard plant
Fatoua pilosa	Hairy crabweed
Fatsia japonica	Japanese aralia
Fatsia polycarpa	Aralia
Ferula assa-faoetida	Asafetida
Ficus benjamina	Whipping fig
Ficus carica	Common fig
Ficus erecta var. *beecheyana*	Milk fig tree
Ficus formosana	Formosan fig tree
Ficus hispida	Rough-leaf stem fig
Ficus microcarpa	Chinese banyan tree
Ficus pedunculosa var. *mearnsii*	Lanyu fig
Ficus pumila var. *awkeotsang*	Creeping fig
Ficus religiesa	Fig
Ficus sarmentosa var. *nipponica*	Alishan fig
Ficus septica Burm.	Angular fruit fig
Ficus superba var. *japonica*	Fig
Ficus virgata	Fig
Ficus wightiana	Fruit fig tree
Flemingia macrophylla	Long leaf wurrus
Flemingia prostrata	Wild hops
Foeniculum vulgare	Florence fennel
Galium echinocarpum	Cleavers
Gardenia angusta var. *kosyunensis*	Cape jasmine

Gardenia jasminoides	Cape jasmine
Gardenia oblongifolia	Gardenia
Gelsemium elegans	Jessamine
Gendarussa vulgaris	White justicia
Gentiana atkinsonii	Gentian
Gentiana arisanensis	Gentian
Gentiana campestris	Gentian
Gentiana flavo-maculata	Gentian
Gentiana lutea	Bitterwort
Gentiana scabrida	Gentian
Gentiana scabrida var. *horaimontana*	Gentian
Geranium nepalense var. *thunbergii*	Geranium
Geranium suzukii	Geranium
Geranium suzukii var. *hayatanum*	Geranium
Glechoma hederacea var. *grandis*	Ground ivy
Glehnia littoralis	Beech silver-top
Glochidion acuminatum	Abacus plant
Glochidion eriocarpum	Abacus plant
Glochidion laceolorium	Large-leaf abacus plant
Glochidion puberum	Glochidion
Glochidion rubrum	Common glochidion
Glochidion zeylanicum	Abacus plant
Glossogyne tenuifolia	Shu Qu Cao (English name not available)
Glycine javanica	Rhodesian kudzu
Glycine tabacina	Rhodesian kudzu
Glycine tomentella	Rhodesian kudzu
Glycosmis citrifolia	Rhodesian kudzu
Glycyrrhiza uralensis	Licorice
Gnaphalium adnatum	Cudweed
Gnaphalium affine	Cudweed

Scientific Name	Common Name
Gnaphalium hypoleucum	Cudweed
Gnaphalium luteoalbum subsp. *affine*	Everlasting
Goldfussia formosanus	Taiwan Ma Lan
Goldfussia psilostachys	Taiwan Ma Lan
Gomphrena globosa	Bachelor's button
Goniothalamus amuyon	Pawpaw
Gonostegia hirta	Stinging nettle
Gonostegia pentandra	Stinging nettle
Goodyera nankoensis	Bachelor's buttons
Goodyera procera	Bottlebrush orchid
Gossampinus malabarica	Cotton
Graptopetalum paraguayense	Ghost plant
Grevillea robusia	Silk oak
Gynostemma pentaphyllum	Sweet tea vine
Gynura bicolor	Velvet plant
Gynura elliptica	Velvet plant
Gynura formosana	Taiwan velvet plant
Gynura japonica var. *flava*	Japanese velvet plant
Habenaira dentata	Fringe orchid
Habenaria repens	Creeping orchid
Haraella retrocalla	Fragrant orchid
Hedychium coronarium	Ginger lily
Hedyotis corymbosa	Carymobse hedyotis
Hedyotis diffusa	Spreading hedyotis
Hedyotis pinifolia	Bluets
Hedyotis uncinella	Meadow hedyotis
Helianthus annuus	Sunflower
Helicteres angustifolia	Shan Zhi Ma (in Chinese)
Heliotropium indicum	India helotrope

Helwingia formosana	Taiwan helwingia
Helwingia japonica ssp. *formosana*	Helwingia
Hemerocallis fulva	Daylily
Hemerocallis longituba	Daylily
Hemiboea bicornuta	Gesnerias
Hemiphragma heterophyllum var. *dentatum*	Figworts
Heterostemma brownii	Chinese climber
Heterotropa hayatanum	Wild ginger
Heterotropa macrantha	Wild ginger
Heterotropa taitonensis	Cellar fungus
Hibiscus esculentus	Hibiscus
Hibiscus mutabilis	Cotton rose
Hibiscus rosa-sinensis	Rose of China
Hibiscus sabdariffa	Roselle
Hibiscus syriacus	Rose of Sharon
Hibiscus taiwanensis	Taiwan cotton rose
Hibiscus tillaceus	Cuban bast
Hippeastrum equestre	Amaryllis
Hippeastrum regina	Amaryllis
Hippobroma longiflora	Oechids
Houttuynia cordata	Fishwort
Hoya carnosa	Wax plant
Humulus scandens	Hops
Hydrangea chinensis	Chinese hydrangea
Hydrangea macrophylla	Big hydrangea
Hydrocotyle asiatica	Tian Hu Sui (English name not available)
Hydrocotyle formosana	Coin penny wort
Hydrocotyle nepaleniss	Coin penny wort
Hydrocotyle sibthorpioides	Coin penny wort
Hylocereus undatus	Night-blooming cereus

Scientific Name	Common Name
Hypericum chinense	Chinese St. John's wort
Hypericum geminiflorum	Creeping St. John's wort
Hypericum japonicum	Japanese St. John's wort
Hypericum patulum	Golden St. John's wort
Hyphea kaoi	Cellar fungus
Hypoestes purpurea	Polka dot plant
Hypolepis tenuifolia	Wax weed
Hypoxis aurea	Star grass
Hyptis rhomboides	Capitate bushmint
Hyptis suaveolens	African potato
Ilex asprella	Holly
Ilex cornuta	Burford's holly
Ilex pubescens	Hairy holly
Ilex rotunda	Chinese holly
Illicium arborescens	Japanese anise
Impatiens balsamina	Garden balsam
Imperata cylindrica var. *major*	Thatch grass
Indigofera longeracemosa	Common indigo
Indigofera suffruticosa	Common indigo
Indigofera tinctoria	Common indigo
Indigofera trifoliata	Common indigo
Indigofera zollingeriana	Common indigo
Ipomoea batatas	Sweet potato
Ipomoea obscura	Sweet potato
Ipomoea pes-caprae ssp. *brasiliensis*	Railroad vine
Ipomoea quamoclit	Cypress vine
Ipomoea stans	Ornamental sweet potato vine
Iris tectorum	Iris
Ixeris chinensis	Rabbit milkweed

Ixeris laevigata var. *oldhamii*	Sword wound weed
Ixeris tamagawaensis	Jian Dao Gu (English name not available)
Ixora chinensis	Long Chuan Hua (English name not available)
Jasminum hemsleyi	Mountain jasmine
Jasminum sambac	Arabian jasmine
Jatropha curcas	Sweet cassava
Juncus effusus var. *decipiens*	Common rush
Juniperus formosana	Taiwan juniper
Justicia gendarussa	Jascobinia
Justicia procumbens	Rat tail willow
Justicia procumbens var. *hayatai*	Brazilian plume
Kadsura japonica	Scarlet kadsura
Kaempferia galanga	Galanga
Kalanchoe crenata	Airplant
Kalanchoe gracillis	Airplant
Kalanchoe pinnata	Airplant
Kalanchoe spathulata	Airplant
Kalanchoe tubiflora	Airplant
Kalimeris indica	Field aster
Kyllinga brevifolia	Short-leaf kyllinga
Lactuca indica	Lettuce
Lagenaria siceraria var. *microcarpa*	Bottle gourd
Lagerstroemia subcostata	Crape myrtle
Lantana camara	Common lantana
Laportea moroides	Wood nettle
Laportea pterostigma	Poisonous wood nettle
Laungusa galanga	Nan Jan (English name not available)
Leea guineensis	Manila leea
Lemnaphyllum microphyllum	Green penny fern
Leonurus artemisia	Chinese mother wort

Scientific Name	Common Name
Leonurus sibiricus f. *albiflora*	Siberian motherwort
Lepidagathis cristata	Pata de gallina
Lepidagathis formosensis	Pata de gallina
Lepidagathis hyalina	Pata de gallina
Lespedeza cuneata	Perennial lespedeza
Leucas chinensis	Chinese leucas
Leucas mollissima var. *chinensis*	Chinese leucas
Ligustrum lucidum	Wax tree
Ligustrum pricei	Privet
Ligustrum sinense	Chinese privet
Lilium formosanum	Taiwan lily
Lilium speciosum	Tiger lily
Limonium sinense	Chinese statice
Lindera communis	Spicebush
Lindera glauca	Spicebush
Lindera okoensis	Spicebush
Lindera strychifolia	Spicebush
Liparis cordifolia	Heart-shaped leaf liparis
Liparis keitaoensis	Heart-shaped leaf liparis
Liparis loeselii	Heart-shaped leaf liparis
Liquidambar formosana	Sweet gum tree
Liriope spicata	Lily turf
Litsea acutivena	Cubebs
Litsea cubeba	Cubebs
Litsea hypophaea	Cubebs
Lobelia chinensis	Chinese lobelia
Lobelia laxiflora	Lobelia
Lobelia nummularia	Lobelia
Lonicera apodonta	Honeysuckle

Lonicera confusa	Orange honeysuckle
Lonicera japonica	Japanese honeysukle
Lonicera japonica var. *sempervillosa*	Japanese honeysukle
Lonicera kawakamii	Honeysuckle
Lonicera macrantha	Honeysuckle
Lonicera shintenensis	Trumpet honeysuckle
Lophatherum gracile	Sasagrass
Loropetalum chinense	Strap flower
Ludwigia octovalvis	Lantern seedbox
Luffa cylindrica	Luffa sponge
Lycium chinense	Matrimony vine
Lycopersicon esculentum	Tomato
Lycopodium cunninghamioides	Club moss
Lycopodium salvinioides	Club moss
Lycopus lucidus var. *formosana*	Bugleweed
Lygodium japonicum	Climbing fern
Lysimachia ardisioides	Loosestrife
Lysimachia davurica	Yellow loosestrife
Lysimachia mauritiana	Loosestrife
Lysimachia simulans	Loosestrife
Macaranga tanarius	Macaraßga
Machilus kusanoi	Wild machilus
Machilus zuihoensis	Wild machilus
Maesa lanceolata	Du Jing Shan (English name not available)
Maesa laxiflora	Du Jing Shan (English name not available)
Maesa perluria var. *formosana*	Kong Xin Hua (English name not available)
Maesa tenera	Kong Xin Hua (English name not available)
Magnolia liliflora	Red magnolia
Mahonia japonica	Japanese mahonia
Mahonia oiwakensis	Oregon grape

Scientific Name	Common Name
Mallotus apelta	Mallotus
Mallotus japonicus	Japanese mallotus
Mallotus paniculatus	Mollotus
Mallotus repandus	Mollotus
Mallotus tiliaefolius	Mollotus
Malvastrum coromandelianum	False mallow
Manihot utilissima	Bitter cassava
Maranta arundinacea	Arrow root
Mariscus cyperinus	Long-headed sedge
Marsilea crenata	Ping (English name not available)
Marsilea minuta	Ping (English name not available)
Marsdenia formosan	Tong Guang San (English name not available)
Maytenus diversifolia	Ci Luo Shi (English name not available)
Maytenus emarginata	Ci Luo Shi (English name not available)
Maytenus serrata	Ci Luo Shi (English name not available)
Medicago polymorpha	Toothed bur clover
Melanolepis multiglandulosa	Molucca mallotus
Melastoma candidum	Common melastoma
Melastoma dodecandrum	Lesser melastoma
Melastoma septemnervium	Lesser melastoma
Melia azedarach	China berry tree
Melicope semecarpifolia	Alani
Melissa officinalis	Lemon balm
Melodinus angustifolius	Taiwan Shan Chen
Mentha canadensis	Canadian mint
Mentha haplocalyx	Pepper mint
Mesona chinensis	Grass jelly
Mesona procumbens	Mesona
Messerschmidia argentea	Silvery messerschmidia

Michelia alba	White champac
Microcos paniculata	Huo Tan Mu (English name not available)
Microglossa pyrifolia	Small everlasting
Mikania cordata	Chinese creeper
Milletia nitida	Milletha
Milletia pachycarpa	Milletha
Milletia speciosa	Showy milletha
Milletia taiwaniana	Milletha
Mimosa pudica	Sensitive plant
Mirabilis jalapa	Marvel of Peru
Miscanthus floridulus	Giant miscanthus
Miscanthus sinensis var. condensatus	Japanese silver grass
Mollugo pentaphylla	Di Ma Huang (English name not available)
Momordica charantia	Bitter melon
Monochoria vaginalis	Ya She Cao (English name not available)
Morinda citrifolia	Noni
Morinda umbellata	Common Indian mulberry
Morus alba	Mulberry tree
Morus australis	Small-leaf mulberry
Mosla punctulata	Chinese mosia
Mucuna macrocarpa	Rusty leaf mucuna
Mucuna nigricans	Velvet bean
Mucuna pruriens	Velvet bean
Muehlenbeckia hastulata	Wire vine
Muehlenbeckia platychodum	Wire vine
Murdannia keisak	Water murdannia
Murdannia loriformis	Water murdannia
Murraya paniculata	Jasmin orange
Musa basjoo var. formosana	Plantain
Musa formosana	Banana

Scientific Name	Common Name
Musa insularimontana	Plantain
Musa paradisiaca	Banana
Musa sapientum	Banana
Mussaenda parviflora	Taihoku mussaenda
Mussaenda pubescens	Splash-of-white
Myrica adenophora	Bayberry
Myrica rubra	Chinese strawberry
Myristica cagayanensis	Nutmeg
Myristica fragrans	Nutmeg
Nandina domestica	Sacred bamboo
Nelumbo nucifera	East Indian lotus
Neolitsea acuminatissima	White bolly gum
Nephrolepis auriculata	Sword fern
Nerium indicum	Indian oleander
Nervilia purpurea	Bird's nest orchid
Nervilia taiwaniana	Bird's nest orchid
Nicotiana tabacum	Tobacco
Nothapodytes foetida	Chin Cui Zi (English name not available)
Nothapodytes nimmoniana	Chin Cui Zi (English name not available)
Nymphaea tetragona	Pigmy water lily
Nymphar shimadai	Water lily
Ocimum basilicum	Basil
Ocimum gratissimum	Sweet basil
Oenanthe javanica	Water celery
Oldenlandia diffusa	Shui Xian Cao (English name not available)
Oldenlandia hedyotidea	Shui Xian Cao (English name not available)
Onychium japonicum	Carrot fern
Ophioglossum vulgatum	Adder's tongue
Ophiopogon japonicus	Japanese lily turf

Opuntia dillenii	Prickly-pear cactus
Oreocnide pedunculata	Cha Gen Zi Ma (English name not available)
Orthosiphon aristatus	Cat's whiskers
Osbeckia chinensis	Tin Xiang Lu (English name not available)
Osmanthus fragrans	Sweet olive
Orthosiphon stamineus	Cat's whiskers
Oxalis corniculata	Wood sorrel
Oxalis corymbosa	Wood sorrel
Paederia cavaleriei	Fevervine
Paederia foetida	Fevervine
Paederia scandens	Chinese fevervine
Paliurus ramosissimus	Ma Jia Zi Ye (English name not available)
Pandanus amaryllifolius	Screwpine
Pandanus odoratissimus var. *sinensis*	Screwpine
Pandanus pygmaeus	Screwpine
Parachampionella filexicaulis	Tidon Sha Lan (English name not available)
Parachampionella flexicaulis	Tidon Sha Lan (English name not available)
Parachampionella rankanensis	Tidon Sha Lan (English name not available)
Paracyclea gracillima	Tu Gang Ji (English name not available)
Paracyclea ochiaiana	Tu Gang Ji (English name not available)
Paris arisanensis	Paris
Paris formosana	Paris
Paris lancifolia	Paris
Paris polyphylla	Himalayan paris
Parthenocissus tricuspidata	Boston ivy
Passiflora foetida var. *hispida*	Weed passion flower
Passiflora suberosa	Black fruit passion
Pedilanthus tithymaloides	Redbird cactus
Pemphis acidula	Reef pemphis
Pericampylus formosanus	Taiwan slat vine

Scientific Name	Common Name
Pericampylus glaucus	Yellow stem fig
Pericampylus trinervatus	Taiwan slat vine
Perilla frutescens	Perilla
Perilla frutescens var. *crispa*	Perilla
Perilla ocymoides	Perilla
Peristrophe japonica	Marble leaf
Peristrophe roxburghiana	Marble leaf
Petasites formosanus	Taiwan coltsfoot
Petasites japonicus	Japanese coltsfoot
Peucedanum formosanum	Taiwan hogfennel
Phellodendron amurense	Cork tree
Phellodendron chinensis	Chinese cork tree
Phellodendron wilsonii	Cork tree
Phoenix dactylifera	Date palm
Phyla nodiflora	Frogfruit
Phyllanthus emblica	Wrinkle fruit leaf flower
Phyllanthus multiflorus	Wrinkle fruit leaf flower
Phyllanthus urinaria	Wrinkle fruit leaf flower
Phyllodium pulchellum	Round brack tick clover
Physalis angulata	Ground cherry
Phytolacca acinosa	Pokeberry
Phytolacca americana	Pokeberry
Phytolacca japonica	Japanese pokeberry
Pieris formosa	Taiwan pieris
Pieris hieracloides	Fetterbush
Pieris taiwanensis	Taiwan pieris
Pilea microphylla	Aluminum plant
Pilea rotundinucula	Aluminum plant
Pinellia pedatisecta	Pinellia

Pinellia ternata	Ternate pinellia
Pinus massoniana	Pine
Pinus taiwanensis	Taiwan pine
Piper arboricola	Pepper
Piper betle	Betel pepper
Piper kadsura	Pepper
Piper kawakamii	Pepper
Piper nigrum	Black pepper
Piper sanctum	Piper
Piper sarmentosum	Piper
Pittosporum pentandrum	Pittosporum
Plantago asiatica	Dooryard weed
Plantago major	Plantain
Platycodon grandiflorum	Kikio root
Plectaanthus amboinicus	Mexican mint
Pleione formosana	Shan Ci Gu (English name not available)
Pluchea indica	Indian pulchea
Plumbago zeylanica	Ceylon leadwort
Plumeria rubra cv. *acutifolia*	Frangipani
Podocarpus macrophyllus var. *nakaii*	Southern yew
Podocarpus nagi	Nagi podocarp
Pogonatherum crinitum	Golden-hair grass
Pogonatherum paniceum	Golden-hair grass
Pogostemon amboinicus	Patchouli
Pogostemon cablin	Patchouli
Pollia secundiflora	Zhu Ye Lian (English name not available)
Polygala aureocauda	Milkwort
Polygala glomerata	Chinese milkwort
Polygonatum falcatum	Solomon's seal
Polygonatum kingianum	Solomon's seal

Scientific Name	Common Name
Polygonatum odoratum	Solomon's seal
Polygonum chinense	Hill buckwheat
Polygonum cuspidatum	Japanese knotweed
Polygonum multiflorum var. *hypoleucum*	Fo-ti (English name not available)
Polygonum paleaceum	Chinese indigo
Polygonum perfoliatum	Chinese indigo
Polygonum plebeium	Joint flowered knotweed
Pometia pinnata	Fiji longan
Portulaca grandiflora	Garden portulaca
Portulaca oleracea	Purslane
Portulaca pilosa	Hairy purslane
Potentilla discolor	Wolf tooth
Potentilla leuconta	Potentilla
Potentilla tugitakensis	Potentilla
Pothos chinensis	Orange leaf pothos
Pouteria obovata	Pouteria
Pouzolzia elegans	Wu Shui Ge (English name not available)
Pouzolzia pentandia	Wu Shui Ge (English name not available)
Pouzolzia zeylanica	Wu Shui Ge (English name not available)
Pratia nummularia	Pratia
Premna crassa	Chou Huang Jing Zi (English name not available)
Premna microphylla	Chou Huang Jing Zi (English name not available)
Premna obtusifolia	Chou Huang Jing Zi (English name not available)
Premna serratifolia	Chou Huang Jing Zi (English name not available)
Prinsepia scandens	Ruiren
Procris laevigata	Perfume herb
Prunella vulgaris	Heal-all
Prunus persica	Peach
Pseudosasa usawai	Tonkin bamboo

Psidium guajave	Guava
Psophocarpus tetragonolobus	God bean
Psychotria rubra	Red psychotria
Pteris ensiformis	Brake
Pteris multifida	Brake
Pteris semipinnata	Panpienchi
Pteris vittata	Brake
Pterocypsela indica	Wild lettuce
Pueraria lobata	Kudzu vine
Pueraria montana	Kudzu
Punica granatum	Pomegranate
Pyracantha fortuneana	Chinese firethorn
Pyrola japonica	Japanese wintergreen
Pyrola morrisonensis	Wintergreen
Pyrrosia adnascens	Felt fern
Pyrrosia petiolosa	Felt fern
Pyrrosia polydactylis	Felt fern
Quisqualis indica	Rangoon creeper
Rabdosia lasiocarpus	Striped plectranthus
Randia spinoa	Spiny randia
Ranunculus japonicus	Japanese radish
Ranunculus sceleratus	Ground mulberry
Rauvolfia verticillata	Devil pepper
Rhamnus formosana	Taiwan buckthorn
Rhinacanthus nasutus	Rhinacanthus
Rhodea japonica	Lily of the valley
Rhododendron simsii	Red azalea
Rhodomyrtus tomentose	Hill gooseberry
Rhoeo spathacea	Purple-leaf spiderwort
Rhus chinensis	Chinese sumac

Scientific Name	Common Name
Rhus javanica var. roxburghiana	Lemonade berry
Rhus microphylla	Sumac
Rhus semialata var. roxburghiana	Sumac
Rhus succedanea	Sumac
Rhus typhina	Staghorn summac
Rhus verniciflua	Sumac
Rhynchoglossum holglossum	Gou Teng Diao (English name not available)
Rhynchosia minima	Lu Huai Hua (English name not available)
Rhynchosia volubilis	Lu Huai Hua (English name not available)
Ribes formosanum	Taiwan black currant
Ribes nigrum	Black currant
Ricinus communis	Castor bean
Rollinia mucosa	Wild cashina
Rosa davurica	Dog rose
Rosa taiwanensis	Taiwan rose
Rotala rotundifolia	Round-leaf rotala
Rubia akane	Indian madder
Rubia lanceolata	Indian madder
Rubia linii	Indian madder
Rubus croceacanthus	Blackberry
Rubus formosensis	Formosan raspberry
Rubus hirsutus	Blackberry
Rubus parvifolius	Japanese raspberry
Ruellia tuberosa	Petunia
Rumex acetosa	Garden sorrel
Rumex crispus	Dock
Rumex japonicus	Japanese curly dock
Ruta graveolens	Common rue
Saccharum officinarum	Sugar cane

Salix warburgii	Willow
Salvia coccinea	Texas sage
Salvia hayatana	Taiwan sage
Salvia japonica	Japanese sage
Salvia plebeia	Cudweed
Sambucus chinensis	Chinese elderberry
Sambucus formosana	Formosan elderberry
Sambucus javanica	Elderberry
Sanguisorba formosana	Taiwan burnet
Sanguisorba minor	Garden burnet
Sanguisorba officinalis	Salad burnet
Sanicula elata	Sanicle
Sanicula petagniodes	Sanicle
Sansevieria trifasciata	Snake plant
Sapindus mukorossi	Soap berry
Sapium sebiferum	Tallow tree
Sarcandra glabra	Jiu Jie Cha (English name not available)
Saurauja oldhamii	Oldham saurauia
Saurauja tristyla var. *oldhamii*	Shu Don Gua (English name not available)
Saururus chinensis	Lizard's tail
Saxifraga stolonifera	Strawberry geranium
Scaevola sericea	Beach naupaka
Scutellaria javanica var. *playfairi*	Skullcap
Schefflera octophylla	Minimum light
Schisandra arisanensis	Schisandra
Schizophragma integrifolium	Bluestem
Scirpus maritimus	San Leng (English name not available)
Scirpus ternatanus	San Leng (English name not available)
Scoparia dulcis	Sweet broom wort
Scrophularia yoshimurae	Figwort

Scientific Name	Common Name
Scurrula ferruginea	Ta Yeh Sang Chih (English name not available)
Scurrula liquidambaricolus	Ta Yeh Sang Chih (English name not available)
Scurrula loniceritolius	Ta Yeh Sang Chih (English name not available)
Scurrula ritozonensis	Ta Yeh Sang Chih (English name not available)
Scutellaria barbata	Barbed skullcap
Scutellaria formosana	Taiwan Skullcap
Scutellaria indica	Skullcap
Scutellaria rivularis	Skullcap
Securinega suffruticosa	Yi Ye Qiu (English name not available)
Securinega virosa	Yi Ye Qiu (English name not available)
Sedum formosanum	Linear stonecrop
Sedum lineare	Stonecrop
Sedum morrisonense	Stonecrop
Sedum sempervivoides	Stonecrop
Selaginella delicatula	Chinese resurrection plant
Selaginella uncinata	Chinese resurrection plant
Semnostachya longespicata	Chung Wei Ma Lan (English name not available)
Senecio nemorensis	Ragwort
Senecio scandens	Ragwort
Serissa foetida	Serissa
Serissa japonica	Serissa
Sesamum indicum	Sesame seed
Setaria italica	Palm grass
Setaria palmifolia	Palm grass
Setaria viridis	Palm grass
Setcreasea purpurea	Purple heart
Severinia buxifolia	Chinese box orange
Sida acuta Burm	Narrow-leaf sida
Sida rhombifolia	Sida

Siegesbeckia orientalis	Common St. Paul's wort
Silene morii	Campion
Silene vulgaris	Campion
Siphonostegia chinensis	Siphonostegia
Smilacina formosana	Taiwan Solomon's seal
Smilax bracteata	Smilax
Smilax china	China root
Solanum abutiloides	Nightshade
Solanum aculeatissimum	Gold silver nightshade
Solanum biflorum	Two-flower nightshade
Solanum capsicastrum	False Jerusalem cherry
Solanum indicum	Indian nightshade
Solanum lyratum	Eggplant
Solanum nigraum	Black nightshade
Solanum undatum	Nightshade
Solanum verbascifolium	Tobacco nightshade
Solidago altissima	Goldenrod
Solidago virgo-aurea	Goldenrod
Sonchus arvensis	Field sow thistle
Sonchus oleraceus	Sow thistle
Sophora flavescens	Sophora
Sophora tomentosa	Sophora
Spilanthes acmella	Gold button
Spilanthes acmella var. *oleracea*	Gold button
Spinacia oleracea	Spinach
Spiraea formosana	Taiwan spirea
Spiraea prunifolia var. *pseudoprunifolia*	Bridal wreath
Spiranthes sinensis	Ladies' tresses
Stachys sieboldii	Betony
Stachytarpheta jamaicensis	Jamaica vervain

Scientific Name	Common Name
Stellaria media	Chick weed
Stemona tuberosa	Stemona
Stephania cephalantha	Stephania
Stephania hispidula	Stephania
Stephania japonica	Japanese stephania
Stephania tetrandra	Stephania
Sterculia lychnophora	Bottle tree
Sterculia nobilis	Noble bottle tree
Stevia rebaudiana	Stevia
Strychnos angustiflora	Strychnine
Swertia randaiensis	Green gentian
Symphytum officinale	Common comfrey
Tabernaemontana amygdalifolia	Crape jasmine
Tabernaemontana divaricata	Crape jasmine
Tabernaemontana pandacaqui	Crape jasmine
Tagetes erecta	Mexican marigold
Taiwania cryptomeriodes	Taiwan fir
Talinum paniculatum	Fame flower
Talinum patens	Fame flower
Talinum triangulare	Fame flower
Tamarix chinensis	Chinese tamarisk
Tamarix juniperina	Juniper tamarisk
Taraxacum formosanum	Taiwan dandelion
Taraxacum mongolicum	Dandelion
Taraxacum officinale	Dandelion
Taxillus levinei	Eusolex
Taxillus matsudai	Eusolex
Taxus mairei	Yew
Ternstroemia gymnanthera	Cleyera

Tetrastigma dentatum	Zou You Cao (English name not available)
Tetrastigma formosanum	Zou You Cao (English name not available)
Tetrastigma umbellatum	Zou You Cao (English name not available)
Tetrapanax papyriferus	Rice paper tree
Teucrium viscidum	Germander
Thalictrum fauriei	Taiwan meadow
Thevetia peruviana	Huang Hua Jia Tao (English name not available)
Thladiantha nudiflora	Chi Pao (English name not available)
Tinospora tuberculata	Jin Guo Lan (English name not available)
Tithonia diversifolia	Mexican sunflower
Toddalia asiatica	Asian toddalia
Toona sinensis	Chinese mahogany
Torenia concolor var. *formosana*	Wishbone plant
Tournefortia sarmentosa	Bearing runners tournefortia
Trachelospermum jasminoides	Star jasmine
Trichosanthes cucumeroides	Snake-gourd
Trichosanthes dioica	Trichosanthes
Trichosanthes homophylla	Trichosanthes
Tricytis formosana	Taiwan toad lily
Tridax procumbens	Lantern tridax
Tripterygium wilfordii	Yellow vine
Tropaeolum majus	Nasturtium
Tubocapsicum anomalum	Nightshade
Turpinia formosana	Smoketree
Tylophora lanyuensis	Lanyu tylophora
Tylophora ovata	Ovate leaf tylophora
Typhonium divaricatum	Li Tou Jian (English name not available)
Uncaria hirsuta	Gambir
Uncaria kawakamii	Cat's claw
Uncaria rhynchophylla	Cat's claw

Scientific Name	Common Name
Uraria crinita	Chinese honeysuckle
Uraria lagopodioides	Honeysuckle
Urena lobata	Aramina
Urena procumbens	Duck foot
Urtica dioica	Stinging nettle
Urtica thunbergiana	Stinging nettle
Vaccinium emarginatum	Blueberry
Vaccinium japonium	Japanese blueberry
Vaccinium myrtillus	Bilberry
Vandellia cordifolia	Blue pig ear
Vandellia crustacea	Blue pig ear
Ventilago leiocarpa	Xue Feng Teng (English name not available)
Veratrum formosanum	Taiwan mountain onion
Verbena officinalis	Vervain
Vernonia cinerea	Ironweed
Vernonia gratiosa	Ironweed
Veronicastrum simadai	Chinese culver's root
Viburnum awabuki	Cramp bark
Viburnum luzonicum	Cramp bark
Viburnum odoratissimum	Cramp bark
Viburnum plicatum var. formosanum	Cramp bark
Vigna angularis	Adzuki bean
Vigna radiata	Mung bean
Vigna umbelbita	Rice bean
Viola betonicifolia	Pansy
Viola confusa	Pansy
Viola diffusa	Pansy
Viola hondoensis	Pansy
Viola inconspicua ssp. nagasakiensis	Pansy

Viola mandshurica	Violet
Viola philippica	Leaf violet
Viola tricolor	Field pansy
Viola verecunda	Violet
Viola yedoensis	Violet
Viscus alniformosanae	Mistletoe
Viscus angulatum	Mistletoe
Viscus multinerve	Mistletoe
Vitex cannabifolia	Chase tree
Vitex negundo	Five-leaf chaste tree
Vitex rotundifolia	Indian privet
Vitis thunbergii	Taiwan wild grape
Wedelia biflora	Wedelia
Wedelia chinensis	Chinese Wedelia
Wendlandia formosana	Liao Ge Wang (English name not available)
Wikstroemia indica	Liao Ge Wang (English name not available)
Xanthium sibiricum	Cocklebur
Xanthium strumarium	Spiny cocklebur
Youngia japonica	Oriental hawksbeard
Zanthoxylum ailanthoides	Ailanthus prickly ash
Zanthoxylum avicennae	Prickly ash
Zanthoxylum dimorphophylla	Prickly ash
Zanthoxylum integrifoliolum	Prickly ash
Zanthoxylum nitidum	Shiny bramble
Zanthoxylum piperitum	Prickly ash
Zanthoxylum pistaciflorum	Prickly ash
Zebrina pendula	Wandering jew
Zephyranthes candida	White zephylily
Zephyranthes carinata	White zephylily
Zingiber kawagoii	Ginger

Scientific Name	Common Name
Zingiber officinale	Ginger
Zingiber rhizoma	Ginger
Zornia diphylla	Ding Gui Cao (English name not available)

Index

H

M

N

O

P